2023 IEEE Regional Symposium on Micro and Nanoelectronics (RSM 2023)

Langkawi, Malaysia
28-30 August 2023

IEEE Catalog Number: CFP2368N-POD
ISBN: 979-8-3503-2369-6

**Copyright © 2023 by the Institute of Electrical and Electronics Engineers, Inc.
All Rights Reserved**

Copyright and Reprint Permissions: Abstracting is permitted with credit to the source. Libraries are permitted to photocopy beyond the limit of U.S. copyright law for private use of patrons those articles in this volume that carry a code at the bottom of the first page, provided the per-copy fee indicated in the code is paid through Copyright Clearance Center, 222 Rosewood Drive, Danvers, MA 01923.

For other copying, reprint or republication permission, write to IEEE Copyrights Manager, IEEE Service Center, 445 Hoes Lane, Piscataway, NJ 08854. All rights reserved.

*** This is a print representation of what appears in the IEEE Digital Library. Some format issues inherent in the e-media version may also appear in this print version.

IEEE Catalog Number: CFP2368N-POD
ISBN (Print-On-Demand):   979-8-3503-2369-6
ISBN (Online):   979-8-3503-2368-9
ISSN: 2639-4650

Additional Copies of This Publication Are Available From:

Curran Associates, Inc
57 Morehouse Lane
Red Hook, NY 12571 USA
Phone: (845) 758-0400
Fax: (845) 758-2633
E-mail: curran@proceedings.com
Web: www.proceedings.com

TABLE OF CONTENTS

Chair message	v
Organizing Committee	vi
List of Reviewers	viii
Keynote 1: Microelectronics: A Concise Overview of The Industry Landscape In Malaysia And In The Emergence of IR 4.0 *Prof. Dato' Ts. Dr. Zaliman Sauli*	xii
Keynote 2: Market and Technology Trends of Advanced Packaging *Dr. Tan Yik Yee*	xiii
Keynote 3: Graphene Nanoballs for Performance Improvement of Thermoelectric Energy Harvester *Prof. Dr. Azrul Azlan Hamzah*	xiv
Performance Analysis of 14nm SOI-based Trigate Gaussian Channel Junctionless FinFET with Punchthrough Stop Layer *Mathangi Ramakrishnan, Nurul Ezaila Alias, Michael Loong Peng Tan, Afiq Hamzah, Yasmin Abdul Wahab and Hanim Hussin*	1
The thermal conductivity of stacked hexagonal Boron Nitride (hBN) and Graphene - A molecular dynamics approach *Dharma Darren Ram, Muhammad Aniq Shazni Mohammad Haniff, Mohd Ambri Mohamed and Abdul Manaf Hashim*	5
Investigating the Performance of Deep Reinforcement Learning-Based MPPT Algorithm under Partial Shading Condition *Yew Weng Ho, Chien Fat Chau, Ahmad Wafi Mahmood Zuhdi, Wan Syakirah Wan Abdullah, Yew Weng Kean and Nowshad Amin*	9
Simulink Model of Noise of Piezoelectric Charge Amplifier *Ghulam Ali and Faisal Mohd-Yasin*	13
Proposal for stochastic resonance in a ferroelectric-graphene transistor *Madhav Ramesh, Amit Verma and Arvind Ajoy*	17
Simulation of Macro-Compact Model of Graphene-based Three-Branch Nano-Junction *Alireza Kalantari, Shaharin Fadzli Bin Abd Rahman and Abdul Manaf Hashim*	21
Surface Defects Originated Photoresponse Study in hBN-ReS2 FETs *Mohd Amir Zulkefli and Muhammad Hilmi Johari*	25
Linear, Efficient and Wideband Emitter Follower Class B Amplifier for Auxiliary Envelope Tracking Supply Modulator *Zubaida Yusoff, Md Mushfiqur Rahman, Farid Zubir and Jahariah Sampe*	28
Fabricating SWCNT thin film via Spray coating and Nitric Acid Vapor Treatment *Arulampalam Kunaraj, Puvaneswaran Chelvanathan, Ahmad Ashrif A. Bakar, Avinash Kumaresan and Iskandar Yahya*	32

Equivalent Circuit Model and Simulation of 2D Asymmetrical PMUT for Non-Destructive Testing — 36

Darven Raj Ponnuthurai

Effect of biasing under illumination on GaAsBi/GaAs multiple quantum wells for solar cell performance — 40

Faezah Harun, Robert D. Richards and John P.R David

The effects of particle sizes of Neodymium Iron Boron microstructure on the magnetic characteristics — 44

Siti Aisyah Binti Ishak, Jumril Yunas and Abdul Manaf Hashim

Evaluation of Cross-Contamination Risk during CMOS Devices Fabrication in an Industrial Silicon Wafer Processing — 47

Mohd Amir Zulkefli, Ismail Umar, Vanita Manaoogaran, Wan Hidayatulhusna Wan Mohamad Rani, Guan Kai Oh and Deyline Samail, Izzuddin Iskandar

NBTI Defects Characterization Using Energy Profiling Simulation Technique — 50

Hanim Hussin, Sharifah Fatmadiana Wan Muhamad Hatta, Norhayati Soin, Yasmin Abdul Wahab, Maizan Muhamad and Nurul Ezaila Alias

Surface Morphology of Fabricated TiO2-Graphene Thin Film by Spin-Coating Technique for pH Sensing Electrode Application — 54

Anis Nabilah Mohd Daud, Aina Syakirah Mohd Masri, NurSyahirah Kamarozaman, Muhammad AlHadi Zulkefle, Zurita Zulkifli and Sukreen Hana Herman

Graphene-Based Hybrid Sensor for the Detection of Cancer Cells Using K-SPR Technology — 58

P. Susthitha Menon, Nur Shahirah Shaari, Vatsala Pithaih, Siti Nasuha Mustaffa, Affa Rozana Abdul Rashid, Vikneswary Ravi Kumar, Nor Haslinda Abd Aziz and Nirmala Kampan

Enhancing Industrial Machine Monitoring with IoT: A Wireless Solution — 62

Maizatul Zolkapli, Ahmad Sabirin Zoolfakar, Rozina Abdul Rani and Yusof Johan

Electrochemical EGFET pH Sensing Performance using ZnO-based Composite Thin Films Sensing Electrode — 66

Zainal Nurbaya, NurSyahirah Kamarozaman, Abdur Rahman, Sukreen Hana Herma and, Zurita Zulkifli

Fabrication of TiO2-PANI Nanostructure using Electrospray for the pH Sensing Electrode — 70

Aina Syakirah Mohd Masri, Zainal Nurbaya, Sukreen Hana Herman, NurSyahirah Kamarozaman and Zurita Zulkifli

Determination of the Aptamer Probe Density by Double Layer and Redox Capacitance of CNT-Based Electrochemical DNA-Aptasensors — 74

Yasmin Abdul Wahab, Mohammad Al Mamun, Mohd Rafie Johan, M. A. Motalib Hossain, Abu Hashem, Nurul Ezaila Alias, Hanim Hussin and Maizan Muhamad

Enhancing Sensitivity of Thermal Biosensors through Vanadium Dioxide (VO2) Thin Films 78

Abdelkader Hassein-Bey, Leila Sabeha Asmaa Hassein-Bey, Slimane Lafane, Samira Abdelli-Messaci and Burhanuddin Yeop Majlis

Fabrication of Flexible and Printable Organic Thin-Film Transistor-based Sensor 82
Fazliyatul Azwa Md Rezali, Norhayati Soin, Siti Nabila Aidit, Sharifah

Fatmadiana Wan Muhamad Hatta

Smoothing Sensor Data in a Controlled IoT Framework with Moving Averages 86

Akmal Mustaffa Zulhakim, Wan Fazlida Hanim Abdullah, Ili Shairah Abdul Halim, Robaiah Mamat, Muhammad Izzat Alif Muslan and Ahmad Zaki Abu Bakar

Morphology and Electrical Properties of Pristine and Composite rice husk ash 90
Nano/ Microparticles thick films for Gas Sensing Applications

Jamila Lamido Sumaila, Dahiru Sani Shu'aibu, Mohd Nizar Hamidon, Zainab Yunusa, Nuradden Magaji, Azlinda Abu Bakar and Suleiman Babani

Effect of the Electrodeposition Cycle of RGO Towards Glucose Detection 94

Muhammad Haziq Bin Ilias, Norhazlin Khairudin, Ahmad Sabirin Zoolfakar, Maizatul Zolkapli, Rozina Abdul Rani, Azrif Manut, Zainiharyati Mohd Zain and Noor Fitrah Abu Bakar

Characterization and Optimization of Ion-Sensitive Field Effect Transistor 98
(ISFET) with Different Gate Dielectric and Thickness

Suhana Mohamed Sultan and Jason Kai Seng Kong

Trade-offs and Optimization: Low Power Approaches for Area, Power 102
Consumption, and Performance in Microprocessor Design

Maizan Muhamad, Hanim Hussin, Abdul Karimi Halim, Yasmin Abdul Wahab and Nur Mahirah Sallehuddin

Design and Implementation of 32 bit SDRAM Memory Controller with Optimized 106
Dynamic Power using ASIC

Toy Zheng Hong, Nurul Ezaila Alias, Michael Loong Peng Tan and Yasmin Abdul Wahab

A Study of the Optimum Input Matching Simulation Networks for Integrated 110
Differential Amplifiers

Moh'd Khier Abdallah Alshamaileh, Lutfi Albasha and Nasir Quadir

Study of Error Amplifiers for Low Power Capacitorless Low Dropout Voltage 114
Regulator using 110 nm CMOS Technology

Julie Roslita Rusli

Chitosan as Natural Binder for Eco-Friendly Printable Conductive Ink 118

Nur Iffah Irdina Maizal Hairi, Aliza Aini Md Ralib, Anis Nurashikin Nordin, Rosminazuin Ab Rahim, Lai Ming Li and, Muhammad Farhan Affendi Mohamad Yunos

Achieving Compact Structure and Good Mechanical Properties of AlN Thin Film through Low Temperature HiPIMS   122

Zulkifli Azman, Nafarizal Nayan, Chin Fhong Soon, Ahmad Shuhaimi, Norain Sahari, Yusmar Palapa Wijaya, Ahmad Nasrull Mohamed and Muhammad Yazid Ahmad

FDTD Simulation for Optical Characteristics Study of Inverted Micro-pyramidal Surface Structure of Black Silicon   126

Md. Yasir Arafat, Yasmin Abdul Wahab, Mohammad Aminul Islam, Sharifah Fatmadiana Wan Muhamad Hatta and Nurul Ezaila Alias

Investigation of the Performance Impact of Active Layer Parameter Variations on Inverted Perovskite Solar Cells Using GPVDM   130

Ahmad Muhajer Abdul Aziz, Muhammad Idzdihar Bin Idris, Zul Atfyi Fauzan Mohammed Napiah, Zarina Baharudin Zamani, Nurbahirah Norddi and, Marzaini Rashid, Subathra Muniandy

Advanced Solar-Powered Seed Sowing Machine with Precision Seeding and Smart Control Features   134

Sadiq Ur Rehman, A. Zaidi Asad, Yasmin Abdul Wahab, Md. Yasir Arafat and Sharifah Fatmadiana Wan Muhamad Hatta

Finite Element Simulation of Single Zinc Oxide Nanorod for Piezoelectric Nanogenerator   138

Muhammad Adhwa Fathullah bin Nor Asmadi, Aliza Aini Md Ralib and Anis Nurashikin Nordin, Norazlina Saidin

Acoustic Streaming in Microchannel as Micro-mixing   142

Anjam Waheed, Farhanulhakim Mohd Razip Wee and Muhamad Ramdzan Buyong

MESSAGE FROM THE CHAIRMAN

Ir. Dr. Hazian Mamat

Assalamualaikum warahmatullahi wabarakatuh.

Dear esteemed Guests, Distinguished Participants, and Honoured Sponsors,

We are delighted to extend our warmest welcome to all of you to the Regional Symposium on Micro and Nanoelectronics (RSM 2023). This year 2023 is the first time after the COVID-19 pandemic we are able to organise a physical event and hopefully the event will run smoothly.

RSM 2023 brings together exceptional participants including esteemed lecturers, industry engineers and researchers from various disciplines, each contributing their expertise to the pursuit of innovative engineering solutions. This vibrant convergence of minds promises to inspire ground-breaking ideas and set new benchmarks for progress int the field of engineering.

On behalf of the organizing committee, we thank you for your active participation in RSM 2023. Your strong continuous support in selecting RSM 2023 as the platform to publish your latest research in semiconductor electronics is greatly appreciated. During the 2-day conference, 38 oral presentations will be delivered across a broad spectrum of technical sessions. These include three keynote speakers which are Dato Prof. Dr. Zaliman Sauli (UNIMAP), Prof. Dr. Azrul Azlan Hamzah (UKM) and Dr. Yik Yee Tan (YOLE Group)

This is the 14th RSM organized by the Electron Devices Chapter of IEEE Malaysia Section and technically co-sponsored by the IEEE Electron Devices Society. Over the last Thirty years, RSM conference scrics has become the prominent international forum on semiconductor electronics embracing all aspects of the semiconductor technology under 4 main clusters which are Devices, Nanophotonic, IC Design and Manufacturing and Material, Process and Products.

As we embark on this remarkable journey of the Regional Symposium on Micro and Nanoelectronics (RSM 2023) we express our profound gratitude to all participants, sponsors and collaborators for their unwavering support in realizing this exceptional event. Together let us pave the way for a resilient, sustainable and transformative future through pioneering solutions. Finally, I hope that RSM 2023 will be successful and enjoyable to all participants.

Thank you and Terima kasih.

Ir. Dr. Hazian Mamat
Chairman
2023 IEEE Regional Symposium on Micro and Nanoelectronics
2023 & 2024 IEEE EDS Malaysia Chapter

ORGANIZING COMMITTEE

Advisor: Prof. Dato' Dr. Burhanuddin Yeop Majlis, and Prof. Dr. Norhayati Soin

Chair : Ir. Dr. Hazian Mamat

Co-Chair: Dr. Maizatul Zolkapli

Technical Chair 1: Prof. Dr. AHM Zahirul Alam

Technical Chair 2: Dr Nurul Ezaila Alias

Technical Chair 3: Dr. Yasmin Abdul Wahab

Secretary : Dr Zubaida Yusoff

Treasurer : Dr Haslina Jaafar

Webmaster: Dr. Aliza Aini Md Ralib

Local Arrangement : Dr Hasnizah Aris

Secretariat Committee

Secretariat leader: Dr. Hanim Hussin

    Dr. Sharifah Fatmadiana Wan Muhammad Hatta

    Dr. Iskandar Yahya

    Dr. Maizan Muhamad

    Dr Suhana Mohamed Sultan

    Dr. Azrif Manut

Committee members:

    Assoc. Prof. Dr. Badariah Bais

    Assoc. Prof. Dr. P Susthitha Menon

    Prof. Dr. Mohd Nizar Hamidon

    Prof. Ir. Dr. Norhayati Soin

    Ir. Dr. Hazian Mamat

    Ir Bernard Lim Kee Weng

    Assoc. Prof. Dr. Rosminazuin Ab Rahim

    Prof. Dr. Nafarizal Nayan

    Assoc. Prof. Ir. Dr. Ahmad Sabirin Zoolfakar

Leader for each cluster:

Cluster 1 (MEMS): Prof. Ir. Dr. Norhayati Soin

Cluster 2 (Nanophotonic): Prof. Dr. Nafarizal Nayan

Cluster 3 (Manufacturing): Ir Bernard Lim Kee Weng

Cluster 4 (Material): Dr Suhana Mohamed Sultan

Cluster 5 (Emerging technology): Dr. Iskandar Yahya

LIST OF REVIEWERS

Name	Affiliation	Country
Affa Rozana Abdul Rashid	USIM	Malaysia
AHM Zahirul Alam	International Islamic University Malaysia	Malaysia
Ahmad Alabqari Ma' Radzi	Universiti Tun Hussein Onn Malaysia	Malaysia
Ahmad Rifqi Md Zain	Institute of Microengineering and Nanoelectronics (IMEN), UKM	Malaysia
Ahmad Sabirin Zoolfakar	Universiti Teknologi MARA	Malaysia
Alireza Ghasempour	University of Applied Science and Technology	USA
Aliza Aini Md Ralib	International Islamic University Malaysia	Malaysia
Amiza Rasmi	TM Research & Development	Malaysia
Azli Yahya	Universiti Teknologi Malaysia	Malaysia
Azrif Manut	Universiti Teknologi MARA Shah Alam	Malaysia
Azura Hamzah	Universiti Teknologi Malaysia	Malaysia
Badariah Bais	Universiti Kebangsaan Malaysia	Malaysia
Badrul Hisham Ahmad	Universiti Teknikal Malaysia Melaka	Malaysia
Chang Fu Dee	Universiti Kebangsaan Malaysia (UKM)	Malaysia
Chau Yuen	Nanyang Technological University	Singapore
China Sonagiri	Institute of Aeronautical Engineering	India
Chutisant Kerdvibulvech	National Institute of Development Administration	Thailand
Dan Ciulin	E-I-A	Switzerland
Dan Milici	University of Suceava	Romania
Datta Chavan	Bharati Vidyapeeth Deemed University College of Engineering, Pune	India
David Forsyth	UTM	United Kingdom
Dilla Duryha	Institute of Microengineering and Nanoelectronics (IMEN), UKM	Malaysia
Domenico Ciuonzo	University of Naples Federico II	Italy
Duu Sheng Ong	Multimedia University	Malaysia
EDS Malaysia Malaysia	Universiti Kebangsaan Malaysia	Malaysia
Ekaterina Pshehotskaya	Moscow State University	Russia
Faizah Abu Bakar	Universiti Malaysia Perlis	Malaysia
Haidawati Nasir	Universiti Kuala Lumpur	Malaysia
Hanim Hussin	University Technology MARA	Malaysia
Harikrishnan Ramiah	Universiti Malaya	Malaysia
Haslina Jaafar	Universiti Putra Malaysia	Malaysia
Heydar Toossian Shandiz	Ferdowsi University of Mashhad	Iran
Hing Keung Lau	Vocational Training Council	Hong Kong
Ibrahim Ahmad	Universiti Tenaga Nasional	Malaysia

LIST OF REVIEWERS

Name	Affiliation	Country
I-Cheng Chang	National DongHwa University	Taiwan
Ir. Hazian Mamat	Mimos Berhad	Malaysia
Iskandar Yahya	Universiti Kebangsaan Malaysia	Malaysia
Ismail Saad	Universiti Malaysia Sabah	Malaysia
Iwan Adhicandra	Bakrie University	Indonesia
John Dennis	Universiti Teknologi PETRONAS	Malaysia
Josip Music	University of Split	Croatia
Jumril Yunas	Universiti Kebangsaan Malaysia	Malaysia
Li Wah Thong	Multimedia University	Malaysia
Li-Cheng Wu	Taiwan Power Research Institute	Taiwan
Maizan Muhamad	Universiti Teknologi MARA	Malaysia
Maizatul Zolkapli	Universiti Teknologi MARA	Malaysia
Marinah Othman	Universiti Sains Islam Malaysia	Malaysia
Mastura Shafinaz Zainal Abidin	Universiti Teknologi Malaysia	Malaysia
Md Ali Rani	Universiti Putra Malaysia	Malaysia
Md Islam	International Islamic University Malaysia	Malaysia
Md Nabil	UNITEN	Malaysia
Md. Akhtaruzzaman Akhtaruzzaman	Universiti Kebangsaan Malaysia	Malaysia
Mehmet Ertugrul	Ataturk University	Turkey
Mohamed Abdelhalim	Arab Academy for Science, Technology & Maritime Transport	Egypt
Mohamed Atef	United Arab Emirates University, AlAin	United Arab Emirates
Mohammad Faiz Liew Abdullah	Universiti Tun Hussein Onn Malaysia (UTHM)	Malaysia
Mohd Amrallah Mustafa	Universiti Putra Malaysia	Malaysia
Mohd Khairuddin Md Arshad	Universiti Malaysia Perlis	Malaysia
Mohd Natashah Norizan	Universiti Malaysia Perlis	Malaysia
Mohd Nazim Mohtar	Universiti Putra Malaysia	Malaysia
Mohd Tafir Mustaffa	Universiti Sains Malaysia	Malaysia
Mohd. Zulhakimi Ab. Razak	Universiti Kebangsaan Malaysia	Malaysia
Montadar Taher	University of Diyala	Iraq
Muhamad Ramdzan Buyong	Universiti Kebangsaan Malaysia	Malaysia
Muhammad Ibrahimy	International Islamic University Malaysia	Malaysia
Muhammad Mokhzaini Azizan	Universiti Sains Islam Malaysia	Malaysia
Muzamir Isa	Universiti Malaysia Perlis	Malaysia

LIST OF REVIEWERS

Name	Affiliation	Country
Nafarizal Nayan	Universiti Tun Hussein Onn Malaysia	Malaysia
Noor Ain Kamsani	Universiti Putra Malaysia	Malaysia
Nor Farahidah Za'bah	International Islamic University Malaysia	Malaysia
Nor Hafizah Ngajikin	Universiti Tun Hussein Onn Malaysia	Malaysia
Norhafizah Burham	Universiti Teknologi MARA	Malaysia
Norhana Arsad	Universiti Kebangsaan Malaysia	Malaysia
Norhayati Soin	University of Malaya	Malaysia
Norhisam Misron	Universiti Putra Malaysia	Malaysia
Nowshad Amin	Universiti Kebangsaan Malaysia	Malaysia
Nurul Ezaila Alias	Universiti Teknologi Malaysia	Malaysia
P. Susthitha Menon	Universiti Kebangsaan Malaysia	Malaysia
Pin-Yu Chen	IBM Research	USA
Poming Lee	NCTU	Taiwan
Pooya Ghani	MAPNA Electric & Control, Engineering & Manufacturing Co. (MECO)	Iran
Puteri Sarah Mohamad Saad	Universiti Teknologi MARA	Malaysia
Rachit Patel	ABESIT Engineering College, Ghaziabad	India
Ratheesh Kumar Meleppat	University of California Davis	USA
Robiah Ahmad	Universiti Teknologi Malaysia	Malaysia
Rosario Morello	University Mediterranea of Reggio Calabria	Italy
Rosaura Palma-Orozco	Instituto Politécnico Nacional	Mexico
Rosminazuin Ab Rahim	International Islamic University Malaysia	Malaysia
S.M.A. Motakabber	International Islamic University Malaysia	Malaysia
Saadah Abdul Rahman	University of Malay	Malaysia
Sabrin Samsudin	Universiti Teknologi Mara	Malaysia
Samir Ladaci	National Polytechnic School of Algiers	Algeria
Satya Nagabhushana Rao Kamisetti	JNTUK	India
Sergey Biryuchinskiy	Vigitek, Inc.	USA
Sew Sun Tiang	Universiti Sains Malaysia	Malaysia
Shaharin Fadzli Abd Rahman	Universiti Teknologi Malaysia	Malaysia
Shahrir Rizal Kasjoo	Universiti Malaysia Perlis	Malaysia
Sharifah Fatmadiana Wan Muhamad Hatta	University of Malaya	Malaysia
Sharifah Md Yasin	Universiti Putra Malaysia	Malaysia
Sheroz Khan	Onaizah College of Engineering	Saudi Arabia
Siti Azlida Ibrahim	Multimedia University	Malaysia

LIST OF REVIEWERS

Name	Affiliation	Country
Siti Ibrahim	International Islamic University Malaysia	Malaysia
Siti Nooraya Mohd Tawil	Universiti Pertahanan Nasional Malaysia	Malaysia
Smain Femmam	University UHA	France
Sotiris Karachontzitis	Independent Authority for Public Revenue	Greece
Sreedharan Pillai Sreelal	Indian Space Research Organization	India
Suhana Mohamed Sultan	Universiti Teknologi Malaysia	Malaysia
Sukreen Hana Herman	Universiti Teknologi MARA	Malaysia
Sulaiman Wadi Harun	Uni Malaya	Malaysia
Syahrul Ashikin Azmi	Universiti Malaysia Perlis	Malaysia
Syarifah Abd. Rahim	Universiti Malaysia Pahang	Malaysia
Teddy Gunawan	International Islamic University Malaysia	Malaysia
Tuan Norjihan Tuan Yaakub	Universiti Teknologi Mara	Malaysia
Ulas Kilic	Ege University	Turkey
Umapathy Eaganathan	School of Technology	Malaysia
Usman Ullah Sheikh Izzat Ullah Sheik	Universiti Teknologi Malaysia	Malaysia
Wan Zuha Wan Hasan	Universiti Putra Malaysia	Malaysia
Wei Wei	Xi'an University of Technology	China
Wira Hidayat bin Mohd Saad	Universiti Teknikal Malaysia Melaka	Malaysia
Xiaoce Feng	Wayne State University	USA
Yaareb Al-Khashab	Ministry of Water Resources/Badush Dam	Iraq
Yasmin Abdul Wahab	Universiti Malaya	Malaysia
Zaharah Johari	Universiti Teknologi Malaysia	Malaysia
Zainal Nurbaya	Universiti Teknologi MARA	Malaysia
Zubaida Yusoff	Multimedia University	Malaysia
Zul Atfyi Fauzan Mohammed Napiah	Universiti Teknikal Malaysia Melaka (UTeM)	Malaysia
Zurita Zulkifli	Universiti Teknologi MARA	Malaysia

KEYNOTE 1

MICROELECTRONICS: A CONCISE OVERVIEW OF THE INDUSTRY LANDSCAPE IN MALAYSIA AND IN THE EMERGENCE OF IR 4.0.

Prof. Dato' Ts. Dr. Zaliman Sauli

Abstract: Microelectronics has been a prominent field for several decades since the introduction of the world's first transistor at Bell Laboratories in 1947. Over the years, significant progress has been made in microelectronics, including the development of the Integrated Circuit (IC), advancements in microfabrication processes, and the utilization of these technologies in the fabrication of MEMS and other advanced small-scale devices.

As the world embraces the era of Industrial Revolution 4.0 (IR 4.0), the importance of microelectronics in driving this revolutionary transformation becomes increasingly crucial. Malaysia has also been actively striving to keep pace with the ever-evolving microelectronics industry and semiconductor supply chain by fostering collaboration between relevant industries, government, and higher education institutions.

This talk aims to explore the evolutionary change of microelectronics in the IR 4.0 era with a glimpse of its application. The need for Microelectronic in the advent of IR 4.0 is highly required to fulfill the nation's aspiration to produce more sophisticated and holistic engineers and technologists.

Prof. Dato' Ts. Dr. Zaliman bin Sauli was born on 9 September 1967 in Kota Bharu, Kelantan. He furthered his studies at Universiti Teknologi Malaysia (UTM) in Physics followed by his Master's in Advanced Semiconductor Materials & Devices at the University of Surrey, United Kingdom and his Doctor of Philosophy (PhD) in Microelectronic Engineering at Universiti Malaysia Perlis (UniMAP).

Prof. Zaliman's career began at MIMOS in 1992 to 2002 where his final post there was as a Wafer Fabrication Product Manager and Training Services Manager. He then joined UniMAP on 09 January 2003 where he has been appointed as the Dean of the Centre for Communication and Entrepreneurship Skills, Director of the Centre for Industrial Collaboration, Dean of the School of Microelectronic Engineering, Director of the Co- Curriculum Centre and Dean of the Centre for Graduate Studies. Prof. Zaliman was appointed as the Deputy Vice Chancellor (Student & Alumni Affairs) of Universiti Malaysia Kelantan for a period of 3 years from 15 March 2018 until 14 March 2021. Beginning on August 09, 2021, Prof. Zaliman has been entrusted by the Ministry of Higher Education (MOHE) as the Vice Chancellor of UniMAP.

Prof. Zaliman has published over 200 WoS/Scopus indexed Journals and Proceedings as lead author and co -author and has also been a lead and co-researcher for 16 research grants. Some of his main research interests are in Wafer Fabrication Process Technologies, Device Characterization, Parametric and Functional Testing, Failure Analysis & Wafer Packaging, MEMS Technologies as well as Solid State & Theoretical Physics.

Currently, he is also serving as a Board Member for the Malaysia Board of Technologist (MBOT) beginning on the 22nd of July 2022 and is also a member of the national Specialized Task Force TVET (STF TVET).

KEYNOTE 2

MARKET AND TECHNOLOGY TRENDS OF ADVANCED PACKAGING

Dr. Tan Yik Yee
(Yole Intelligence, Malaysia)

Abstract: Advanced packaging has been rapidly growing in recent years, driven by the increasing demand in high-performance computing, artificial intelligence, and autonomous driving. It is getting traction from semiconductor industry as more than Moore solution to enhance system performance to enable higher device performance, increase bandwidth, offer lower latency, and lower power consumption. This keynote will give an overview of the market and technology trend in the advanced packaging. Other than that, the presentation will highlight the emerging trend of chiplet and how it drives the advanced packaging to attain heterogeneous integration. Advanced packaging players and their innovation direction and commercial product will be briefly discussed. Last, the presentation will highlight the importance of IC substrate in advanced packaging supply chain of semiconductor.

Dr. Tan Yik-Yee is a Senior Technology & Market Analyst, Semiconductor Packaging & Assembly at Yole Intelligent, within the Semiconductor, Memory & Computing division. Based in Malaysia, Yik Yee follows the semiconductor packaging industry and its evolution. Based on her technical expertise and market knowledge, she develops technology & market reports and is engaged in dedicated custom projects. Prior to Yole, Yik Yee Tan worked as a failure analyst and interconnect principal at Infineon Technologies (Malaysia) and later as an open innovation senior manager at Onsemi (Malaysia). While at Onsemi, Yik Yee was deeply involved in numerous innovative advanced packaging projects. Yik Yee Tan holds a Ph.D. in Engineering from Multimedia University (MMU, Malaysia).

KEYNOTE 3

GRAPHENE NANOBALLS FOR PERFORMANCE IMPROVEMENT OF THERMOELECTRIC ENERGY HARVESTER

Prof. Dr. Azrul Azlan Hamzah

Abstract: Renewable energy has been the center of attention in sustainable energy research for the past decade, as it is in line with the Sustainable Development Goals (SDG) of the United Nations. It supports SDG 7: affordable and clean energy, SDG 11: sustainable cities and communities, and SDG 13: climate action. Among the renewable energy sources, thermoelectric energy (TE) harvester stands out as a clean and environmental friendly energy source as it directly converts waste heat into electrical energy. Upon successful implementation, TE harvester would greatly reduce world's dependency on fossil fuel, promotes clean energy conversion and supply for domestic and industrial use, while reducing global carbon footprint and greenhouse effect. In this context, our prototype increases the thermoelectric conversion efficiency of a TE harvester by infusing graphene nanoballs into bismuth telluride (Bi_2Te_3) thermoelectric generator (TEG). The graphene nanoballs increase the total ZT value of the Bi_2Te_3/graphene nanoballs composite, resulting in a better performance TEG. In our laboratory prototype, the ZT value increased by 22.7%, which plausibly increases TEG efficiency from the typical 8% to 11%, thus pushing this Bi_2Te_3/graphene nanoballs TEG into a commercially viable product.

Azrul Azlan Hamzah (Senior Member, IEEE) received the B.S. degree in manufacturing engineering from the University of California (UC) at Berkeley, USA, in 2000, and the Ph.D. degree in microelectromechanical systems from Universiti Kebangsaan Malaysia in 2008. He is currently pursuing the M.S. degree in clinical microbiology and infectious diseases with the University of Edinburgh, U.K. He is also a Professor of microelectromechanical systems (MEMS) and the former Director of the Institute of Microengineering and Nanoelectronics (IMEN), Universiti Kebangsaan Malaysia. His research interests include bioMEMS, biomedical devices, artificial kidney, biosensor, and microenergy. He leads several projects funded by the Ministry of Science, Technology and Innovation Malaysia (MOSTI) and the Ministry of Higher Education Malaysia (MoHE) in the development of MEMS and bioMEMS devices. He is a Senior Member of the Electron Devices Society of IEEE Malaysia Section.

PROCEEDINGS

2023 IEEE REGIONAL SYMPOSIUM ON MICRO AND NOELECTRONICS (RSM)

28 – 29 AUGUST 2023 Langkawi, Malaysia

RSM 2023

Organized by:

PUBLICATION CONTACT

AHM Zahirul Alam
Faculty of Engineering
International Islamic University Malaysia
Jalan Gombak, 53100 Kuala Lumpur
Malaysia
Tel: +6 03 6421 4529
Email: zahirulalam@iium.edu.my
web: https://zahirulalam.staffat.iium.edu.my/

Performance Analysis of 14nm SOI-based Trigate Gaussian Channel Junctionless FinFET with Punchthrough Stop Layer

Mathangi Ramakrishnan
Faculty of Electrical Engineering,
Universiti Teknologi Malaysia,
81310 Johor Bahru, Malaysia
rmathangi@graduate.utm.my

N. Ezaila Alias*
Faculty of Electrical Engineering,
Universiti Teknologi Malaysia,
81310 Johor Bahru, Malaysia
ezaila@fke.utm.my

M. L. Peng Tan
Faculty of Electrical Engineering,
Universiti Teknologi Malaysia,
81310 Johor Bahru, Malaysia
michael@utm.my

Afiq Hamzah
Faculty of Electrical Engineering,
Universiti Teknologi Malaysia,
81310 Johor Bahru, Malaysia
mafiq@fke.utm.my

Yasmin Abdul Wahab
Nanotechnology & Catalysis Research
Centre,
Universiti Malaya,
50603 Kuala Lumpur, Malaysia
yasminaw@um.edu.my

Hanim Hussin
School of Electrical Engineering,
College of Engineering,
Universiti Teknologi MARA,
40450 Shah Alam, Malaysia
hanimh@uitm.edu.my

Abstract—In this paper, 14nm Silicon-On-Insulator-based Gaussian Channel Junctionless FinFET is presented. The gate length of 14nm is considered along with an Equivalent Oxide Thickness (EOT) of 1nm, 5nm as fin width, and the work function of the gate metal is 4.75eV. The device architecture has a non-uniform doping profile (Gaussian distribution) across the fin's thickness. The results show that the I_{on}=101.5μA/μm and I_{on}/I_{off} is 3.2x10^7, DIBL=25.3 mV/V and Subthreshold Swing (SS) = 63.88 mV/dec are obtained. Thus, the Gaussian Channel-based FinFET architecture can provide optimum results for Junctionless-based FinFET devices. Further, to limit the parasitic leakage current in SOI-based FinFETs, possible solutions such as the Punch-Through Stop layer are also examined in this work, and about 36.8% of leakage current is reduced.

Keywords— SOI FinFET, Gaussian Channel, Junctionless FinFET, PTS layer.

I. INTRODUCTION

As MOSFET size decreases, the impact of Short Channel Effects (SCEs) becomes more pronounced, negatively affecting device performance. Various solutions have been proposed to tackle these issues, including Silicon On Insulator (SOI) technology, and Multi Gates FET architectures [1-2]. In the nanoscale regime, the fabrication of transistors becomes challenging due to the complexity added by junctions and the required thermal conditions. To address these difficulties, Junctionless Transistors have been introduced as an intriguing concept. These transistors eliminate the need for p-n junctions and offer a simplified fabrication process [3-5]. Junctionless Transistors operate through bulk conduction, which distinguishes them from other transistors like accumulation and inversion mode FETs that rely on surface conduction. This bulk conduction mode enhances performance in Subthreshold and reduces surface scattering effects [6]. The multi-gate Junctionless Transistors structures such as Double Gate FETs [7] and Tri-Gate FETs have been developed in which the Tri-Gate architecture has better controllability of gate, reduced DIBL, Subthreshold Swing and increased ON current [8].

The use of high-k spacers has been shown to decrease the fringing field, leading to an improvement in the OFF-state current without impacting the ON current [9-10]. Additionally, Nawaz et al. investigated the impact of quantum capacitance on Junctionless FinFETs, specifically focusing on the threshold voltage variability caused by Random Dopant Fluctuations [11].

The JL-FinFET offers several inherent advantages, but achieving high uniform doping in the device layer is a critical challenge during fabrication. This complexity becomes even more pronounced in non-planar architectures like FinFET, where the doping must be applied in a three-dimensional pattern around the fin region, resulting in non-uniform doping [12]. The use of a Gaussian doping profile is a common approach, where other doping distributions can be derived by adjusting specific parameters of the Gaussian distribution as needed [13]. Previous studies have explored the performance and operation of Junctionless concept-based FinFETs with heavily doped structures. Recently, Gaussian channel doping profile in Junctionless Transistors are proposed [14-15].

In this work, a Gaussian Channel-based Junctionless FinFET structure simulation in a 14nm technology is presented. The device structure is simulated and analyzed using Silvaco TCAD simulation tool, and the IV characteristic is compared to the uniformly doped Junctionless FinFET structure. This paper presents a thorough investigation into Gaussian doping Junctionless FinFETs, aiming to provide a solution for reducing the device's leakage current by introducing a Punch-Trough Stop (PTS) layer.

II. DEVICE STRUCTURE AND SIMULATION

A. Device Validation

The 3D schematic view of the proposed n-type Junctionless FinFET architecture used for simulation is shown in Fig 1. The detailed device parameters list is provided in Table 1. The source and the drain (S/D region) have the same n-type doping concentration of 4 x 10^{19} cm^{-3}

for all the cases. In uniform doping distribution, the channel has the same doping concentration of 4 x 10^{19} cm^{-3}, similar to the S/D regions. However, the S/D regions are highly n-doped, and the channel has non-uniform n-type doping of 4 x 10^{19} cm^{-3} across the thickness of the fin.

TABLE I.   SIMULATION DEVICE PARAMETERS OF THE 14NM SOI GAUSSIAN CHANNEL JUNCTIONLESS FINFET

Device parameters	[15]	This work
Gate Length, L_g	14nm	14nm
Fin Width, F_w	8nm	5nm
Fin Height, F_h	N/A	42nm
EOT	1nm	1nm
S/D Doping, N_{sd}	2.5 x 10^{19} cm^{-3}	4.0 x 10^{19} cm^{-3}
V_{DS}	0.70V	0.70V
Spacer width	(2-22) nm	10nm

Junctionless FinFETs were simulated using the Silvaco TCAD simulation tool. The simulation incorporated the Bohr Quantum Potential model to account for quantization effects and improve result accuracy. The density gradient model was included in the quantum transport equation. For highly doped JLFinFETs, the Band Gap Narrowing (BGN) model accurately modelled the bipolar current gain. Additionally, the Band-To-Band-Tunneling (BTBT) model and Shockley Read Hall (SRH) and Auger models were used to consider band-to-band tunnelling and recombination effects. Lastly, the Lombardi CVT model was introduced in the simulation to account for carrier mobility.

For fair validation, the device is carefully validated using the experimental results for 14nm Conventional Inversion Mode FinFET structure using the 3D TCAD simulator [15]. Precise calibration can be achieved by implementing the exact device parameters and dimensions and doping in TCAD simulation to match current-voltage (I-V curves). After which, the mobility models can be varied to achieve the electrical characteristics of the proposed work. The simulated IV characteristics of both n- and p-type FinFET exhibit good compliance with the experimental results of work in [15] as depicted in Fig. 1 by validating the results of the device simulations.

Fig. 2 shows the net doping concentration of 14nm Inversion-Mode FinFET, Uniformly Doped Junctionless FinFET and Gaussian Channel Junctionless FinFET. The IM-FinFET has p-n junctions, resulting in a Gaussian doping profile with a junction. In Uniformly doped Junctionless FinFETs, the doping profile is uniform throughout the device structure. However, in Gaussian Channel Junctionless FinFETs (GC-JLFinFET), the doping concentration peaks at the fin's sidewalls and gradually decreases towards the centre. The decrement of doping concentration at the channel centre leads to the increment of electron mobility during the ON state (V_{gs}=0V, V_{ds}=0.7V). Additionally, the depletion of electrons at the fin sidewalls is reduced when the device is in the OFF state (V_{gs}=0V, V_{ds}=0.7V).

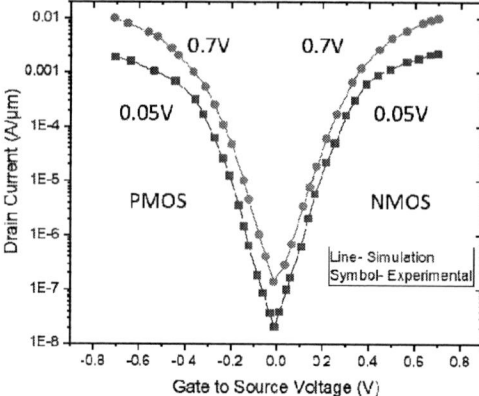

Fig. 1.   Validation of the simulation with the experimental data for 14nm IM-FinFET structure [15].

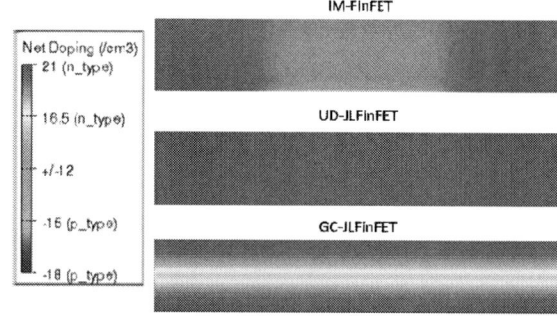

Fig. 2.   Cross-Sectional view of the channel in Inversion mode FinFET, Uniformly Doped Junctionless FinFET and Gaussian Channel Junctionless FinFET mode at V_{gs}=1V, V_{ds}=0.7V.

The spread of doping across the channel also reduces electron crowding, which is a major issue in FETs as it causes impurity scattering due to Random Dopant Fluctuation (RDF) in the device. Hence, the 14nm SOI-based GC-JLFinFET exhibits reduced impurity scattering caused by RDF and shows improved electrical and electrostatic performance compared to conventional FinFET structures.

B. Device Performance Analysis

The simulated data indicate that the Gaussian Channel JL-FinFET performs better with a lower gate dielectric constant in the Subthreshold. Various dielectric materials with dielectric constant such as k=3.9, 7.5,11 and 22 (SiO$_2$, Si$_3$N$_4$, HfSiO$_4$ and HfO$_2$) are considered for the analysis. As the Junctionless device operates in bulk conduction mode, the gate leakage current will not show much effect due to the presence of low-k dielectric material such as SiO$_2$. The GC-JLFinFET device shows an exemplary gate leakage conduct when it is in OFF condition (i.e., 5.676x10^{-18}A/μm) with Silicon dioxide of k=3.9 as the gate dielectric. Therefore, combining the low-k gate dielectric constant with the high-k spacer dielectric in the 14nm GC-JLFinFET holds great promise for achieving improved performance. The specific values can be found in Table 2. The I_{on} extraction is based on V_{gs}=0.4V; V_{ds}=0.65V and I_{off} extraction is based on V_{gs}=0V; V_{ds}=0.65V. Drain-induced Barrier Lowering (DIBL) is defined as the difference in threshold voltage (V_{th}) at linear

979-8-3503-2369-6/23 $31.00 © 2023 IEEE

and saturation regions over the difference in drain voltage (V_{DS}) at linear and saturation regions. Subthreshold swing (SS) is defined as the change in gate bias (V_G) required to change the subthreshold drain current by one decade at the steepest slope on log plot I_D-V_G.

TABLE II. PERFORMANCE ANALYSIS OF DIFFERENT DOPING CONFIGURATIONS OF 14NM JUNCTIONLESS FINFET.

Device parameter	Gaussian Channel JL-FinFET [21]	Uniformly JL-FinFET simulation	Gaussian Channel JL-FinFET simulation
Ion (A/µm)	4.84×10^{-6}	1.02×10^{-6}	1.78×10^{-5}
Ioff (A/µm)	N/A	4.57×10^{-10}	5.54×10^{-10}
Ion/Ioff	1.1×10^{7}	2.2×10^{5}	3.2×10^{7}
DIBL (mV/V)	53.6	54.52	25.3
SS (mV/dec)	77.7	70.2	63.88
Transconductance, g_m (S/µm)	N/A	1.28×10^{5}	3.62×10^{5}

C. Leakage Current Suppression and Analysis of Punch-Through Stop Layer

It was found that even with the optimal parameters, there is no significant reduction in the leakage current of the device. The OFF current is crucial for the performance of the device when it is utilized in digital or analog circuits. Therefore, this section will explore effective methods to mitigate the leakage current when the device is in the OFF state. One potential solution to address this issue is the implementation of a Punch-Through Stop (PTS) layer, and its properties will be examined to evaluate its impact on the suggested device.

Punch-through leakage current causes high leakage current in bulk or SOI FinFETs. This mainly occurs in the region below the channel, and it is considered a part of S/D leakage because of the drain voltage, V_d. This issue can be cancelled out by placing a PTS layer with highly counter-doped impurity underneath the channel. The layer with p-type doping ranges from 10^{13} to 10^{18} cm^{-3} in SOI FinFET, as shown in Fig. 3.

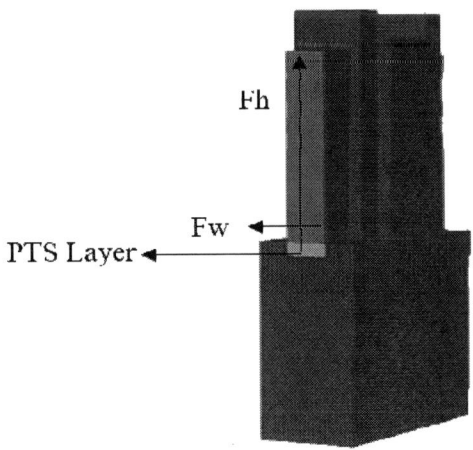

Fig. 3. The schematic diagram of 14nm SOI Junctionless FinFET with Punch-Through-Stop Layer beneath the bottom of the fin.

I_{on} and I_{off} for 14nm SOI GC-JLFinFET with varying PTS doping concentrations were plotted in Fig. 4. It can be

seen that the PTS layer does not have any impact on the ON current of the device and suppresses the leakage by 36.8% in Gaussian Channel Junctionless FinFET. Likewise, the ON-OFF current ratio also increased from 4.3×10^3 to 2.4×10^6 for the proposed device structure. Thus a commendable electrostatic performance is obtained by adjusting the PTS doping concentration.

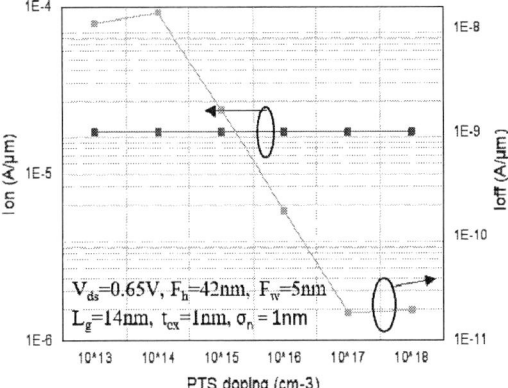

Fig. 4. The plot of I_{on} and I_{off} of 14nm SOI GC-JLFinFET with varying PTS doping concentration.

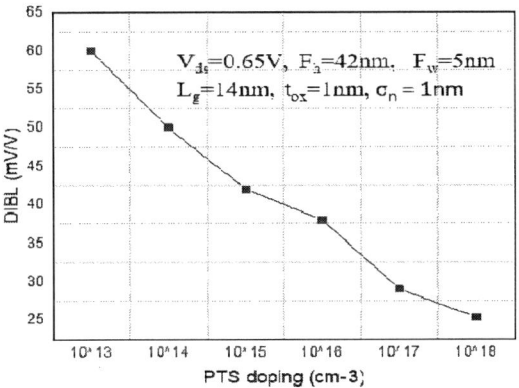

(a)

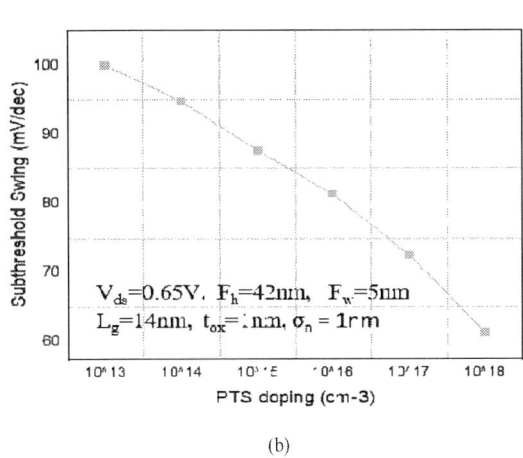

(b)

Fig. 5. The change in (a) DIBL and (b) Subthreshold Swing with respect to the doping concentration of PTS layer.

The change in DIBL and SS with respect to the doping concentration of the PTS layer is illustrated in Fig. 5 (a) and (b). The SS is reduced by 42.67% for the proposed structure when the doping increases from 10^{13} to 10^{18} cm^{-3}. Thus, it is possible to achieve enhanced short-channel effects with a rising doping concentration of the PTS layer, and the corresponding performance values are mentioned in Table 3. The I_{on} extraction is based on V_{gs}=0.4V; V_{ds}=0.65V and I_{off} extraction is based on V_{gs}=0V; V_{ds}=0.65V.

TABLE III. PERFORMANCE ANALYSIS OF DIFFERENT DOPING CONFIGURATIONS OF PUNCH-THROUGH-STOP-LAYER FOR VARIOUS ELECTRICAL PARAMETERS FOR 14NM SOI GC-JLFINFET STRUCTURE.

Device Parameters / Doping Configuration	10^{13} cm^{-3}	10^{14} cm^{-3}	10^{15} cm^{-3}
I_{on} (A/μm)	1.78 x 10^{-5}	1.78 x 10^{-5}	1.78 x 10^{-5}
I_{off} (A/μm)	1.39 x 10^{-8}	1.97 x 10^{-8}	1.64 x 10^{-9}
I_{on} / I_{off}	1.60 x 10^{3}	2.36 x 10^{3}	1.09 x 10^{4}
DIBL (mV/V)	60.11	50.32	42
SS (mV/dec)	100	94.8	87.6
Device Parameters / Doping Configuration	10^{16} cm^{-3}	10^{17} cm^{-3}	10^{18} cm^{-3}
I_{on} (A/μm)	1.78 x 10^{-5}	1.78 x 10^{-5}	1.78 x 10^{-5}
I_{off} (A/μm)	1.76 x 10^{-10}	1.85 x 10^{-11}	1.96 x 10^{-11}
I_{on} / I_{off}	1.01 x 10^{5}	0.97 x 10^{6}	1.21 x 10^{6}
DIBL (mV/V)	38.63	29	23.3
SS (mV/dec)	81.32	72.5	61.32
I_{on} (A/μm)	1.78 x 10^{-5}	1.78 x 10^{-5}	1.78 x 10^{-5}

III. CONCLUSION

This paper investigated the Gaussian doped channel through 3-D simulations. The possible solutions for lowering the leakage current were also analyzed by implementing Punch Through Stop Layer beneath the channel for the proposed 14nm SOI-based Gaussian Channel-based Junctionless FinFET. The proposed design showed better efficiency in short channel characteristics, namely DIBL = 25.3 mV/V and SS = 63.88 mV/dec. The simulated results also show that the Gaussian Channel JL-FinFET exhibited enhanced ON current along with improved Short Channel Characteristics and an overall improvement in the device's performance by 23.8%. It can be concluded that the presented architecture can be used in low-power digital circuit applications with the optimum design parameters. Therefore, the presented GC-JLFinFET structure is optimized for potent device architecture in the sub-20 nm regime.

ACKNOWLEDGMENT

Authors would like to acknowledge the financial support under the UTM Fundamental Research Grant Project No. Q.J130000.3823.22H52. Also, thanks to the Research Management Center (RMC) of Universiti Teknologi Malaysia (UTM) for providing an excellent research environment in which to complete this work.

REFERENCES

[1] Ferain I, Colinge CA, Colinge JP. Multi-gate transistors as the future of classical metal–oxide–semiconductor field-effect transistors. Nature.; 479(7373):310-6, Nov 2011.

[2] Fossum JG, Zhou Z, Mathew L, Nguyen BY. SOI versus bulk-silicon nanoscale FinFETs. Solid-State Electronics.1;54(2):86-9, Feb 2010

[3] Liu X, Wu M, Jin X, Chuai R, Lee JH, Lee JH. The optimal design of 15 nm gate-length junctionless SOI FinFETs for reducing leakage current. Semiconductor science and technology. 16;28(10):105013, Aug 2013.

[4] Lee CW, Ferain I, Afzalian A, Yan R, Akhavan ND, Razavi P, Colinge JP. Performance estimation of junctionless multigate transistors. Solid-State Electronics. 1;54(2):97-103, Feb 2010.

[5] Ionescu AM. Nanowire transistors made easy. Nature nanotechnology.;5(3):178-9, Mar 2010

[6] Ichii M, Ishida R, Tsuchiya H, Kamakura Y, Mori N, Ogawa M. Computational study of effects of surface roughness and impurity scattering in Si double-gate junctionless transistors. IEEE Transactions on Electron Devices 17; 62(4):1255-61, Feb 2015.

[7] Liu X, Wu M, Jin X, Chuai R, Lee JH. Simulation study on deep nanoscale short channel junctionless SOI FinFETs with triple-gate or double-gate structures. Journal of Computational Electronics, 13(2):509-14, Jun 2014.

[8] Doria RT, Pavanello MA, Trevisoli RD, de Souza M, Lee CW, Ferain I, Akhavan ND, Yan R, Razavi P, Yu R, Kranti A. Junctionless multiple-gate transistors for analog applications. IEEE Transactions on Electron Devices, 27; 58(8):2511-9, Jun 2011.

[9] Sule MA, Ramakrishnan M, Alias NE, Paraman N, Johari Z, Hamzah A, Tan ML, Sheikh UU. Impact of Device Parameter Variation on the Electrical Characteristic of N-type Junctionless Nanowire Transistor with High-k Dielectrics. Indonesian Journal of Electrical Engineering and Informatics (IJEEI), 30; 8(2):400-8, Jun 2020.

[10] Alias NE, Sule MA, Tan ML, Hamzah A, Saidu KA, Mohammed S, Aminu TK, Shehu A. Electrical characterization of n-type cylindrical gate all around nanowire junctionless transistor with SiO2 and high-k dielectrics. In 2020 IEEE International Conference on Semiconductor Electronics (ICSE) 28 (pp. 13-16). IEEE], Jul 2020.

[11] Nawaz SM, Mallik A. Role of quantum capacitance on the random dopant fluctuation induced threshold voltage variability in junctionless InGaAs FinFETs. Solid-State Electronics.1;171:107862, Sep 2020

[12] Parihar MS, Kranti A. Revisiting the doping requirement for low power junctionless MOSFETs. Semiconductor Science and Technology. 28; 29(7):075006, Apr 2014.

[13] Dubey S, Tiwari PK, Jit S. A two-dimensional model for the potential distribution and threshold voltage of short-channel double-gate metal-oxide-semiconductor field-effect transistors with a vertical Gaussian-like doping profile. Journal of Applied Physics. 1; 108(3):034518, Aug 2010.

[14] Mondal P, Ghosh B, Bal P. Planar junctionless transistor with non-uniform channel doping. Applied Physics Letters.1; 102(13):133505, Apr 2013.

[15] Sze SM, Li Y, Ng KK. Physics of semiconductor devices. John Wiley & Sons; 19, Mar 2021.

Thermal Conductivity of Stacked Hexagonal Boron Nitride (hBN) and Graphene – A Molecular Dynamics Approach

Dharma Darren Ram
Institute of Microengineering and Nanoelectronics
Universiti Kebangsaan Malaysia
43600, Bangi, Selangor
P116360@siswa.ukm.edu.my

Muhammad Aniq Shazni Mohammad Haniff*
Institute of Microengineering and Nanoelectronics
Universiti Kebangsaan Malaysia
43600, Bangi, Selangor
*aniqshazni@ukm.edu.my

Abdul Manaf bin Hashim
Malaysia−Japan International Institute Of Technology
Universiti Teknologi Malaysia
81310 Skudai, Johor
abdmanaf@utm.my

Mohd Ambri Mohamed
Institute of Microengineering and Nanoelectronics
Universiti Kebangsaan Malaysia
43600, Bangi, Selangor
ambri@ukm.edu.my

Abstract— **Developing solutions to keep up with the needs of increased power output of contemporary microprocessors is an ongoing challenge in the electronics industry. As such, thermal interface materials, which act as a filler to smooth out the contact imperfections between heat source and heat sink have been an important area of research. Two–dimensional (2D) materials may be a solution to having a material that has high thermal conductivity, flexibility, and a long service life. Although highly thermally conductive, the electrical conductivity graphene makes it unsuitable for use directly adjacent to the active layer in electronics. Hexagonal boron nitride (hBN) has attracted attention for use as an insulating layer due to its structural similarity to graphene with a lattice mismatch of only 1.8%. In this research, equilibrium molecular dynamics (EMD) via the Green–Kubo (GK) method is used to calculate the thermal conductivity of a hexagonal boron nitride/graphene (hBN/Gr) heterostructure. It is thought that replacing the secondary hBN layer would increase the thermal conductivity of the structure.**

Keywords—graphene, hexagonal boron nitride, van der Waals stacked structure, Green–Kubo method

I. INTRODUCTION

The latest generation of microchips has become smaller yet more powerful, leading to increased cooling requirements. Server grade CPUs, for example, now have TDPs (thermal design power) of 360 – 400 W and with this increased power consumption comes increased heat load to be dissipated [1]. When two materials come into contact, surface roughness leads to imperfect contact that hinders efficient heat transfer. Thermal interface materials (TIMs) are used to fill these gaps and enhance heat conductivity [2, 3]. Selecting the right TIM involves considering various factors most crucially, the material's thermal conductivity, as it determines how effectively heat can be transferred across the interface. Electrical conductivity is also important because the TIM must contact the microchip's active layer without interfering with its function. Additionally, the TIM should be reliable over the expected service life of the component, which can be on the order of decades [4]. The state of the art for TIM technology includes thermal greases (0.9 W/m·K – 5 W/m·K), phase change materials (20 W/m·K) [5], and thermal pads (0.24 W/m·K – 0.56 W/m·K) [6, 7]. While functional, they fall short of the potential offered by 2D materials with thermal conductivity values in the range of 2000 – 5000 W/m·K [8].

In this study, we use molecular dynamics (MD) to investigate the thermal transport in the proposed hBN/Gr structure where replacing the secondary hBN layer with graphene would enhance the structure's thermal conductivity [29]. The method chosen is equilibrium molecular dynamics (EMD) via the Green–Kubo (GK) due to its relative insensitivity to size effects which enable smaller simulation domains, and thus lowered computational power to be used compared to non-equilibrium methods. This work also aims to explain the disparity in thermal conductivity values in the context of the phononic coupling between the individual layers in BL-Gr, BL-hBN and hBN/Gr. Dispersion curves obtained from density functional theory (DFT) studies are used to compare the behaviour of phonons of the stacked structure against that of the isolated layers. Significant changes in the low frequency and low wave–vector regions are noticeable and are attributed to layer interactions. It was found that, SL-Gr, SL-hBN, BL-Gr, BL-hBN, and hBN/Gr presented thermal conductivity values of 2402.03, 51.94, 1480.28, 210.75, and 586.93 – 669.37 W/m·K, respectively.

II. SIMULATION SET UP AND METHODS

Classical molecular dynamics is used to predict the thermal conductivity of the heterostructure. This method calculates the time–dependent position of every individual atom in the simulation box using Newton's 2nd Law and predefined interatomic potentials. This research uses EMD through the GK method which calculates the thermal conductivity, λ, by integrating the heat flux vectors over time [9, 10].

$$\lambda = \frac{V}{k_B T^2} \int_{-\infty}^{\infty} \{J(t) \cdot J(0)\} \, dt \qquad (1)$$

The in–plane thermal conductivity of SL-Gr is calculated by averaging the thermal conductivity in the x and y directions (λ_{xx} and λ_{yy}) using periodic boundary conditions. This calculation involves the system volume (V), the

Boltzmann constant (k_B), and the heat flux ($J(t)$) in the x, y, or z direction. Periodic boundary conditions are used in the x and y directions. Sufficient convergence for thermal conductivity values using the Green–Kubo method required a large simulation time; thus, this study uses 2.5 ns which has been observed to achieve convergence.

The simulations were conducted using the Large-scale Atomic/Molecular Massively Parallel Simulator (LAMMPS) [11]. EMD simulations were performed to calculate the thermal conductivity of the samples. The equations of motion were integrated using the Verlet velocity algorithm. Initial velocities were randomized using an initial temperature of 300 K. The data files were created using the Open Visualizer Tool [12] (OVITO) and Visual Molecular Dynamics (VMD) [13]. The intermolecular potentials used to describe the interactions are Tersoff [14] and AIREBO [15] for intralayer interactions and ILP.BNCH [16–18] for interlayer interactions. The Coulombic interactions between boron and nitrogen were accounted for by the coul/shield parameter with a cut–off distance of 16 Å, to accurately simulate the behaviour of hBN. The x and y directions were set as the in–plane direction with periodic boundary conditions, while the z direction was set as the normal with fixed boundary conditions. The timestep was set to 0.5 fs. The system was equilibrated twice. Initially, in the isothermal–isobaric ensemble for 1.25 ns, then in the canonical ensemble for 2.5 ns. This is achieved using a Nosé–Hoover thermostat [19, 20]. Finally, the system is run in the NVE ensemble to compute the thermal conductivity at 300 K. The stacking configuration is shown in Fig. 2. Bernal stacking (AB) was chosen as it more realistically represents real world stacking especially if no specific attention is focused to alignment of the grains.

The thermal conductivity of all the samples were computed using the Green–Kubo method as described in the previous section. The values for the thermal conductivity presented in the table at the end of the section are obtained by evaluating the average of 5 independent simulations once the heat flux autocorrelation function converges to relatively constant values. Fig. 1 shows the graph of thermal conductivity convergence. The red section is averaged over to obtain the thermal conductivity used.

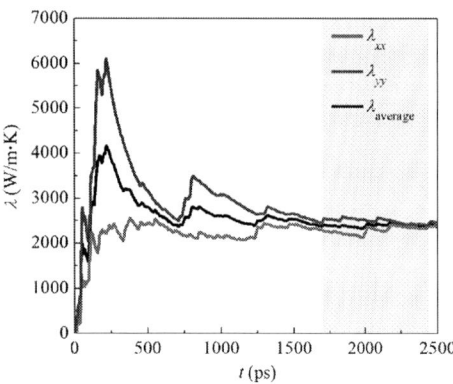

Fig. 1: Convergence of 50 × 50 Å models of SL-Gr. The method was repeated for all other samples. The yellow region is the values averaged over to obtain average thermal conductivity.

III. RESULTS AND DISCUSSION

The small size would allow some phonon frequencies, especially large wavelength phonons to pass multiple times through the material without scattering, which would impact the thermal conductivity. Small domain sizes would also cause fewer scattering events leading to high incidences of phonons travelling ballistically which would cause thermal conductivity to inflate to values much higher than what could be observed in reality [21]. The 50 × 50 Å square sheet was chosen because the thermal conductivity difference was negligible, at the cost of significantly higher processing time.

The individual phonon modes (longitudinal, transverse, and flexural) each contribute to the thermal conductivity of the material. The trend of decreasing thermal conductivity with increasing number of layers is due to the suppression of the flexural acoustic phonon mode which is thought to be the main contributor to the overall thermal conductivity [22]. Fig. 2(a, b) shows the Bernal (AB) stacking configuration, which was used because unless special care is taken to align the atoms, the atoms will more commonly fall into this position. Another factor influencing the thermal conductivity is phonon scattering at the interfacial boundary [23]. Fig. 2(c) shows the interfacial boundary for the hBN/Gr structure. As the phonon traverses the interlayer boundary, it may experience one or more scattering effects, such as specular reflection or diffuse scattering which would cause the phonon wave to lose coherence and affect the thermal conductivity.

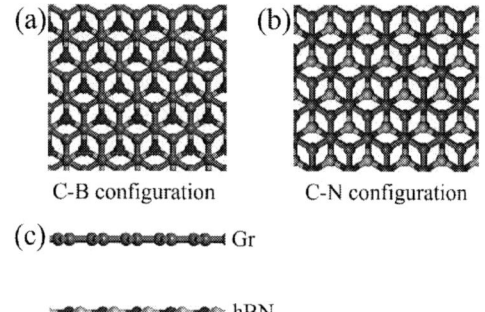

Fig. 2: The material boundary for the stacked bilayer. Schematic representation of the AB stacked hBN/Gr heterostructure with (a) C-B and (b) C-N configuration. (c) Interfacial boundary between layers in the zigzag direction.

Some recent studies investigate the thermal transport in the SL-Gr and multilayer graphene (ML-Gr) using the exact solution of the Boltzman transport equation (BTE). The effect of the number of layers was shown to saturate at around $n = 10$ to the value of the bulk material. However, the effects of adding individual layers have not been studied, especially the few−layer ($1 < n < 3$) configuration. In this section, we calculate the dispersion relations by means of DFT and explain the behaviour of the stacked structure. For the current investigation, we employ an ultrasoft pseudopotential within the framework of the generalized gradient approximation (GGA) to simulate the lattice dynamics of hBN. The calculations are performed using a theoretical lattice parameter of $a = 2.504$ Å, which closely matches the experimental value at a temperature of 300 K. From Fig. 3(a, b), it can be observed that the effects of the additional layer are absent in the longitudinal (LA) and transverse acoustic

(TA) modes, as well as all three (longitudinal (LO), transverse (TO), and flexural (ZO)) optical modes. However, the flexural acoustic (ZA) modes in the bilayer samples are impacted, which can be attributed to the secondary layer (in red). An additional mode in the low frequency (small wavevector region) corresponds to an out–of–phase vibration [26]. Previous work has suggested that there exists a cut–off frequency below which the coupling with the substrate phonons and cross–plane phonons become strong in the supported material (SL-Gr in the reference, however it can be assumed that hBN would behave similarly) [27, 28]. Consequently, the energy of the phonons is released to the secondary layer, leading to a significant decrease in the involvement of these low–frequency phonons in heat transfer. Fig. 3(d) shows a threshold frequency of 86 cm^{-1} for the ZA phonon modes in the bilayer structure. This threshold frequency is identified by the region where the individual phonon modes converge. Considering that the ZA phonons reach a maximum frequency of 313 cm^{-1} at the M point, this threshold frequency encompasses 30% of the permissible ZA phonon modes in the direction from G to M which is in agreement with the measured experimental value of BL-hBN ($\lambda = 484$ W/m·K) [32] against the theoretical value of pristine SL-hBN ($\lambda = 550$ W/m·K) (see Table 1) [34].

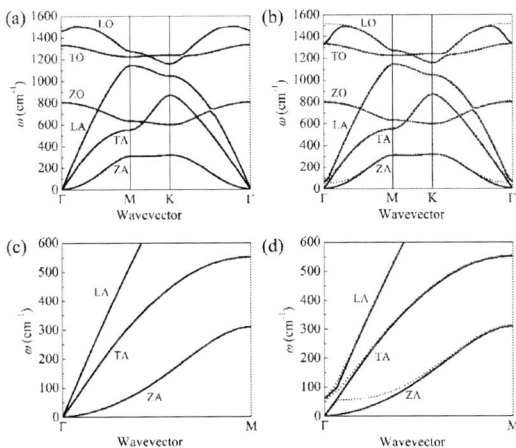

Fig. 3: Phonon dispersion curves of (a, c) SL-hBN and (b, d) BL-hBN in the high symmetry directions.

Previous phononic studies have shown that using graphene and hBN together could result in the improvement of the thermal conductivity of hBN [29] while keeping graphene electrically insulated and able to be used in nanoscale electronics. Notably, the obtained SL-hBN thermal conductivity values of $5 - 52$ W/m·K do not agree with the current literature values as opposed to the obtained values of hBN/Gr and BL-hBN. It is believed that this discrepancy is caused by unoptimized Tersoff parameters used similar to Matsunaga et al. [36, 37]. Furthermore, Jo et al. demonstrated that the thermal conductivity of multilayered hBN decreases with increasing number of layers (360 to 250 W/m·K for 11 to 5 layers) (see Table 1) [32]. Linearly extrapolating this data results in a thermal conductivity of 195 W/m·K for BL-hBN which agrees well with the obtained thermal conductivity. It can be assumed that the correction factors in the inter- and intra-molecular used successfully optimize the parameters leading to results which are in line with expectations. The

thermal conductivity calculations of hBN/Gr present a value of $586.93 - 669.37$ W/m·K, which is between that of SL-hBN ($550 - 650$ W/m·K) and SL-Gr (550 W/m·K) [34, 35]. The decrease in thermal conductivity from that of SL-Gr can be attributed to the interfacial coupling through the van der Waals (vdW) bonding between the layers. The peaks of the LA modes at the K point (983 cm^{-1}) also suggest that graphene has a higher thermal conductivity. The difference in the Debye temperature (1813 K for SL-Gr and 1700 K for SL-hBN) implies that the phonon group velocity in hBN is lower than that of graphene as shown in Fig. 4(a, b). Although the thermal conductivity would be dominated by the contribution of graphene, the difference is not extremely profound due to the similar lattice constants and unit cell mass.

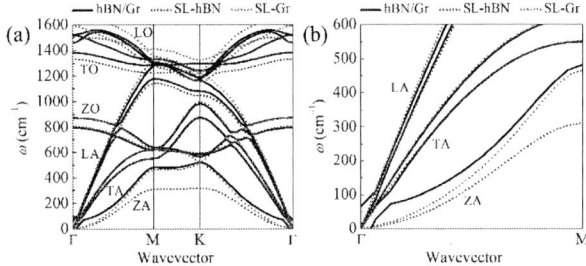

Fig. 4: The phonon dispersion of (a, b) the vdW stacked hBN/Gr heterostructure, with the dispersion of the constituent SL-Gr (blue dots) and SL-hBN (red dots) plotted alongside.

Table 1. Thermal conductivity as measured in this work and literature.

Mat.	Properties		
	Dimensions (Å)	Potentials (inter/intra)	λ (W/m·K)
SL-Gr	10×10	Tersoff	2570.26
	50×50		2402.03
	100×100		2434.93
SL-hBN	10×10		5.75
	50×50		51.94
	100×100		49.76
BL-Gr	50×50	Tersoff/ ILP.BNCH	1480.28
BL-hBN	50×50	Tersoff/ ILP.BNCH	210.75
h-BN/Gr (C–B)	50×50	Tersoff/ ILP.BNCH	586.93
h-BN/Gr (C–N)	50×50	Tersoff/ ILP.BNCH	669.37
SL-hBN [30]	1 µm × 1 µm	Tersoff/ L-J	600
BL-hBN [31]	$l = 3$ µm $w = 3.3$ µm $h = 0.666$ nm	Experimental	484±141
ML-hBN [32]	11-layer (6.7 µm ×5 µm) 5-layer (6.6 µm ×7.5 µm)	Experimental	360 250
hBN/Gr [29]	-	DFT	751

IV. CONCLUSION

In this study, a molecular dynamics approach was undertaken to study the effects of domain size and vdW stacking on graphene and hBN layers. Isolated graphene and hBN layers were probed using the Green–Kubo method of heat current autocorrelation functions to calculate the thermal conductivity. It is noteworthy that the thermal conductivity of SL-hBN is significantly lower than experimental values, which implies that the potentials used were improperly calibrated. It was shown that the addition of a graphene sheet above the hBN sheet resulted in an increased value of thermal conductivity, thus proving the hypothesis theoretically. It can also be concluded that the thermal conductivity values will likely change if supported on different materials due to interfacial interactions. Overall, the calculated thermal conductivity agreed with previous literature whereby $\lambda_{BL\text{-}hBN} < \lambda_{hBN/Gr} < \lambda_{SL\text{-}Gr}$. The heterostructure proposed in this study offers a promising design for creating devices that enable effective control and management of heat.

V. ACKNOWLEDGEMENTS

Research reported in this publication was supported by the Ministry of Higher Education under the Fundamental Research Grant Scheme (FRGS/1/2020/STG05/UKM/02/9). The authors would also like to thank Dr. Simon Gravelle of LIPhy, University Grenoble Alpes for his help with LAMMPS.

REFERENCES

[1] AMD. "AMD EPYC™ 9654P." https://www.amd.com/en/product/12251 (accessed 9/1/2023).

[2] A. Bejan and A. D. Kraus, Heat transfer handbook. John Wiley & Sons, 2003.

[3] C. V. Madhusudana and C. Madhusudana, Thermal contact conductance. Springer, 1996.

[4] R. Mahajan, C. Chiu, and R. Prasher, "Thermal interface materials: a brief review of design characteristics and materials," Electronics Cooling, vol. 10, no. 1, p. 10, 2004.

[5] F. Sarvar, D. C. Whalley, and P. P. Conway, "Thermal interface materials–A review of the state of the art," in 2006 1st electronic systemintegration technology conference, 2006, vol. 2: IEEE, pp. 1292–1302.

[6] Q. Mu, S. Feng, and G. Diao, "Thermal conductivity of silicone rubber filled with ZnO," Polymer Composites, vol. 28, no. 2, pp. 125–130, 2007.

[7] L. C. Sim, S. Ramanan, H. Ismail, K. Seetharamu, and T. Goh, "Thermal characterization of Al_2O_3 and ZnO reinforced silicone rubber as thermal pads for heat dissipation purposes," Thermochim. Acta, vol. 430, no. 1–2, pp. 155–165, 2005.

[8] A. A. Balandin et al., "Superior thermal conductivity of single–layer graphene," Nano Lett., vol. 8, no. 3, pp. 902–907, 2008.

[9] M. S. Green, "Markoff random processes and the statistical mechanics of time‑dependent phenomena. II. Irreversible processes in fluids," The Journal of Chemical Physics, vol. 22, no. 3, pp. 398–413, 1954.

[10] R. Kubo, "Statistical–mechanical theory of irreversible processes. I. General theory and simple applications to magnetic and conduction problems," J. Phys. Soc. Jpn., vol. 12, no. 6, pp. 570–586, 1957.

[11] S. Plimpton, "Fast parallel algorithms for short–range molecular dynamics," Journal of Computational Physics, vol. 117, no. 1, pp. 1–19, 1995.

[12] A. Stukowski, "Visualization and analysis of atomistic simulation data with OVITO–the Open Visualization Tool," Modell. Simul. Mater. Sci. Eng., vol. 18, no. 1, p. 015012, 2009.

[13] W. Humphrey, A. Dalke, and K. Schulten, "VMD: visual molecular dynamics," Journal of molecular graphics, vol. 14, no. 1, pp. 33–38, 1996.

[14] J. Tersoff, "Empirical interatomic potential for silicon with improve elastic properties," Phys. Rev. B, vol. 38, no. 14, p. 9902, 1988.

[15] S. J. Stuart, A. B. Tutein, and J. A. Harrison, "A reactive potential for hydrocarbons with intermolecular interactions," The Journal of Chemical Physics, vol. 112, no. 14, pp. 6472–6486, 2000.

[16] T. Maaravi, I. Leven, I. Azuri, L. Kronik, and O. Hod, "Interlayer potential for homogeneous graphene and hexagonal boron nitride systems: reparametrization for many–body dispersion effects," The Journal of Physical Chemistry C, vol. 121, no. 41, pp. 22826–22835, 2017.

[17] I. Leven, T. Maaravi, I. Azuri, L. Kronik, and O. Hod, "Interlayer potential for graphene/h–BN heterostructures," Journal of Chemical Theory and Computation, vol. 12, no. 6, pp. 2896–2905, 2016.

[18] I. Leven, I. Azuri, L. Kronik, and O. Hod, "Inter–layer potential for hexagonal boron nitride," The Journal of Chemical Physics, vol. 140, no. 10, p. 104106, 2014.

[19] S. Nosé, "A molecular dynamics method for simulations in the canonical ensemble," Molecular Physics, vol. 52, no. 2, pp. 255–268, 1984.

[20] W. G. Hoover, "Canonical dynamics: Equilibrium phase–space distributions," Phys. Rev. A, vol. 31, no. 3, p. 1695, 1985.

[21] D. Singh, J. Y. Murthy, and T. S. Fisher, "On the accuracy of classical and long wavelength approximations for phonon transport in graphene," J. Appl. Phys., vol. 110, no. 11, p. 113510, 2011.

[22] L. Lindsay, D. A. Broido, and N. Mingo, "Flexural phonons and thermal transport in multilayer graphene and graphite," Phys. Rev. B, vol. 83, no. 23, p. 235428, 2011.

[23] Q. Yu et al., "Control and characterization of individual grains and grain boundaries in graphene grown by chemical vapour deposition," Nature Mat., vol. 10, no. 6, pp. 443–449, 2011.

[24] R. Kubo, M. Yokota, and S. Nakajima, "Statistical–mechanical theory of irreversible processes. II. Response to thermal disturbance," J. Phys. Soc. Jpn., vol. 12, no. 11, pp. 1203–1211, 1957.

[25] R. Zwanzig, "Time–correlation functions and transport coefficients in statistical mechanics," Annual Review of Physical Chemistry, vol. 16, no. 1, pp. 67–102, 19

[26] B. D. Kong, S. Paul, M. B. Nardelli, and K. W. Kim, "First–principles analysis of lattice thermal conductivity in monolayer and bilayer graphene," Phys. Rev. B, vol. 80, no. 3, p. 033406, 2009.

[27] P. Klemens, "Theory of thermal conduction in thin ceramic films," International Journal of Thermophysics, vol. 22, pp. 265–275, 2001.

[28] D. Nika, S. Ghosh, E. Pokatilov, and A. Balandin, "Lattice thermal conductivity of graphene flakes: Comparison with bulk graphite," Appl. Phys. Lett., vol. 94, no. 20, p. 203103, 2009.

[29] H. Gholivand and N. Donmezer, "Phonon mean free path in few layer graphene, hexagonal boron nitride, and composite bilayer h-BN/graphene," IEEE Transactions on Nanotechnology, vol. 16, no. 5, pp. 752-758, 2017.

[30] L. Lindsay and D. A. Broido, "Enhanced thermal conductivity and isotope effect in single-layer hexagonal boron nitride," Phys. Rev. B, vol. 84, no. 15, p. 155421, 2011.

[31] C. Wang, J. Guo, L. Dong, A. Aiyiti, X. Xu, and B. Li, "Superior thermal conductivity in suspended bilayer hexagonal boron nitride," Sci. Rep., vol. 6, p. 25334, May 4 2016, doi: 10.1038/srep25334.

[32] I. Jo et al., "Thermal conductivity and phonon transport in suspended few-layer hexagonal boron nitride," Nano Lett., vol. 13, no. 2, pp. 550-554, 2013.

[33] M. S. Alborzi and A. Rajabpour, "Thermal transport in van der Waals graphene/boron-nitride structure: a molecular dynamics study," The European Physical Journal Plus, vol. 136, no. 9, p. 959, 2021.

[34] M. H. Rahman et al., "Phonon thermal conductivity of the stanene/hBN van der Waals heterostructure," Phys. Chem. Chem. Phys., vol. 23, no. 18, pp. 11028-11038, 2021.

[35] X. Wu and Q. Han, "Thermal conductivity of monolayer hexagonal boron nitride: From defective to amorphous," Comput. Mater. Sci., vol. 184, p. 109938, 2020.

[36] K. Matsunaga, C. Fisher, and H. Matsubara, "Tersoff potential parameters for simulating cubic boron carbonitrides," Japanese Journal of Applied Physics, vol. 39, no. 1A, p. L48, 2000.

[37] B. Mortazavi and Y. Rémond, "Investigation of tensile response and thermal conductivity of boron-nitride nanosheets using molecular dynamics simulations," Physica E: Low-dimensional Systems and Nanostructures, vol. 44, no. 9, pp. 1846-1852, 2012

Investigating the Performance of Deep Reinforcement Learning-Based MPPT Algorithm under Partial Shading Condition

Weng Ho Yew
Institute of Sustainable Energy (ISE)
Universiti Tenaga Nasional,
Kajang, Selangor, Malaysia
whyew97@hotmail.com

Chien Fat Chau
Institute of Sustainable Energy (ISE)
Universiti Tenaga Nasional,
Kajang, Selangor, Malaysia
cchienfat@uniten.edu.my

Ahmad Wafi Mahmood Zuhdi
Institute of Sustainable Energy (ISE)
Universiti Tenaga Nasional,
Kajang, Selangor, Malaysia
wafi@uniten.edu.my

Wan Syakirah Wan Abdullah
TNB Renewables Sdn. Bhd.
Tenaga Nasional Berhad
Petaling Jaya, Malaysia
syakirahwa@tnb.com.my

Weng Kean Yew
School of Engineering & Physical Sciences
Heriot-Watt University
Putrajaya, Malaysia
W.Yew@hw.ac.uk

Nowshad Amin
Institute of Sustainable Energy (ISE)
Universiti Tenaga Nasional
Kajang, Selangor, Malaysia
nowshad@uniten.edu.my

Abstract— For renewable energy systems to operate as efficiently and as effectively as possible, maximum power point tracking (MPPT) controllers are essential. They make it possible to precisely and dynamically track the peak output of solar panels or wind turbines, ensuring that the system will be stable and reliable even in the face of changing environmental factors. Recently, more robust algorithms based on deep reinforcement learning (DRL) have been proposed. These DRL-based algorithms optimize the local and global maximum power point (MPP) using deep Q-learning and deep deterministic policy gradient (DDPG). In this study, MATLAB models of a DRL-based MPPT algorithm were developed, tested, and compared to simulation based on two established MPPT algorithms – the Particle Swarm Optimization (PSO), and the Perturb and Observe (P&O). The simulations were conducted under various conditions, including standard test conditions (STC), and partial shading conditions (PSC). Simulation results demonstrate that at STC, both the DRL-based MPPT and PSO algorithm tracks the steady-state power at 0.02 seconds, outperforming the traditional P&O technique of 0.08 seconds. However, the PSO algorithm manages to track 1.18% more power than DRL MPPT at PSC. Despite the limitations of training the DRL, it shows a promising method for addressing MPPT issues under PSC.

Keywords – deep reinforcement learning, energy, maximum power point tracking (MPPT), partial shading conditions (PSC), particle swarm optimization (PSO), perturb and observe (P&O), off-grid PV.

I. Introduction

The rapid depletion of conventional energy sources like gas, coal, and fossil fuels is a growing concern worldwide, necessitating the search for alternative sources of energy to prevent energy shortages. Among the many potential renewable energy sources, solar energy has emerged as one of the viable solutions. Photovoltaic (PV) system harvesting solar energy has thus gained increasing popularity in many countries due to their efficiency, cost-effectiveness, and ability to generate clean energy while reducing CO_2

emissions. However, partial shading remains a significant issue affecting the efficiency of PV arrays. Partial shading occurs when weather conditions cause shading across the PV arrays, leading to a reduction in energy production. To address this issue, a maximum power point tracking (MPPT) algorithm is often used to track the maximum power of the PV arrays. However, the use of bypass diodes can cause multiple peaks on the power curve [1], making it challenging to distinguish between the global maximum power point (MPP) and the local MPP. An intelligent and efficient MPPT algorithm can distinguish between these points, allowing it to obtain the most power under partial shading conditions [1].

Conventional MPPT methods are commonly used in the industry due to their simplicity and easy implementation. These methods can be categorized into direct power control and indirect power control. Direct power control methods include the classic Perturb and Observe (P&O) method and the Incremental Conductance (IC) method, which compares the power of the PV panel with the MPP and adjusts the duty cycle accordingly [2]. Meanwhile, indirect power control methods include the Fixed Voltage and the Fractional Open Circuit Voltage methods, which indirectly control the power by comparing the voltage of the PV panel with the reference voltage, using the error between the two voltages to produce a new voltage [3]. In [2], the P&O method is demonstrated to be able to track the power efficiently under uniform solar irradiance. However, it fails to track the global MPP when there are fast changes in PV power, which typically happens during partial shading conditions (PSC) [4]. Another drawback of the P&O method is that it has a slow tracking speed, weak convergence speed, and high oscillation. Modified P&O methods have been reported which aimed to reduce the oscillation and quicken the tracking speed [5], [6].

Soft computing MPPT methods are more complex than conventional methods, as they use advanced logic and have a higher computational burden. This review [7] summarizes a number of popular artificial intelligence (AI)-based MPPT techniques such as fuzzy logic (FL) [8], artificial neural network (ANN) [9], Genetic Algorithm (GA) [10], and Swarm Intelligence (SI) [11]. The main advantage of these methods is their ability to efficiently solve non-linear problems and obtain the global MPP under partial shading conditions with fast-tracking accuracy and convergence speed. In [11]–[13], the PSO algorithm is proven to be more efficient than the conventional P&O with a faster tracking speed. However, these methods are costly to implement and are model-based, which means they have specific parameters according to the PV system environment and cannot automatically adapt to changes in the system.

One of the most recent research MPPT methods that are being studied is the deep reinforcement learning (DRL) MPPT control [14]. DRL-based MPPT integrates the reinforcement learning technique with the deep learning aspects of the charge controller. The main advantage of DRL-based MPPT compared to previous methods is that it does not require system identification.

Hence, this paper aims to investigate and compare the performance of a DRL-based MPPT algorithm with two of the more established MPPT algorithms, namely the Particle Swarm Optimization (PSO) and the Perturb and Observe (P&O). Several hybrid methods which have been reported recently with improved tracking speed, efficiency, oscillation, and convergence speed by a small margin were not included in this comparison due to the complexity of simulating them.

II. MODELLING OF THE PROPOSED SYSTEM

In this section, the architecture of a DRL model is presented, along with its structure and the neural network method employed in deep learning training. The approach discussed in this section is then implemented using MATLAB/Simulink. Finally, the hardware specification which can impact the DRL training process is discussed.

A. Design of the DRL-based MPPT control

Deep reinforcement learning (DRL) is an advanced form of reinforcement learning that integrates deep learning into the training process of the agent. The DRL uses a Markov Decision Process (MDP) model to design the overall architecture of the control system. The block diagram of the MDP model is shown in Fig. 1.

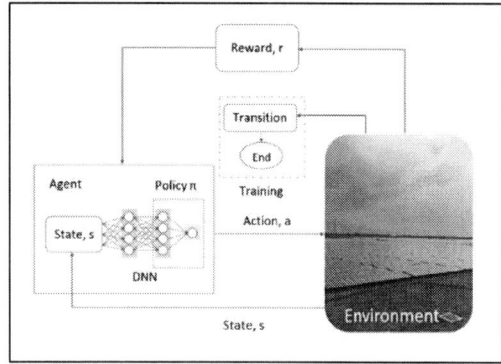

Fig. 1.   MDP model of the DRL PV system

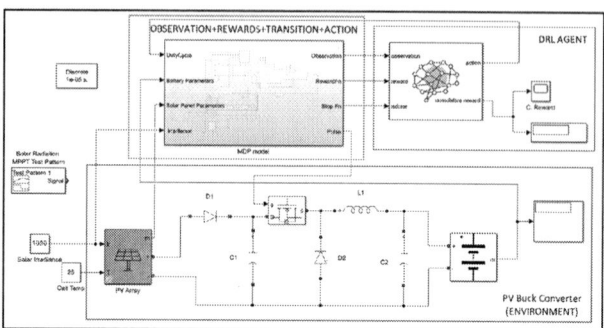

Fig. 2.   MDP model in MATLAB/Simulink

The model consists of a deep learning agent that observes multiple inputs/states, S from the environment. For this paper, the environment represents an off-grid charge controller whereby the solar panel charges the battery. The states, on the other hand, are the PV panel's voltage (V_{PV}), the PV panel's current (I_{PV}), the duty cycle of the MOSFET (D), and the maximum power of the solar panel (P_{MPP}). These inputs/states are required as it is often used in the reward function, r.

The agent would use the aforementioned states and go through a deep learning policy called Deep Deterministic Policy Gradient (DDPG) algorithm to produce an action of a random duty cycle for the PV system. The new duty cycle would then indirectly produce a new PV panel power, P_{PV}, which will result in a new reward that is calculated from the reward function set. The reward functions that were used in this paper are shown in equations (1) to (4)

$$r = r_1 + r_2 + r_3 \tag{1}$$

$$r_1 = \begin{cases} \frac{P_{x+1}}{P_{MPP}} & if \; \Delta P \geq \delta_1 \\ 0 & if \; \Delta P < \delta_1 \end{cases} \tag{2}$$

$$r_2 = \begin{cases} 5 & if \; n \geq 0.99 \\ 4 & if \; n \geq 0.95 \\ 3 & if \; n \geq 0.90 \\ 2 & if \; n \geq 0.85 \\ 1 & if \; n \geq 0.80 \\ 0 & if \; n \geq 0.50 \\ -1 & if \; n \geq 0.40 \\ -2 & if \; n \geq 0.30 \\ -3 & if \; n \geq 0.20 \\ -4 & if \; n \geq 0.10 \end{cases} \quad where \; n = \frac{P_{PV}}{P_{MPP}} \tag{3}$$

$$r_3 = \begin{cases} 0 & if \; 0.01 \leq D \leq 0.99 \\ -1 & otherwise \end{cases} \tag{4}$$

Reward functions serve as a measure of the agent's performance, indicating whether it is improving, maintaining, or failing. The rewards are computed using three distinct functions, which are summed up at the end. The first function, r_1, assigns a positive reward if the change in power ΔP is more than a certain amount δ_1, while providing zero rewards otherwise. The second function, r_2, is designed to be more dynamic, assigning higher rewards as the produced power approaches the expected maximum power, and vice versa. Lastly, the third function, r_3, applies a negative reward if the calculated duty cycle is either 0, 1, or greater than 1.

During agent training, a transition function is utilized to indicate the completion of each episode. This allows the training of the agent to be repeated while recording the cumulative rewards until either the stopping criterion or the transition function is triggered. In this study, the stopping criterion is defined as a maximum of 200 steps per episode. However, if the transition function, indicating that the newly produced power exceeds the maximum power point, is triggered first, it prematurely ends the current episode. Once the stopping criterion is reached, a new episode begins, and the previous cumulative reward is recorded. The number of episodes for training can be specified, but training can be stopped prematurely if the obtained rewards meet the desired outcome. Using the trained agent, the photovoltaic (PV) environment can be simulated, and the results can be obtained.

B. Neural Network of the DRL

For the DRL, the neural network architecture is based on a DDPG policy. The neural network usually compromises of two neural networks which are the actor network and critic network. The actor network inputs the environment's current state and produces a continuous action. On the other hand, the critic network generates a Q-value that indicates the anticipated future reward based on the inputs of the current state and activity. For this paper, the networks are created based on the number of layers recommended by MathWorks®. The structure of the critic network and actor network are shown in Fig. 3.

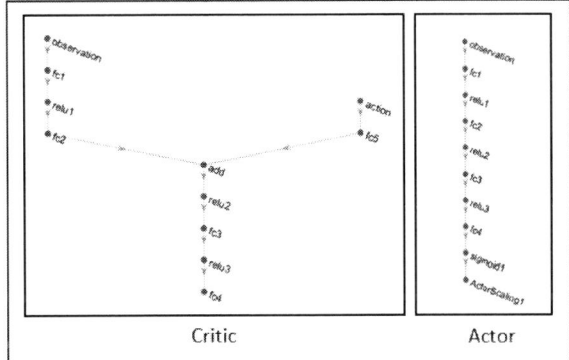

Fig. 3. Structure of the Critic and Actor neural network

The critic network takes in the current observation and action as input to produce the Q-value which represents the expected future rewards. Meanwhile, the actor network takes in the current observation of the environment which outputs a probability distribution over the possible actions. The hidden layers consist of rectified linear unit (ReLU) layers and fully connected layers. The ReLU layer is a non-linear function that leaves positive inputs unaltered while replacing all negative inputs with zero. The fully connected layer is a network that helps to learn complex relationships between the input and output data. Lastly, the probability distribution for the probable actions of 0 and 1 is provided by the sigmoid layer.

III. RESULT AND DISCUSSION

In this section, the simulation for the MATLAB model of the DRL-based MPPT control after training is tested and the results are discussed. Then, the DRL simulation is compared with the simulation of two other algorithms, which are obtained using the Particle Swarm Optimization (PSO) MPPT algorithm and the Perturb and Observe (P&O) MPPT algorithm. The simulations are done in standard test conditions (STC) and partial shading conditions (PSC) to show the effectiveness of the algorithm in each condition. For this paper, a panel count of 3 with each panel having a power rating of 213.15W in parallel configuration was selected. As a result, the maximum power that can be obtained is 639.45W.

A. Outcome of DRL Training

The training process was conducted using the Reinforcement Learning toolbox in MATLAB, and the corresponding results are presented in Fig. 4. The environment and agent setup are illustrated in Fig. 2 and 3, respectively. From Fig. 4, it is evident that the average episode rewards gradually converge to zero over the course of 500 training episodes which shows that the agent is slowly learning.

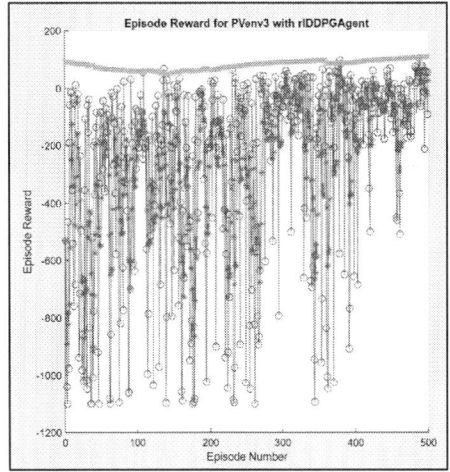

Fig. 4. Training Result of the DRL agent

B. Performance at Standard Test Condition (STC)

Under STC (Irradiance, $G = 1000W/m^2$, and temperature, $T = 25°C$), the simulations are done for 1 second and the results are recorded. Fig. 5 shows the output of each simulation for the trained DRL agent, PSO algorithm, and P&O algorithm respectively under STC.

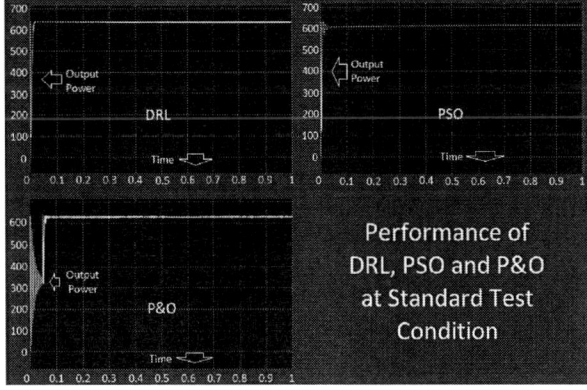

Fig. 5. Tracked power of each algorithm under STC

The DRL and PSO algorithm manages to track the steady-state power in 0.02s whereas the P&O algorithm takes approximately 0.08s. In terms of steady-state oscillation, it can be observed that the P&O algorithm has the worst result as it has a thicker graph. All three algorithms are capable of achieving the maximum power point of 639.45W from the solar panel.

979-8-3503-2369-6/23 $31.00 © 2023 IEEE

C. Performance at Partial Shading Condition (PSC)

Firstly, an ideal partial shading model with the irradiance of 350, 500, and 200 W/m² is simulated in MATLAB/Simulink, and the P-V curve is plotted. The P-V curve in Fig. 6 helps to validate the results that are obtained from each algorithm. From Fig. 6, the global maximum power point (GMPP) and the local maximum power point can be differentiated.

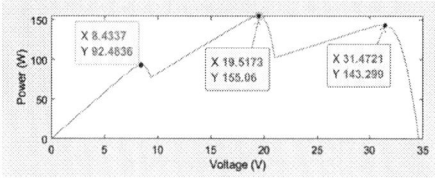

Fig. 6. P-V curve of the PV system at 350, 500, 200 W/m²

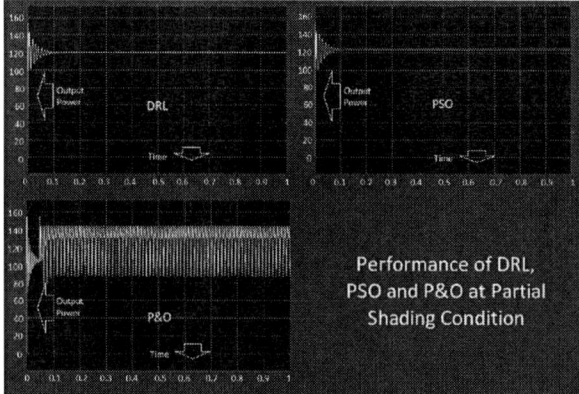

Fig. 7. Tracked power of each algorithm under PSC (350, 500, 200 W/m²)

From the graphs in Fig. 7, it can be seen that the output power of the DRL and PSO algorithms stabilize at 120 W and 123 W, respectively, whereas the P&O algorithm kept fluctuating from 85W to 142W. This shows that the DRL and PSO algorithms can detect the MPP at PSC. However, the value obtained isn't the GMPP as shown in Fig. 6 for both DRL and PSO algorithms.

Among the three algorithms, the PSO MPPT appears as the best for both standard test condition (STC) and partial shading condition (PSC), which manage to obtain the maximum power point (MPP) with the lowest oscillation around the steady state and fastest tracking speed. For the trained DRL-based MPPT, the algorithm manages to perform well during STC whereby it manages to track the MPP with fast tracking speed and low oscillation. During PSC, the algorithm manages to perform as well as the PSO algorithm with a 1.18% differential. As for the P&O MPPT, the algorithm manages to obtain the MPP during STC, however, it has high oscillation and a slower tracking speed of 0.08s. During PSC, the algorithms fail to stabilize on the MPP and kept fluctuating.

IV. CONCLUSION

In conclusion, this paper demonstrated that the DRL algorithm manages to track the MPP significantly better than the traditional method, P&O, with faster tracking speed and stable oscillation. Against a more modern soft computing algorithm, the PSO algorithm, it is proven that the DRL algorithm

performs at a similar level at STC. Therefore, more studies will be required in the future to fine-tune all the parameters to improve the tracking ability of the DRL algorithms. Besides, the computational power of the hardware used in training the DRL is quite crucial, and longer training can be done in future studies. Lastly, the DRL algorithm can be further experimented on a real-time prototype to validate the results.

ACKNOWLEDGMENT

The authors wish to thank Tenaga Nasional Berhad (TNB) Seeding Fund (U-TE-RD-18-01) for supporting this research.

REFERENCES

[1] J. C. Teo, R. H. G. Tan, V. H. Mok, V. K. Ramachandaramurthy, and C. Tan, "Impact of partial shading on the P-V characteristics and the maximum power of a photovoltaic string," *Energies*, vol. 11, no. 7, 2018, doi: 10.3390/en11071860.

[2] K. Harini and S. Syama, "Simulation and analysis of incremental conductance and Perturb and Observe MPPT with DC-DC converter topology for PV array," *Proc. 2015 IEEE Int. Conf. Electr. Comput. Commun. Technol. ICECCT 2015*, pp. 1–5, 2015, doi: 10.1109/ICECCT.2015.7225989.

[3] J. Ahmad, "A fractional open circuit voltage based maximum power point tracker for photovoltaic arrays," *ICSTE 2010 - 2010 2nd Int. Conf. Softw. Technol. Eng. Proc.*, vol. 1, pp. V1-247-V1-250, 2010, doi: 10.1109/ICSTE.2010.5608868.

[4] M. S. Nkambule, A. N. Hasan, and A. Ali, "MPPT under Partial Shading Conditions based on Perturb & Observe and Incremental Conductance," *2019 11th Int. Conf. Electr. Electron. Eng.*, 2019.

[5] P. E. SARIKA, J. Jacob, S. Mohammed, and S. Paul, "A Novel Hybrid Maximum Power Point Tracking Technique With Zero Oscillation Based On P&O Algorithm," *Int. J. Renew. Energy Res.*, vol. 10, no. 4, pp. 1960–1971, 2020.

[6] M. Kamran, M. Mudassar, M. R. Fazal, M. U. Asghar, M. Bilal, and R. Asghar, "Implementation of improved Perturb & Observe MPPT technique with confined search space for standalone photovoltaic system," *J. King Saud Univ. - Eng. Sci.*, vol. 32, no. 7, pp. 432–441, 2020, doi: 10.1016/j.jksues.2018.04.006.

[7] K. Y. Yap, C. R. Sarimuthu, and J. M. Y. Lim, "Artificial Intelligence Based MPPT Techniques for Solar Power System: A review," *J. Mod. Power Syst. Clean Energy*, vol. 8, no. 6, pp. 1043–1059, 2020, doi: 10.35833/MPCE.2020.000159.

[8] S. K. Sonam and P. Harika, "A Fuzzy Logic Based MPPT for Solar Power Generation," *Iciccs*, pp. 1182–1186, 2017.

[9] R. B. Roy and N. Amin, "Performance Analysis and Comparison of Three ANN Algorithms for MPPT Energy Harvesting in Solar PV System," vol. XX, pp. 1–17, 2017.

[10] M. Ben Smida and A. Sakly, "Genetic based algorithm for maximum power point tracking (MPPT) for grid-connected PV systems operating under partial shaded conditions," *Proc. 2015 7th Int. Conf. Model. Identify. Control. ICMIC 2015*, no. Icmic, pp. 1–6, 2016, doi: 10.1109/ICMIC.2015.7409433.

[11] F. M. Oliveira, S. A. O. Da Silva, F. R. Durand, and L. P. Sampaio, "Application of PSO method for maximum power point extraction in photovoltaic systems under partial shading conditions," *2015 IEEE 13th Brazilian Power Electron. Conf. 1st South. Power Electron. Conf. COBEP/SPEC 2016*, vol. 1000, no. 1, 2015, doi: 10.1109/COBEP.2015.7420175.

[12] Z. Cheng, H. Zhou, and H. Yang, "Comparison between the Conventional Methods and PSO Based MPPT Algorithm for Photovoltaic Systems," *Int. J. Electr. Robot. Electron. Commun. Eng.*, vol. 8, no. 4, pp. 887–892, 2014, doi: 10.1109/CCDC.2010.5498097.

[13] K. Ishaque, Z. Salam, M. Amjad, and S. Mekhilef, "An improved particle swarm optimization (PSO)-based MPPT for PV with reduced steady-state oscillation," *IEEE Trans. Power Electron.*, vol. 27, no. 8, pp. 3627–3638, 2012, doi: 10.1109/TPEL.2012.2185713.

[14] B. C. Phan, Y. C. Lai, and C. E. Lin, "A deep reinforcement learning-based MPPT control for PV systems under partial shading condition," *Sensors (Switzerland)*, vol. 20, no. 11, 2020, doi: 10.3390/s20113039.

Simulink Model of Noise of Piezoelectric Charge Amplifier

Ghulam Ali
School of Engineering and Built Environment
Griffith University
Nathan, Queensland, Australia
ghulam.ali@griffithuni.edu.au

Faisal Mohd-Yasin
School of Engineering and Built Environment
Griffith University
Nathan, Queensland, Australia
f.mohd-yasin@griffith.edu.au

Abstract—**This paper models noise of piezoelectric charge accelerometer. First, input and output state-space equations are derived based on mass-spring-damper system. After that, Simulink model of accelerometer with mechanical and electrical noise components is developed. The device and piezoelectric parameters are taken from commercial products. Simulation results demonstrate impact of noise on output of this sensor, and electrical noise is more dominant than mechanical noise. The Simulink model with noise sources is easy to use and helpful in characterizing the performances of any piezoelectric accelerometers. The model can also be adapted to other piezoelectric-based inertial transducers.**

Keywords—Piezoelectric, Accelerometer, MEMS, Noise, Simulink

I. INTRODUCTION

Piezoelectric (PE) effect is the ability of certain materials to produce electrical charges in responses to the mechanical stresses. It is a reversible transduction mechanism [1]. That is, PE transducer can be configured as a sensor using the direct-effect or as an actuator using the converse-effect [2]. One of major commercial MEMS products is an accelerometer. Capacitive-based accelerometer dominates the markets due to its low-cost and easy integration with CMOS technologies, Albeit the higher cost, PE accelerometers are also in demand as it offers ultra-high sensitivity and wider dynamic range of temperature and frequency. PE accelerometer can be divided into two categories, namely Integrated Electronic Piezo-Electric (IEPE) and charge accelerometer.

Random noise is the intrinsic property of any MEMS. It limits the device's precision, accuracy, and reliability [3]. MEMS that produces small signals are more vulnerable to the impact of its noise. Furthermore, noise also become more critical at low frequencies. Thermal noise is one of the common types. Any dissipative process that is coupled to a thermal reservoir has thermal noise fluctuations [4]. There are two functional parts of a PE transducer that generate thermal noise sources [5]. First, the mechanical resistance generates mechanical thermal noise. This resistance is due to a mechanical damped harmonic oscillator and PE material produce charge on its terminal in proportion to mechanical motion [6]. The second source of thermal noise is the losses in the PE element, which is known as dielectric loss, loss tangent, or loss angle.

Based on the literature search, several groups report on the modelling and measurement of noise of PE accelerometers [7-12]. There is still a lack of simulation tool that allows the MEMS designers to investigate the influence of electrical and mechanical noise components towards the precision of the devices, hence, this work. After deriving the state-space equations, we develop a Simulink model with electrical and mechanical noise sources that enable MEMS designers to simulate the performances of their prototypes.

The composition of this paper is as follows. First, mathematical modelling of the PE charge accelerometer is discussed in section II using a mass-spring-damper system. After that, a Simulink model with noise components is presented. Section III presents the simulation results of the operation of the accelerometer, as well as the impact of noise on the output of the device. Finally, section IV concludes this paper.

II. METHOD

A. Mathematical Modelling

A piezoelectric charge accelerometer contains a seismic mass M and PE transducer. The mass spring equivalent diagram is shown in Fig. 1. The spring constant k depicts the effective rigidity of sensor, and the damping coefficient b represents the sensor's frictional resistance. The mass M exerts a force on the PE transducer whenever the sensor is exposed to an acceleration. The acceleration-measurement a is often related to static acceleration due to gravity (g). The classical second order differential equation for the system in Fig. 1 is given as follow [5, 13].

$$-Ma = M \frac{d^2x(t)}{dt^2} + b \frac{dx(t)}{dt} + kx(t) \tag{1}$$

where $x(t)$ is the displacement relative to the base. Let we define some state variables as below.

$$X_1 = x(t) \tag{2}$$

$$X_2 = \frac{dx(t)}{dt} = \dot{X}_1 \tag{3}$$

$$\dot{X}_2 = \frac{d^2x(t)}{dt^2} = \ddot{X}_1 \tag{4}$$

Thus, equation (1) is rewritten as

$$Ma = M \dot{X}_2 + b X_2 + k X_1 \tag{5}$$

or $$\dot{X}_2 = -\frac{k}{M} X_1 - \frac{b}{M} X_2 - a \tag{6}$$

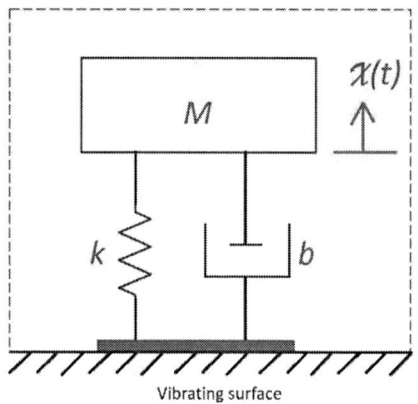

Fig. 1. Mass-spring equivalent of the accelerometer

The accelerometer produces a charge output q that is proportional to the accelerating force F. If d_{ij} represents piezoelectric constant, then q is given as [14]:

$$q = d_{ij}F = d_{ij}Ma = d_{ij}kx = Y \qquad (7)$$

Hence, the Input state-space equation is:

$$\begin{bmatrix} \dot{x}_1 \\ \dot{x}_2 \end{bmatrix} = \begin{bmatrix} 0 & 1 \\ -\frac{k}{M} & -\frac{b}{M} \end{bmatrix} \begin{bmatrix} x_1 \\ x_2 \end{bmatrix} + \begin{bmatrix} 0 \\ -1 \end{bmatrix} a \qquad (8)$$

Accordingly, the Output state-space equation is:

$$[Y] = [d_{ij}k \quad 0] \begin{bmatrix} x_1 \\ x_2 \end{bmatrix} + [0]a \qquad (9)$$

We have the following matrices from (8) and (9):

$$A = \begin{bmatrix} 0 & 1 \\ \frac{k}{M} & -\frac{b}{M} \end{bmatrix}$$

$$B = \begin{bmatrix} 0 \\ -1 \end{bmatrix}$$

$$C = [d_{ij}k \quad 0]$$

$$D = [0] \qquad (10)$$

The transfer function from A, B, C, and D matrices can be written in s-domain (i.e. Laplace domain) as follows.

$$G_{PE}(s) = C[sI - A]^{-1}B + D = \frac{-d_{ij}k}{s^2 + 2\omega_o\xi s + \omega_o^2} \qquad (11)$$

where $\omega_o = \sqrt{k/M}$ is the resonant frequency (rad/s) and $\xi = b/2M\omega_o$ is the damping ratio. The equivalent thermal noise acceleration spectral density (g/√Hz) is given as [5]:

$$a_{nPE} = \sqrt{4k_bT\left(\frac{0.01\omega_o}{MQ} + \frac{\eta C_{PE}}{\omega Q_{PE}^2}\right)} \qquad (12)$$

where k_b is the Boltzmann constant, Q is the quality factor, η is the dissipation factor, C_{PE} is PE transducer's electrical capacitance, T is the temperature and ω is the radian frequency . The Q_{PE} is the charge sensitivity of the PE material. The first part within the square root of equation (12) represents square of equivalent mechanical thermal noise (EMN) acceleration spectral density, while the second part represents square of equivalent electrical thermal noise (EEN) acceleration spectral density.

B. Simulink Modeling

The Simulink model is shown in Fig. 2. First, we build the operating model of the accelerometer without the noise sources using A, B, C, and D matrices as given in equation (10), as well as integrator and summers modules. Some of the device parameters are taken from commercial product Bruel & Kjaer model 4384 [15] and the d_{15} PE properties from [16]. The values of parameters used in equation (10) to equation (12) are summarized in Table 1. Input acceleration (referred as Input 'g' in Fig 2 in unit of g) is given as the step function. The charge output is defined as "q" in Fig. 2.

In order to perform the noise analysis, the total noise acceleration (referred to as "TN acc (g)" in Fig. 2) is added to the Input 'g'. "TN acc (g)" is composed of two components, namely EEN and EMN acceleration spectral densities. We use "user define function" module in Simulink to implement both noise sources from equation (12). As EEN is frequency dependant, the inverse of the input t reflects the frequency f. Both noise sources are then added and square-rooted. Finally, it multiplied with the accelerometer's square-rooted BW to produce the total noise acceleration. This value in g is converted to m/s² by multiplying it with g = 9.806 m/s².

It should be noted that charge accelerometer produces small amount of negative charges (q), which then require a charge amplifier for practical measurement. However, charge amplifier circuit is not included in this model because we want to focus on the noise sources that are intrinsic to the device.

TABLE I.   PARAMETERS USED FOR SIMULINK MODEL

Parameter	Value
Resonant frequency (in Hertz)	30 KHz
Mass M	10 gram
Piezoelectric constant d_{15}	420 pC/N
Damping coefficient b	0.25 Ns/m
Dissipation factor η	0.015
Quality factor Q	100
Capacitance C_{PE}	1100 pF
Boltzmann constant K_b	1.38 X 10^{23} J/K
Temperature T	300 K
Charge sensitivity Q_{PE}	41.2 pC/g

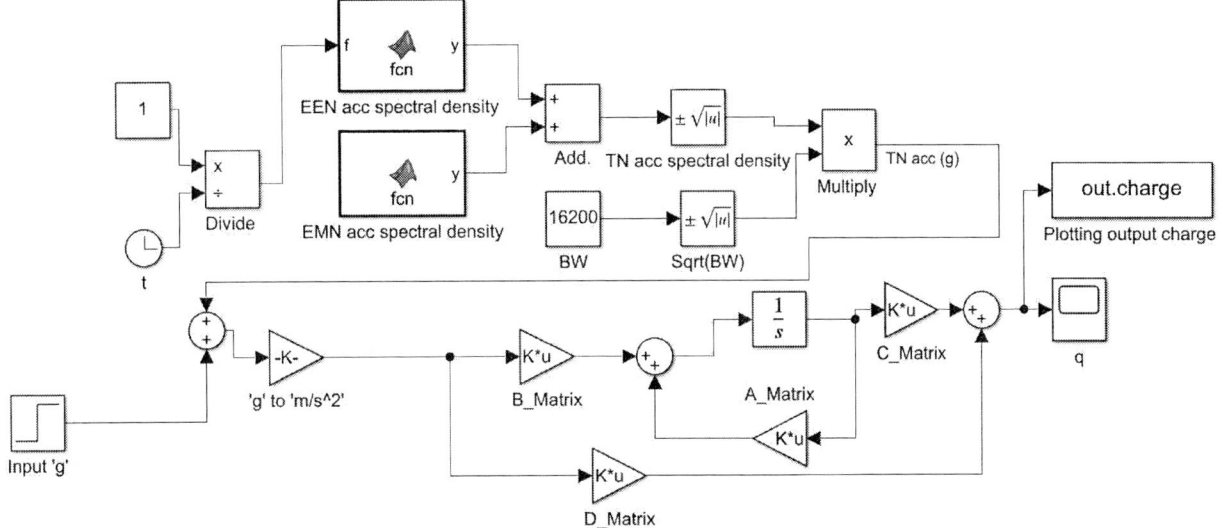

Fig. 2. Simulink model of the charge accelerometer with electrical and mechanical noise sources

III. RESULTS AND DISCUSSION

The Bode plot of the transfer function of equation (11) is shown in Fig. 3. A significant change in amplitude and phase response is evident at resonant frequency of 30 *KHz*. This demonstrates the accuracy of the model in representing the resonant behaviour of the device. From the amplitude response, the bandwidth (BW) at ±3 dB level is approximately 16.2 *KHz*. This BW is needed to calculate the total noise acceleration (TN acc (g)).

Fig. 4 shows the acceleration spectral densities of the EEN (blue color) and EMN (red color). There are two notable points. First, the amplitude of EEN is much higher than EMN, indicating electrical noise's dominance. Second, EEN is frequency dependant, while EMN is not.

Next, we show the impact of EEN and EMN on the output charge q. A small input amplitude is chosen as most PE accelerometers are needed in applications with micro g level measurements e.g. for seismic detection [17]. Fig. 5 shows the output charge versus time plot of the charge accelerometer with 1*ug* input. There are two points. First, the system stabilises after small settling time. Second, the charge output with noise (red color) is higher in amplitude than the one without noise (blue color), indicating the addition of EEN and EMN to the system.

IV. CONCLUSION

A Simulink model of charge accelerometer with noise components is presented in this paper. First, the correct operation of the accelerometer without noise is demonstrated. Then, the behaviours of the electrical and mechanical thermal noise acceleration spectral densities are plotted. Finally, the influence of input noise on the charge output is shown. This is a worrying discovery, as some sensitive applications such as seismic measurement require the accelerometers to operate with small amplitudes.

We establish in this paper that the proposed Simulink model will be very useful in determining the output charge of any PE accelerometers. Furthermore, the methodology and noise analysis could also be applied to other PE-based inertial devices such as pressure sensor [18], microphones

etc. For this purpose, we provide the Simulink and MATLAB files as supplementary information, so it could easily be modified by MEMS designers to suit their purposes.

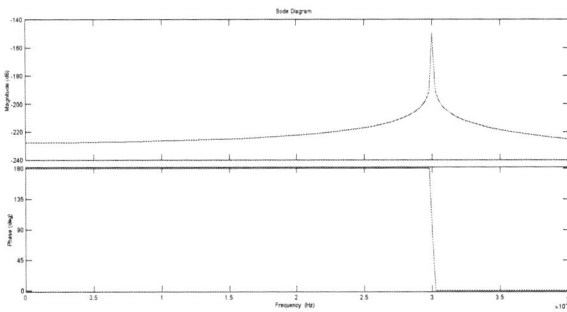

Fig. 3. The Bode plot of the transfer function $G_{PE}(s)$

REFERENCES

[1] X. Le, Q. Shi, P. Vachon, E. J. Ng, and C. Lee, "Piezoelectric MEMS—Evolution from sensing technology to diversified applications in the 5G/Internet of Things (IoT) era," Journal of Micromechanics and Microengineering, vol. 32, no. 1, p. 014005, 2021.

[2] S. Tadigadapa and K. Mateti, "Piezoelectric MEMS sensors: state-of-the-art and perspectives," Measurement Science and technology, vol. 20, no. 9, p. 092001, 2009.

[3] F. Mohd-Yasin and D. J. Nagel, "Noise as diagnostic tool for quality and reliability of MEMS," Sensors, vol. 21, no. 4, p. 1510, 2021.

[4] S. D. Senturia, Microsystem design. Springer Science & Business Media, 2005.

[5] F. Levinzon, Piezoelectric accelerometers with integral electronics. Springer, 2015.

[6] F. A. Levinzon, "Noise of piezoelectric accelerometer with integral FET amplifier," IEEE sensors journal, vol. 5, no. 6, pp. 1235-1242, 2005.

[7] R. De Reus, J. O. Gulløv, and P. R. Scheeper, "Fabrication and characterization of a piezoelectric accelerometer," Journal of Micromechanics and Microengineering, vol. 9, no. 2, p. 123, 1999.

[8] D. L. DeVoe and A. P. Pisano, "A fully surface-micromachined piezoelectric accelerometer," in Proceedings of International Solid State Sensors and Actuators Conference (Transducers' 97), 1997, vol. 2: IEEE, pp. 1205-1208.

979-8-3503-2369-6/23 $31.00 © 2023 IEEE   15

[9] F. Gerfers, M. Kohlstadt, H. Bar, M.-Y. He, Y. Manoli, and L.-P. Wang, "Sub-μg ultra-low-noise MEMS accelerometers based on CMOS-compatible piezoelectric AlN thin films," in TRANSDUCERS 2007-2007 International Solid-State Sensors, Actuators and Microsystems Conference, 2007: IEEE, pp. 1191-1194.

[10] N. N. Hewa-Kasakarage, D. Kim, M. L. Kuntzman, and N. A. Hall, "Micromachined piezoelectric accelerometers via epitaxial silicon cantilevers and bulk silicon proof masses," Journal of microelectromechanical systems, vol. 22, no. 6, pp. 1438-1446, 2013.

[11] F. A. Levinzon, "Fundamental noise limit of piezoelectric accelerometer," IEEE Sensors Journal, vol. 4, no. 1, pp. 108-111, 2004.

[12] B. Hu, Y. Liu, B. Lin, G. Wu, W. Liu, and C. Sun, "A novel trapezoidal ScAlN/AlN-based MEMS piezoelectric accelerometer," IEEE Sensors Journal, vol. 21, no. 19, pp. 21277-21284, 2021.

[13] T. B. Gabrielson, "Mechanical-thermal noise in micromachined acoustic and vibration sensors," IEEE transactions on Electron Devices, vol. 40, no. 5, pp. 903-909, 1993.

[14] Y. Shi, S. Jiang, Y. Liu, Y. Wang, and P. Qi, "Design and Optimization of a Triangular Shear Piezoelectric Acceleration Sensor for Microseismic Monitoring," Geofluids, vol. 2022, 2022.

[15] B. Kjaer. "Product data." https://www.bksv.com/-/media/literature/Product-Data/bp2041.ashx (accessed 13 June 2023).

[16] CTS. "Data Sheet." https://www.ferropermpiezoceramics.com/wp-content/uploads/2021/10/Datasheet-soft-pz23.pdf (accessed 13 June 2023).

[17] D. Instruments. "Data Sheet." https://www.djbinstruments.com/app/djb/files-module/local/datasheets/A-800.pdf (accessed 22 June 2023.

[18] N. Marsi, B.Y. Majlis, A.A. Hamzah AA and F. Mohd-Yasin, "Comparison of mechanical deflection and maximum stress of 3C SiC-and si-based pressure sensor diaphragms for extreme environment", In 2012 10th IEEE International Conference on Semiconductor Electronics (ICSE2012), 2012, IEEE, pp. 186-190)

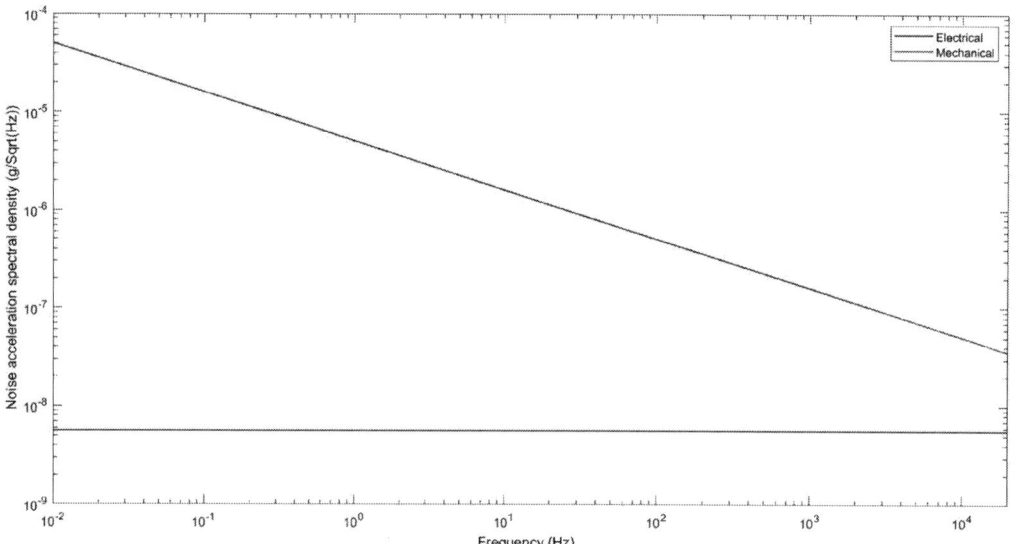

Fig. 4. Electrical and mechanical noise acceleration spectral densities

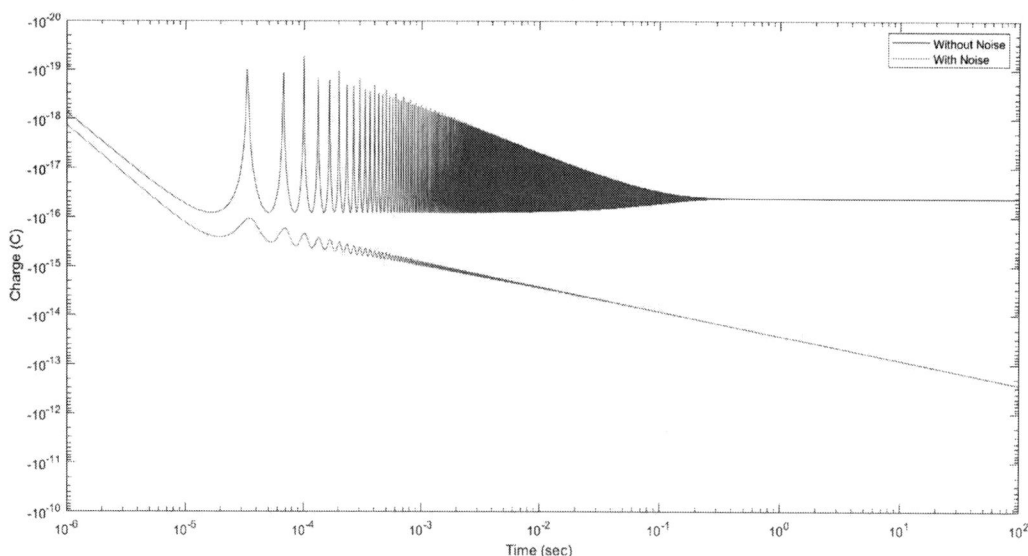

Fig. 5. Output charge vs time plot without and with noise added

Proposal for stochastic resonance in a ferroelectric-graphene transistor

Madhav Ramesh
*School of Electrical and
and Computer Engineering
Cornell University
Ithaca, USA*
mr974@cornell.edu

Amit Verma
*Dept. of Electrical Engineering
Indian Institute of Technology Kanpur
Kanpur, India*
amitkver@iitk.ac.in

Arvind Ajoy
*Dept. of Electrical Engineering
Indian Institute of Technology Palakkad
Palakkad, India*
arvindajoy@iitpkd.ac.in

Abstract—**Stochastic Resonance (SR) is a phenomenon involving non-linear systems, where the addition of noise improves some figure of measurement of the system – for example, power, signal-to-noise ratio, bit error rate. Ferroelectric materials possess a double-well energy landscape, which should enable them to be used for bi-stability based SR. However, measuring the polarization of a ferroelectric is difficult. This work proposes a design for a ferroelectric-graphene transistor. The drain current, which can be measured easily, then tracks the polarization of the ferroelectric. The choice of graphene as the channel material should ensure that the double-well nature of the energy landscape of the ferroelectric is preserved in the transistor. The choice of an n-type wide bandgap (WBG) contact should eliminate ambipolar conduction in the graphene transistor. Via numerical simulations, it is shown that the drain current of the transistor demonstrates SR while detecting weak but periodic signals. This work motivates experimental investigation of SR in ferroelectrics and ferroelectric-graphene transistors.**

Index Terms—**Ferroelectric double well, stochastic resonance, graphene transistor**

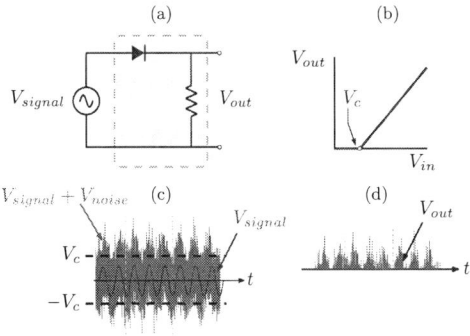

Fig. 1. Example of threshold based stochastic resonance (SR). (a) Non-linear circuit with an ideal diode. (b) $V_{out} - V_{in}$ characteristics. (c) Subthreshold signal and added white noise. (d) V_{signal} alone will give zero output. However, addition of white noise causes a non-zero V_{out} with envelope similar to V_{signal}.

I. INTRODUCTION

Stochastic Resonance (SR) describes the improvement in the figure of merit of a non-linear measurement system due to the addition of an optimal amount of noise. It is a counter-intuitive phenomenon, and is usually facilitated by the non-linearity of the system. SR was first proposed by Benzi et al. [1] to explain the quasi-periodic variation in atmospheric temperature over large time periods. It is interesting to note that SR has been discovered in biological systems as well – for example, the nervous systems of crayfish and paddlefish [2], [3] and the human auditory system [4].

Recently, several engineering applications of SR have emerged, such as a visual aid [5], a low power photodetector [6] and weak-signal detectors [7], [8]. The above applications are based on devices/systems whose non-linearity can be represented by a detection threshold $\pm V_c$, shown for example in Fig. 1. In such systems, the amplitude of the weak signal alone is lower than V_c, resulting in a zero output. However, the addition of an optimum amount of noise causes the weak input to cross the threshold, thereby enabling detection. Since the detector itself is biased in its *off* (below-threshold) state, such systems can be designed to consume very low power.

Devices/systems with a non-linearity resulting from bistable behaviour (such as a Schmitt trigger comparator [9]) demonstrate more robust SR. Ferroelectrics are a class of materials which have an inherent, bistable polarization-voltage $(P - V)$ hysteresis as shown in Fig. 2(a). This bistability arises from the double-well energy landscape (free energy vs. polarization) of the ferroelectric [10]. Harnessing this inherent bistability of ferroelectrics could result in highly-scaled SR detectors. Let V_c represent the coercive voltage of the ferroelectric. Without noise, sub-critical signals $(V_o < V_c)$ cannot switch the polarization of the ferroelectric. However, provided an optimum amount of noise (Fig. 2(b,c)) is added, even sub-critical signals can cause the polarization to switch. Since this switching is synchronized with the signal, such systems can enable robust communication strategies in noisy environments, as shown in Fig. 2(d) for the case of frequency modulated transmission using frequency shift keying (FSK). The additional robustness in this type of SR is due to bistable systems having only two possible output states, similar in spirit to the advantages of

AA thanks SERB (Science and Engineering Research Board, Government of India) for support through MTR/2021/000823 and CRG/2022/008128. AV thanks SERB Early Career Research Award (Grant No. ECR/2018/001076) for supporting the SURGE internship of MR.

979-8-3503-2369-6/23 $31.00 © 2023 IEEE

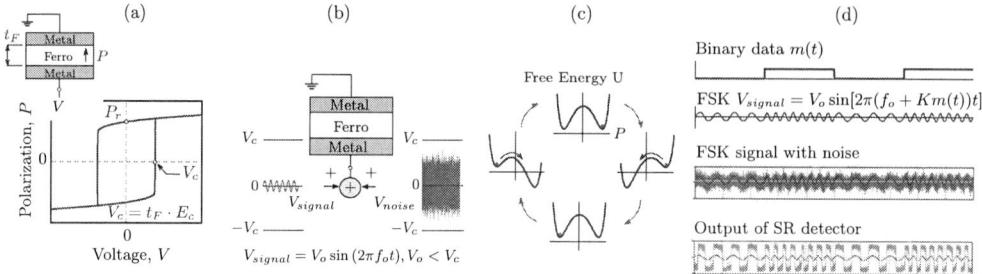

Fig. 2. Mechanism of bistablity based SR, shown using our simulations for a ferroelectric capacitor. (a) $P - V$ Hysteresis loop (b) Noise added to a weak periodic signal $V_o < V_c$ (c) The weak signal modifies the double well landscape of the ferroelectric, but does not eliminate the energy barrier. Noise cause the system to "jump" over the barrier and facilitates switching at the same frequency as the signal. (d) FSK communication in a noisy environment can benefit from bistable SR - the output of the SR detector robustly captures the frequency content of V_{signal}.

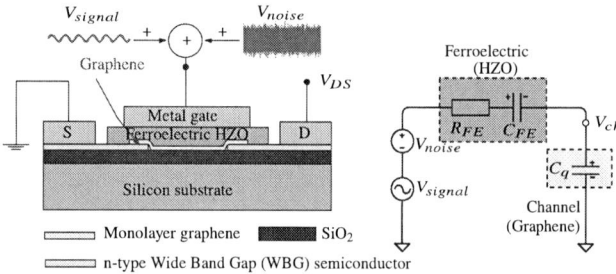

Fig. 3. Proposed ferroelectric-graphene transistor. The circuit diagram helps to determine the effective energy landscape and drain current.

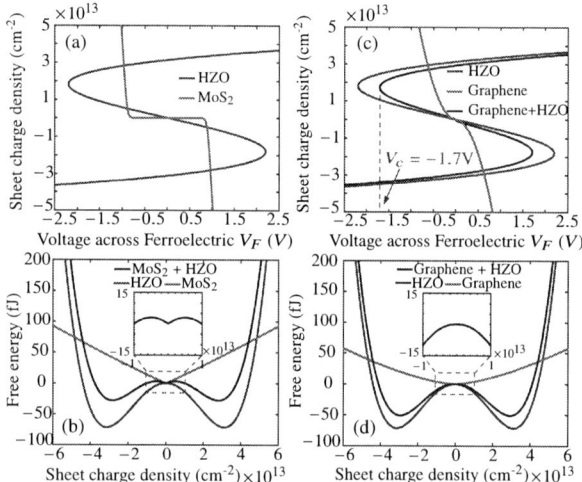

Fig. 4. Load-lines and the corresponding energy landscapes for MoS_2 (a,b) and graphene (c,d) in series with HZO. Note the double well landscape of the latter. The coercive voltage of the graphene-HZO combination is marked in (c).

digital communication over analog communication.

Measuring the polarization of a ferroelectric capacitor is non-trivial. Common techniques include Sawyer-tower circuits, current integration circuits using operational amplifiers [11] and optical methods [12]. However, all of these require either active circuit elements and other electrical components, or involve complex calculations. The aim of this work is to propose a transistor structure that can directly demonstrate SR in its current, due to a weak periodic voltage and noise at its gate terminal. For this, we propose a novel ferroelectric-graphene transistor (shown in Fig. 3) including two main innovations : (1) the channel is to be made from monolayer graphene (2) the contacts are to be made from a wide bandgap n-type material like SiC or MoS_2. The reasons for these design elements will be described in future sections. Due to its design, the drain current tracks the polarization of the ferroelectric. Via numerical simulations, we show that this structure should be able to detect weak signals through bistable SR.

II. Design of Ferroelectric-Graphene Transistor

A. Graphene as channel

As discussed in Fig. 2(c), a ferroelectric material (with thickness t_F) possesses a double well energy landscape given by $U = \alpha P^2 + \beta P^4 - PE$, with α, β being material properties and $E = V/t_F$ the electric field. This landscape generates the bistable behaviour seen in the hysteresis loop in Fig. 2(a). In order to enable SR, it is critical that the double well landscape is preserved when the ferroelectric is incorporated into a transistor. Our calculations reveal that the density of states (DOS) of the transistor channel determines the shape of the energy landscape in a ferroelectric transistor. The effect of finite graphene DOS on the channel charge can be represented by a quantum capacitance C_q as shown in the equivalent circuit of Fig. 3. The ferroelectric and the quantum capacitance of the channel are then equivalent to two non-linear capacitors in series. Using load-line analysis, an expression for the energy of this system is obtained [13]. The points where the load-lines intersect with the ferroelectric curve correspond to extrema in the energy landscape, as shown in Fig. 4. Note the load-line has been shifted (say using an appropriate metal workfunction) to make the energy landscape symmetric at zero DC gate voltage.

We assume the ferroelectric to be $Hf_{1-x}Zr_xO_2$ (HZO) with parameters as listed in Table I. Lets first consider a 2D channel material that has a bandgap, for example MoS_2 as shown in Fig. 4(a,b). Note the flat region that corresponds

TABLE I
HZO PARAMETERS FOR SIMULATIONS

Parameter	Value
Thickness t_F (nm)	20
α (mF^{-1})	-2.858×10^9
β (m^{5}F^{-1}C^{-2})	5.716×10^{11}
Resistivity ρ ($\Omega - m$)	30
Temperature T (K)	300
Area A_F (μm^2)	1

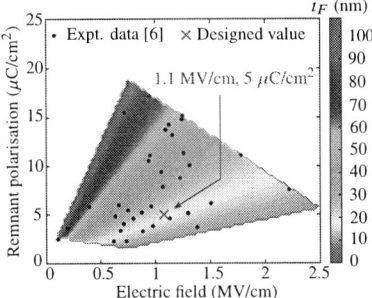

Fig. 5. Contour plot showing the minimum thickness t_F^* for a double well landscape, given different values of critical field (E_c) and remnant polarization (P_r) for ferroelectric HZO.

to the energy gap in MoS$_2$. Hence the load lines intersect at three points, corresponding to this heterostructure inevitably having five extrema (i.e three energy wells). This analysis is transferrable to any other semiconductor (2D or 3D) material with a bandgap as well. However, monolayer graphene does not have a bandgap. As a result, provided the thickness of the ferroelectric $t_F > t_F^* \equiv 1/2\alpha m$, the symmetry of the double-well energy landscape of the ferroelectric is essentially unaffected (see Fig. 4(c,d)). Only the value of the coercive voltage is altered. Here m is the slope of the graphene load-line near the origin.

Note α, β of a ferroelectric are related to its critical electric field (E_c) and remnant polarization (P_r), as $\alpha = -3\sqrt{3}E_c/4P_r$ and $\beta = 3\sqrt{3}E_c/8P_r^3$. Fig. 5 computes the effect of $\{E_c, P_r\}$ of HZO on the minimum thickness t_F^*

that ensures a double well landscape for a HZO-Graphene heterostructure. Note that our choice of t_F, α, β in Table I (corresponding to 20 nm, 1.1 MV/cm, 5 μC/cm^2) is well within the range of experimentally reported values of $\{E_c, P_r\}$ in Ref. [14].

B. Contact design for unipolar transport

If our transistor were to have ohmic contacts for both electrons and holes, then current flows from Drain→Source irrespective of whether the polarization state of the ferroelectric in positive or negative, making it impossible to observe SR (average current will be zero). Hence, we need to block the current due to one set of carriers (say holes). We propose using n-type wide bandgap (WBG) semiconducting contacts, similar to Ref. [15], [16] to suppress ambipolar conduction in the transistor, as shown in Fig. 6. When there is zero charge in the graphene channel and $V_{DS} = 0$, the Fermi levels E_{FS}, E_{FD} in the source, drain align with the Dirac point of the channel. The orange shading depicts the electron distribution as a function of energy. Note that the DOS in graphene is zero at the Dirac point, and grows linearly for energies above or below the Dirac point [17]. When the graphene channel is electron rich (charge -Q as in Fig. 6(b)), the Dirac point lies (say) Δ below E_{FS}. Assuming V_{DS} to be small, E_{FD} also lies above the Dirac point. The DOS in the channel near the conduction band edge of the source contact is high. The current is determined by the product of the DOS in the channel and the electron distribution in the source, depicted by the green inset. Hence, in the case of electron rich graphene, a high current flows from Drain→Source. For the same amount of hole charge (+Q) in the graphene channel (Fig. 6(c)), the Dirac point lies Δ above E_{FS}. Clearly, this alignment leads to a very small DOS in channel, and correspondingly low current from Drain→Source. The combination of n-type and wide bandgap of the contacts ensures that there is no current flow from the valence band of the contacts. With the above considerations, we thus assume the ideal case, where the small current in the case of hole rich graphene is approximately zero, and can be neglected.

C. Calculation of current due to stochastic input

The transistor model involves solving the following stochastic differential equation

$$\rho t_F \frac{dP}{dt} = V_{signal} + V_{noise} - (2\alpha P + 4\beta P^3) - V_{ch} \quad (1)$$

where V_{ch} is the channel voltage [17]. V_{signal} and V_{noise} represent the signal and noise (assumed white) voltages respectively. The method for solving Eq.(1) is similar to that available in literature [18]. Using a ballistic current model [19] and under the assumption of unipolar transport, we compute the current due to a weak periodic input in the presence of noise. We set the input signal $V_{signal} = 0.4\cos(2\pi \times 100\text{kHz} \times t)$V, with amplitude smaller than the coercive voltage of the Graphene-Ferroelectric heterostructure (~ 1.7V). Hence no switching is expected. However, using

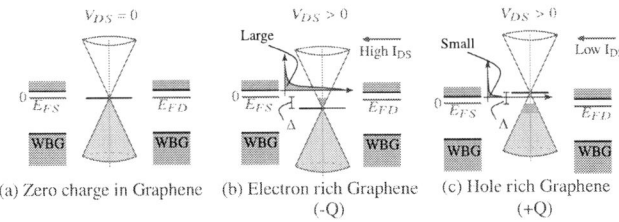

(a) Zero charge in Graphene   (b) Electron rich Graphene (-Q)   (c) Hole rich Graphene (+Q)

Fig. 6. Band-diagram to illustrate how the hole current in the device is suppressed. (a) Alignment when there is no charge in graphene. (b,c) Alignment for electron rich and hole rich graphene. The green region shows the product of the DOS in the channel with the electron distribution in the source, which is significantly lower in (c) as compared to (b), resulting in negligible current in the case of hole rich graphene. Note V_{DS} is assumed to be small in (b,c)

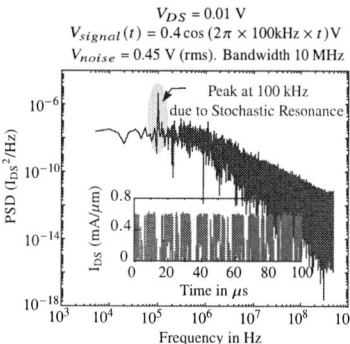

Fig. 7. SR using a current read-out. (Inset) Current I_{DS} switching over $100\mu s$. The power spectral density (PSD) plot shows a peak at 100kHz, which is the frequency of the signal to be detected.

V_{noise} at 0.45 V_{rms} (limited to a 10 MHz bandwidth), the current switches quasi-periodically due to SR. This corresponds to a peak in the power spectral density at 100 kHz, as shown in Fig. 7.

III. CONCLUSION

In conclusion, we have proposed a ferroelectric-graphene transistor capable of demonstrating bistability based stochastic resonance. Our simulations suggest that the periodicity in the weak signal is captured and a direct current read-out can be obtained. The drain current of the proposed transistor tracks the polarization of the ferroelectric, which eliminates the need for other electrical or optical measurement techniques. We also demonstrate how graphene is superior to other semiconductors for the purpose of SR by virtue of its zero bandgap.

ACKNOWLEDGMENT

The authors gratefully acknowledge the use of the CHAN-DRA High Performance Computing cluster at IIT Palakkad.

REFERENCES

[1] R. Benzi, G. Parisi, A. Sutera, and A. Vulpiani, "Stochastic resonance in climatic change," *Tellus*, vol. 34, no. 1, pp. 10–16, Feb. 1982.

[2] J. K. Douglass, L. Wilkens, E. Pantazelou, and F. Moss, "Noise enhancement of information transfer in crayfish mechanoreceptors by stochastic resonance," *Nature*, vol. 365, no. 6444, pp. 337–340, Sep. 1993.

[3] D. F. Russell, L. A. Wilkens, and F. Moss, "Use of behavioural stochastic resonance by paddle fish for feeding," *Nature*, vol. 402, no. 6759, pp. 291–294, 1999.

[4] A. Schilling, K. Tziridis, H. Schulze, and P. Krauss, "The Stochastic Resonance model of auditory perception: A unified explanation of tinnitus development, Zwicker tone illusion, and residual inhibition," *Prog. Brain Res.*, vol. 262, pp. 139–157, 2021.

[5] E. Itzcovich, M. Riani, and W. G. Sannita, "Stochastic resonance improves vision in the severely impaired," *Sci. Rep.*, vol. 7, no. 1, pp. 1–8, 2017.

[6] A. Dodda, A. Oberoi, A. Sebastian, T. H. Choudhury, J. M. Redwing, and S. Das, "Stochastic resonance in MoS$_2$ photodetector," *Nat. Commun.*, vol. 11, no. 1, pp. 1–11, 2020.

[7] S. Arai, W. Tamura, T. Yamazato, H. Hatano, M. Saito, H. Tanaka, and Y. Tadokoro, "Circuit Experiment of Photodiode-type Visible Light Communication Using the Stochastic Resonance Generated by Interfering Light Noise," in *Proc. Int. Symp. Circuits and Systems (ISCAS)*. IEEE, May 2021, pp. 1–5.

[8] Y. Hakamata, Y. Ohno, K. Maehashi, S. Kasai, K. Inoue, and K. Matsumoto, "Enhancement of weak-signal response based on stochastic resonance in carbon nanotube field-effect transistors," *J. Appl. Phys.*, vol. 108, no. 10, p. 104313, 2010.

[9] V. Melnikov, "Schmitt trigger: A solvable model of stochastic resonance," *Phys. Rev. E*, vol. 48, no. 4, p. 2481, 1993.

[10] M. Hoffmann, F. P. Fengler, M. Herzig, T. Mittmann, B. Max, U. Schroeder, R. Negrea, P. Lucian, S. Slesazeck, and T. Mikolajick, "Unveiling the double-well energy landscape in a ferroelectric layer," *Nature*, vol. 565, no. 7740, pp. 464–467, 2019.

[11] M. Stewart, M. G. Cain, and D. Hall, *Ferroelectric hysteresis measurement and analysis*. National Physical Laboratory Teddington, 1999.

[12] S. A. Denev, T. T. Lummen, E. Barnes, A. Kumar, and V. Gopalan, "Probing ferroelectrics using optical second harmonic generation," *J. Am. Ceram. Soc.*, vol. 94, no. 9, pp. 2699–2727, 2011.

[13] R. K. Jana, A. Ajoy, G. Snider, and D. Jena, "Transistor switches using active piezoelectric gate barriers," *IEEE J. Explor. Solid-State Comput. Devices*, vol. 1, pp. 35–42, 2015.

[14] M. Kobayashi, "A perspective on steep-subthreshold-slope negative-capacitance field-effect transistor," *Appl. Phys. Exp.*, vol. 11, no. 11, p. 110101, 2018.

[15] Y. Nagahisa and E. Tokumitsu, "Suppression of Hole Current in Graphene Transistors with N-Type Doped SiC Source/Drain Regions," in *Mat. Sci. Forum*, vol. 717. Trans. Tech. Publ., 2012, pp. 679–682.

[16] Y. Nagahisa, Y. Harada, and E. Tokumitsu, "Unipolar behavior in graphene-channel field-effect-transistors with n-type doped SiC source/drain regions," *Appl. Phys. Lett.*, vol. 103, no. 22, p. 223503, 2013.

[17] T. Fang, A. Konar, H. Xing, and D. Jena, "Carrier statistics and quantum capacitance of graphene sheets and ribbons," *Appl. Phys. Lett.*, vol. 91, no. 9, p. 092109, 2007.

[18] M. Ramesh, A. Verma, and A. Ajoy, "Kramers escape problem for white noise driven switching in ferroelectrics," *arXiv preprint arXiv:2112.01373*, 2021.

[19] S. O. Koswatta, A. Valdes-Garcia, M. B. Steiner, Y.-M. Lin, and P. Avouris, "Ultimate RF performance potential of carbon electronics," *IEEE Trans. Microw. Theory Tech*, vol. 59, no. 10, pp. 2739–2750, 2011.

Simulation of Macro-Compact Model of Graphene-based Three-Branch Nano-Junction

Alireza Kalantari
Malaysia-Japan International Institute of Technology
Universiti Teknologi Malaysia (UTM)
Kuala Lumpur, Malaysia
a.kalantari87@gmail.com

Shaharin Fadzli Abd Rahman
Faculty of electrical engineering
Universiti Teknologi Malaysia (UTM)
Johor, Malaysia
shaharinfadzli@utm.my

Abdul Manaf Hashim
Malaysia-Japan International Institute of Technology (MJIIT)
Universiti Teknologi Malaysia (UTM)
Kuala Lumpur, Malaysia
abdmanaf@utm.my

Abstract— Gated-graphene three-branch nano-junction (G-GTBJ) has been investigated as a promising ballistic device for various applications. Device modelling of G-GTBJ is beneficial for investigating its basic operation both in single device and in circuit level. Simulation of device design model using a dedicated simulator such as TCAD simulator is only practical in evaluating the operation of a single device. This paper evaluated a macro-compact model equivalent circuit and the simulation of the G-GTBJ was done using a general electronic circuit simulator. For validation of the simulation, the obtained results were compared with the reported work that used a TCAD simulator. The investigated macro model produces the characteristics that are in good agreement with the TCAD-based simulation work. The dependences of G-GTBJ's characteristics on carrier mobility, empirical parameter of Fsat, temperature, branch length and width were analysed. Branch length and carrier mobility showed significant effects on the simulated characteristics. The macro model was also used to demonstrate the operation of G-GTBJ based rectifier and logic gate circuit. The proposed approach seems to offer much simpler solution for the investigation of other G-GTBJ based device and logic circuit, and is expected to be beneficial for larger circuit-level simulation.

Keywords—graphene, three-branch nano-junction, macro-compact model, device simulation.

I. INTRODUCTION

A high-quality graphene possesses ultra-high carrier mobility at room temperature which enables ballistic conduction even in a submicron channel [1]. Up to date, operations of various graphene-based ballistic devices have been demonstrated [2-7]. The mobility of graphene can be even more enhanced by combining it with other materials and producing heterostructures [8-10]. Three-branch nano-junction (TBJ) is one of the ballistic devices that is utilizing graphene as its channel [11-13]. A nonlinear transfer characteristic due to the ballistic transport of the carrier could be clearly observed in graphene based TBJ (GTBJ) even at room temperature operation. As for the GTBJ's applications, applications such as rectifier, frequency doubler and logic gates have been demonstrated [14-18]. Owing to the ballistic transport, the high frequency operation of the GTBJ's devices can be expected [19]. The basic TBJ operation in ballistic transport regime can be described using Landauer–Buttiker formalism [20, 21]. Troccolo *et al.* combined this formalism with Monte Carlo transport approach to model and optimise

ballistic rectifiers operation [22]. A Finite Element Method (FEM) modelling based on drift-diffusion transport was presented by P. Butti *et al.* By using the model, the effects of temperature and branch width to the rectification efficiency were investigated [23]. The operation of the GTBJ in the drift-diffusion transport was also simulated using a Technology Computer-Aided Design (TCAD) simulation, namely Silvaco [24, 25].

This work presents an alternative approach to model the operation of a gated GTBJ (G-GTBJ). The macro-compact model used in this work is made from graphene Field-Effect Transistor (GFET). Here, a Verilog-A-based GFET model is utilized. The transistor Verilog-A model includes the device physical parameters. Thus, the simulation allows the investigation of such parameters to the TBJ's nonlinear operation. The simulation was done using a general electronic circuit simulator. The investigated macro-compact model is expected to be suitable for the use of circuit-level analysis. It is expected to be beneficial especially when designing a more complicated GTBJ circuit's application. To verify the reliability and accuracy of the proposed GTBJ model, the obtained results are compared with the outcome using TCAD simulator.

II. EXPERIMENTAL

Figure 1(a) shows a back-gated-GTBJ structure. In its typical operation, an input voltage is applied at right and left branches while a voltage at the central branch is taken as the output voltage. The heavily doped silicon substrate functions as a back-gate that modulates the current in the input branches (i.e. right and left branches). The current modulation via gating in the input branches is analogous to the gate control in a GFET. Therefore, a G-GTBJ can be modelled by an equivalent circuit shown in Figure 1(b). In the typical TBJ operation, the output voltage is measured as an open-circuit voltage. Hence, the central branch of TBJ can be replaced with a zero-resistance wire probe. The right and left branches of the graphene transistor are applied with identical gate voltage provided by the back gate.

For GFET used in this macro model, a Verilog-A GFET model presented by Shaloo Rakheja *et al.* [26] was used. The device model is a physics-based ambipolar-virtual-source model defined using the parameters listed in Table 1. To validate the functionalities of the investigated G-GTBJ macro-model, TCAD simulation results reported in ref. [25] are used for the benchmarking. In that report, the length from the end of left branch to the end of right branch of the GFET is 1.3 μm. In our equivalent circuit, the gate length (L) of both GFETs is set to be half of the total channel length (i.e. Lg = 650 nm). The width (W) of the branches is 300 nm. The dielectric constant and thickness of silicon dioxide is set to 3.1 and 285 nm, respectively. The other parameters in the Verilog-A GFET model which are not mentioned in this paper are fixed at the default values throughout the simulation. The GTBJ equivalent circuit was simulated using an open-source circuit simulator called Quite Universal Circuit Simulator (QUCS). The DC simulation was done to produce current-voltage characteristics. Transient simulation was also performed to demonstrate the application of TBJ as rectifier and digital logic gates.

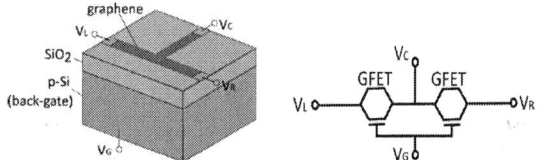

Fig. 1. (a) gated-GTBJ device structure and (b) macro-compact model equivalent circuit

Table I. Parameters used in the GFET model

Model Parameters	Parameters used in the GFET model	
	Value	*Taken from ref [25]?*
Channel width, W	300 nm	Yes
Gate length, L_g	650 nm	Yes
Gate capacitance, C_g	1.21×10^{-8} F/cm^2	Yes
Mobility, μ	5300 cm^2/V$_s$	No
Empirical parameter for Fsat, β	1.8	-
Minimum background doping, Qmin	2.2×10^{-4} C/m^2	-

III. RESULTS AND DISCUSSION

Figure 2(a) is the current-voltage curve used for the benchmarking. It shows the modulation of an input branch current (I_{RL}) under gate voltage (V_G) sweeping from -8 to 8 V. The voltages at the right branch (V_R) and left branch (V_L) are applied in a push-pull mode, where $V_R = - V_L = V_{in} = 0.1$ V. The central branch was kept in open-circuit condition. The simulated curve shows a typical transfer characteristic of a GFET. The ambipolar behaviour of the graphene channel with Dirac voltage at 0 V could be confirmed. The positive curve slope at positive V_G region corresponds to the n-type conductivity. On the other hand, the negative slope curve corresponds to the p-type conductivity. GFET model parameters, namely ΔV, Q_{min}, μ and β, were manually adjusted until the proposed G-GTBJ equivalent circuit produced a characteristic that adequately fits the reported current-voltage curve. The result of the manual current fitting is shown by the dashed line in Figure 2(a). The benchmarking curve shows a gradual current change near the Dirac voltage (i.e. zero V_G region). For the curve fitting, ΔV is set to 1 V to produce such behaviour at the subthreshold region. The obtained μ value is 5,300 cm^2/Vs, which is higher than the

mobility stated in ref. [25]. The change of V_C against the sweeping V_G is shown in Figure 2(b). In general, the obtained result shows a similar curve with the result from the TCAD-simulation [25]. When V_G is a negative value the V_C value becomes positive. The V_C switches to negative when V_G becomes positive. The device's operation could be explained based on the voltage division between right and left branches. If the resistance of the right and left branches is represented by R_R and R_L, respectively, the V_C can be written as follows.

$$V_C = (V_{in}) ((R_L - R_R) / (R_L + R_R)) \qquad (1)$$

The branch resistance depends on the gate-to-source voltage (V_{GS}) applied to the GFETs. When negative V_G is applied, the V_{GS} of the right-branch GFET is more negative than that at the left-branch. The more negative V_{GS} leads to greater hole-conduction at the right branch and subsequently lesser R_R. As $R_R < R_L$, based on equation (1), a positive V_C value can be expected. Similar analysis can be applied for the case of positive V_G.

Although the produced curve showed consistent trend with the TCAD simulation result, the obtained V_C values significantly deviated from the benchmarking results. Similar observation could be made when simulating the GTBJ's output-input voltage characteristics shown in Figure 2(c). In the output-input voltage characteristics, V_R and V_L are variable inputs (V_{in}) and V_C is measured as output, while V_G is fixed at certain value. The inputs are applied in a push-pull fashion. The general shape of the simulated curve and its changes against V_G are in good agreement with the reported experimental and simulation works [12, 17, 25]. In the case of positive V_G, where the electron conduction is dominant in the branch (i.e., n-type conduction), a bell-shaped input-output voltage curve is observed. When the V_G becomes negative, hole conduction is dominant (i.e., p-type conduction), and the input-output voltage characteristic is a V-shaped curve. The V_C can be approximately expressed using a quadratic equation, $V_C \sim \alpha V_{in}^2$, α is associated to the curvature of the curve and controlled by V_G. α is inversely proportional to $V_G - V_{Dirac}$ (i.e., $\alpha \propto 1/(V_G - V_{dirac})$) [12, 13]. As the V_G becomes closer to the Dirac voltage, the magnitude increases significantly.

When different V_R and V_L are applied, the V_C under the electron conduction mode (i.e. positive V_G) will be closer to the smallest value between the two inputs. In push-pull voltage application, V_C will always be negative, and bell-shaped voltage curve can be obtained. On the other hand, at the hole conduction mode (i.e., negative V_G), the V_C will be closer to the highest value between the inputs. As a result, a V-shaped curve is obtained. Similar device operation could be confirmed even under push-fixed input voltage application. In the push-fixed voltage application, the left branch is grounded (i.e. $V_L = 0$ V) while the V_R is varied. Figure 2(d) shows the simulation result. At $V_R = 2V$ and positive V_G, V_C value becomes closer to 0 V. When V_G changes to negative, the V_C becomes closer to 2 V.

979-8-3503-2369-6/23 $31.00 © 2023 IEEE

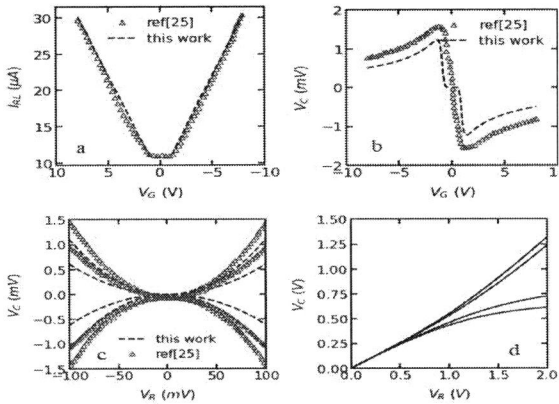

Fig 2. (a) Modulation of branch current by V_G, (b) Dependence of V_C on V_G, (c) output-input voltage characteristics under push-pull application mode, and (d) output-input voltage characteristics under push-fixed application mode.

Results shown in Figure 2 confirm the functionality of the proposed device model for producing the common electrical characteristics of a gated GTBJ's. However, due the difference of approach, this simulation technique using the equivalent circuit cannot be regarded as a direct replacement of the other simulation method such as TCAD simulator. The accuracy of the proposed simulation approach depends on the accuracy of the used GFET model

Effect of GFET model parameters to the output-input voltage curve

Figure 3 shows the dependence of gated GTBJ's output-input voltage characteristics on the branch width, branch length and temperature. No changes could be observed when the branch width and temperature were varied. This is not consistent with the reported simulation and experimental works [6, 23, 25, 27]. As for the effect of branch length, the curvature of V- and bell-shaped increased as the length is reduced.

To justify the simulation result shown in Figure 3, the following mechanism of device operation is considered. In the case of the gated GTBJ operated in drift-diffusion transport regime, the output voltage can be solved using eq. (1). The branch resistance is associated to the GFET's output resistance that controlled the bias condition. At the linear operating region, the GFET possesses low output resistance. When the GFET enters the saturation region, the output resistance increases significantly due to the current saturation. Such change in the bias condition between the right and left branch will lead to the different channel resistance and subsequently results in the V- and bell-shaped nonlinear curve. It can be predicted that the GTBJ which reaches the saturation operating regime at lower branch voltage will have sharper nonlinear curve. The branch length dependence shown in Figure 3(b) supports the presented mechanism. Current of the shorter GFET saturates at lower bias due to higher electric field.

The parameters in the GFET Verilog-a model are user-defined values that are not affected by other parameters. In the GFET compact model, the relationship between the carrier mobility and temperature is not considered and it must manually adjusted according to temperature and other relevant parameters. The carrier mobility is known to be a significant factor that can tune the electrical characteristics of

the GTBJ. The dependence on the mobility is shown in Figure 4(a). The curvature of the V- and bell-shaped curves increased at the higher carrier mobility. In the GFET channel with higher mobility, the carrier velocity can reach the saturated velocity condition even under low applied voltage.

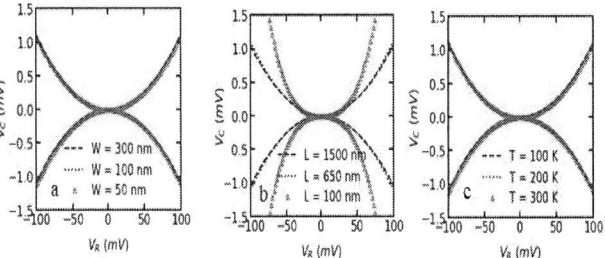

Fig 3. Dependence of output-input voltage characteristics on (a) channel width, (b) gate length, and (c) temperature defined in the GFET model.

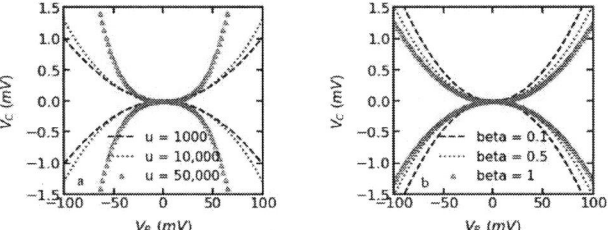

Fig 4. Dependence of output-input voltage characteristics on (a) carrier mobility and (b) empirical parameter for Fsat

To reproduce the branch width dependence reported in the experimental works, other considerations are required. P. Butti *et al.* analyzed the effect of fringing field to GTBJ's characteristic [23]. In the used GFET model, there is empirical parameter denoted as β that defines the transition from the linear to saturation operation regions. Figure 4(b) shows the effect of β value to the curve. As shown in Figure 4(b), the curvature increases when the β value decreases. This is consistent with the presented device mechanism.

Demonstration of TBJ application

By using the equivalent circuit, the circuit application of the GTBJ can be simulated and analyzed. The GTBJ's application as rectifier is shown in Figure 5. For the demonstration of rectifier operation, sinusoidal signals with a peak voltage of 0.1 V and frequency of 1 kHz were inserted at the left and right branches. The input application is also in push-pull mode where the signal at left- and right-branches has a voltage of 0.1 V and a phase difference of 180°. As shown in Figure 5, the produce Vc is a fluctuated positive DC signal. Dividing Vc to input voltage, the rectification efficiency was calculated to be 1.5 %.

Characteristics shown in Figure 2 suggest the potential of using the GTBJ as AND, and OR logic gates. A n-type GTBJ behaves as an AND logic gate because the high-state output is achieved only when both V_R and V_L are in high-state. Otherwise, the output will always in low state. As for p-type GTBJ, the device operates as OR gate. The output is high state when at least one of the inputs are in high state. Figure 6 shows the simulation of the GTBJ equivalent circuit for OR and AND logic gate operation.

979-8-3503-2369-6/23 $31.00 © 2023 IEEE   23

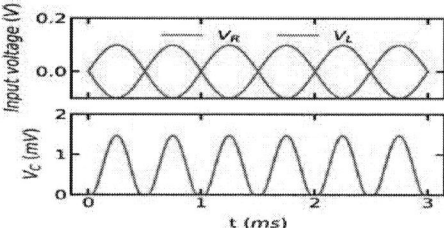

Figure 5. Simulation of rectifier operation of the G-GTBJ

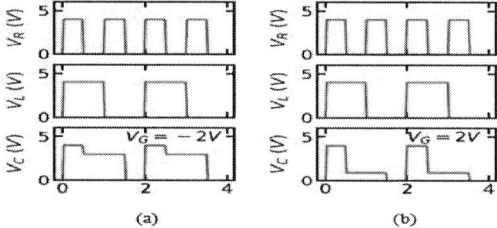

Figure 6. Timing diagram of G-GTBJ operation as (a) OR and (b) AND logic gates.

IV. CONCLUSION

A macro-compact equivalent circuit consisting of GFET was proposed to model the operation of a G-GTBJ. The typical electrical characteristics of a G-GTBJ and its gate modulation were successfully simulated using a general electronic circuit simulator. The results show good agreement with the reported experimental results, and consistent with the simulation work done using a T-CAD simulator. For the demonstration of device applications, G-GTBJ rectifier and digital OR and AND logic gates were presented. The proposed simulation approach offers simple and low-cost alternative to the simulations that utilize dedicated simulator

ACKNOWLEDGMENT

This work was financially supported by the Universiti Teknologi Malaysia through UTM-FR research grant Vot: 22H16.

REFERENCES

[1] A.S. Mayorov et al., "Micrometer-scale ballistic transport in encapsulated graphene at room temperature" Nano Letters, Vol. 11(6), pp. 2396-2399, 2011.

[2] G. Nemnes, T. Mitran, and D. Dragoman, "Ballistic transport in graphene Y-junctions in transverse electric field. Nanotechnology", Vol. 29(35), pp. 355202, 2018.

[3] V.H. Nguyen et al., "Optimum design for the ballistic diode based on graphene field-effect transistors", NPJ 2D Materials and Applications, Vol. 5(1), pp. 1-8, 2021.

[4] X.F. Peng et al., "Ballistic thermal transport in multi-terminal graphene junctions", Computational Materials Science, Vol. 77, pp. 440-444, 2013.

[5] M.S. Mobarakeh et al., "A novel graphene tunnelling field effect transistor (GTFET) using bandgap engineering". Superlattices and Microstructures, Vol. 100, pp. 1221-1229, 2016.

[6] J. Brownless, J. Zhang, and A. Song, "Graphene ballistic rectifiers: Theory and geometry dependence", Carbon, Vol. 168, pp. 201-208, 2020.

[7] A.K. Singh et al., "Graphene based ballistic rectifiers", Carbon, Vol. 84, pp. 124-129, 2015.

[8] C.R. Dean et al., "Boron nitride substrates for high-quality graphene electronics", Nature Nanotechnology, Vol. 5(10), pp. 722-726, 2010.

[9] L, Banszerus et al. "Ultrahigh-mobility graphene devices from chemical vapor deposition on reusable copper", Science Advances, Vol. 1(6), pp. e1500222, 2015.

[10] A. Kalantari et al. "Dry transfer process of single-layer graphene on multi-layer hexagonal boron nitride for high quality heterostructure" Materials Science Forum, Vol 1055, pp. 171-178, 2022.

[11] B. Händel et al., "Electrical gating and rectification in graphene three-terminal junctions". Applied Surface Science, Vol. 291. pp. 87-92, 2014.

[12] W. Kim, et al., "Nonlinear behavior of three-terminal graphene junctions at room temperature. Nanotechnology", Vol. 23(11), pp. 115201, 2012.

[13] S. F. A. Rahman, S. Kasai, and A. M. Hashim, "Room temperature nonlinear operation of a graphene-based three-branch nanojunction device with chemical doping". Applied Physics Letters, Vol. 100(19), pp. 193116, 2012.

[14] W. Kim et al., "All-graphene three-terminal-junction field-effect devices as rectifiers and inverters", ACS Nano, Vol. 9(6), pp. 5666-5674, 2015.

[15] R. Zhu et al., "Gate tunable nonlinear rectification effects in three-terminal graphene nanojunctions. Nanoscale", Vol. 6(9), pp. 4527-4531, 2014.

[16] R. Göckeritz, J. Pezoldt, and F. Schwierz, "Epitaxial graphene three-terminal junctions", Applied Physics Letters, Vol. 99(17), pp. 173111, 2011.

[17] X. Yin, S. Kasai, "Graphene‐based three‐branch nano‐junction (TBJ) logic inverter", Physica Status Solidi (c), Vol. 10(11), pp. 1485-1488, 2013.

[18] K, Prakash et al., "Thermoelectric rectification in a graphene-based triangular ballistic rectifier (G-TBR)", Journal of Computational Electronics, Vol. 20(6), pp. 2308-2316, 2021.

[19] K, Prakash et al., "Thermoelectric rectification in graphene based Y-junction", Micro and Nanostructures, Vol.167, pp. 207242, 2022.

[20] F, Araújo et al., "Gate potential-controlled current switching in graphene Y-junctions", Journal of Physics: Condensed Matter,Vol. 33(37). pp. 375501, 2021.

[21] H. Xu, "Electrical properties of three-terminal ballistic junctions", Applied Physics Letters, Vol. 78(14), pp. 2064-2066, 2001.

[22] D, Truccolo et al., "Modeling and optimization of graphene ballistic rectifiers", Solid-State Electronics, Vol. 194, pp. 108314, 2022.

[23] P. Butti et al., "Finite element simulations of graphene based three-terminal nanojunction rectifiers", Journal of Applied Physics, Vol. 114(3), pp. 033710, 2013.

[24] K, Prakash et al. "Drift diffusion modelling of three branch junction (TBR) based nano-rectifier", IEEE 14th Nanotechnology Materials and Devices Conference (NMDC), pp. 1-4, 2019.

[25] A. Garg, N. Jain, and A.K. Singh, "Modeling and simulation of a graphene-based three-terminal junction rectifier", Journal of Computational Electronics, Vol. 17(2), pp. 562-570, 2018.

[26] S. Rakheja et al., "An ambipolar virtual-source-based charge-current compact model for nanoscale graphene transistors", IEEE Transactions on Nanotechnology, Vol. 13(5), pp. 1005-1013, 2014.

[27] A, Jacobsen et al., "Rectification in three-terminal graphene junctions", Applied Physics Letters, Vol. 97(3), pp. 032110, 2010

Surface Defects Originated Photoresponse Study in hBN-ReS$_2$ FETs

Amir Zulkefli
Quality Reliability Engineering,
Infineon Technologies (Kulim) Sdn. Bhd.,
Kulim, Malaysia.
MohdAmir.Zulkefli@infineon.com

Muhammad Hilmi Johari
Institute of Microengineering and Nanoelectronics (IMEN),
University Kebangsaan Malaysia (UKM), Bangi, Malaysia.
P111586@siswa.ukm.edu.my

Abstract—**Electrical and optical transport properties of materials can be significantly affected by the imperfections in the crystal lattice. For instance, dislocations, defects, or grain boundaries. This paper presents a study on the electrical and optoelectrical properties of hBN-ReS$_2$ flakes field-effect transistors (FETs). The transistors demonstrate n-type behavior with an impressive electron mobility of 2.41 cm2/Vs. Additionally, the research highlights their remarkable photosensitive properties, exhibiting a photo-responsivity (Rλ) of 5.5 A/W and an external quantum efficiency (EQE) of 1240% when subjected to broad light irradiation at a wavelength (λ) of 550 nm. Under dark condition, there is a definite threshold voltage shift by 9.2 V between the device under vacuum and ambient air environment which is due to charge transfer takes place to the surface defects. Under light irradiation, photo-generated carriers (hole) get traps in those surface and bulk defects, which shows high photocurrent with 51.1% increment of I$_{ds}$ when compared to under air environment. There results here represent a crucial route in designing high performance two-dimensional optoelectronics device applications.**

Keywords—***MoS$_2$, Photocurrent, Surface Defects, Field-Effect Transistors***

I. INTRODUCTION

Semiconducting transition metal dichalcogenides (TMDs) with a layered crystal structure exhibit unique electrical [1] and optical properties [2] which hold immerse potential for optoelectronic device application. While efforts were focused on graphene, such graphene-based devices can be impeded its applications due to its zero-band gap that typically possesses a low on-off ratio [3]. To date, a new member of TMDs family, rhenium disulfide (ReS$_2$) has fascinated significant research attention due to its electrical and optoelectrical properties. Compared to ordinary hexagonal layered dichalcogenides like MoS$_2$, MoSe$_2$, WSe$_2$, and WS$_2$, ReS$_2$ forms a stable distorted 1-T structure with triclinic symmetry due to an additional valence electron in rhenium [4]. Although ReS$_2$ has been investigated as a candidate material for photo-response FETs, there remains intense attention in enhancing the performance of the device. Besides, a substantial variation in performance despite similar device layouts is rather irregular and requires further investigation.

Here, we will specify surface and bulk defects contribution apart from interface defects towards electrical and optoelectrical properties of the device. To do so, we introduced 2D-dielectric, hexagonal boron nitride (hBN) as

an insulator layer rather than SiO$_2$ to nullify interface defect of the device (Fig. 1(a)), and thus, clarify the physical mechanisms that give rise to photoconductivity and photogain at ON-state in the device. This high photogain at ON-state is because of the Fermi level around the conduction band, and all the surface defect states near to conduction band are filled, only those near to valence band are empty as shown in Fig. 1(b).

II. RESULTS AND DISCUSSIONS

A. Fabrication and Characterizations of hBN-ReS$_2$ devices

In our work, we investigated back-gated FETs with hBN-ReS$_2$ channels (Fig. 1(c)). The devices were fabricated by dry-transfer ReS$_2$ on top of as-exfoliated hBN on SiO$_2$/Si substrate with d_{ox} = 300 nm. Electron beam lithography and metal evaporation techniques were employed to produce the source-drain contact electrodes (3 nm Cr, 80 nm Au). After fabrication, the electrical measurements were performed at room temperature using a semiconductor analyzer.

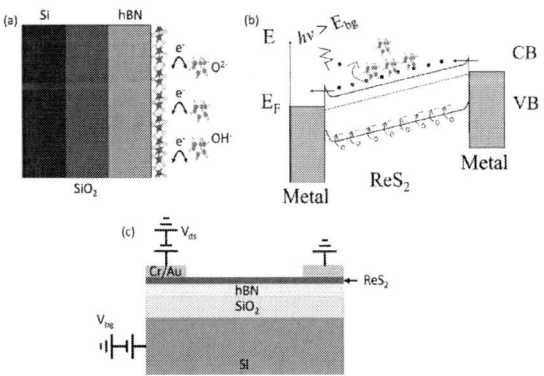

Fig. 1. (a) Schematic diagram of hBN/ReS$_2$, (b) The band diagram of ReS$_2$/metal junction (c) Schematic diagram of back-gated FETs with hBN-ReS$_2$ channels.

B. Electrical properties under dark and green-light irradiation

Fig. 2(a) shows the transfer characteristic curve (source-drain current (I$_{ds}$) vs. back gate voltage (V$_{bg}$) with a log-scale plot) of a hBN-ReS$_2$ device (dark and vacuum environment). We fixed the source-drain voltage (V$_{ds}$) at 2.0 V. The device showed excellent n-type properties, where the off-state current was ~pA, and the on/off current ratio reached 10^7. The threshold voltage (V$_{th}$) and the carrier mobility (μ) which were extracted from Equation 1, were -9.91 V and

979-8-3503-2369-6/23 $31.00 © 2023 IEEE

1.95 cm²/Vs, respectively, indicating natural n-doping / high density of electron in the device. In addition, the n-doping may be ascribed to the S vacancies and the impurities in the ReS₂ flakes.

$$I_{ds} = \left(\frac{W}{L}\right) \mu\, C_i \left(V_g - V_{th}\right) V_{ds} \qquad (1)$$

Based on Equation 1, L and W are the channel length and width, respectively. C_i is the capacitance per unit area between the gate and the channel.

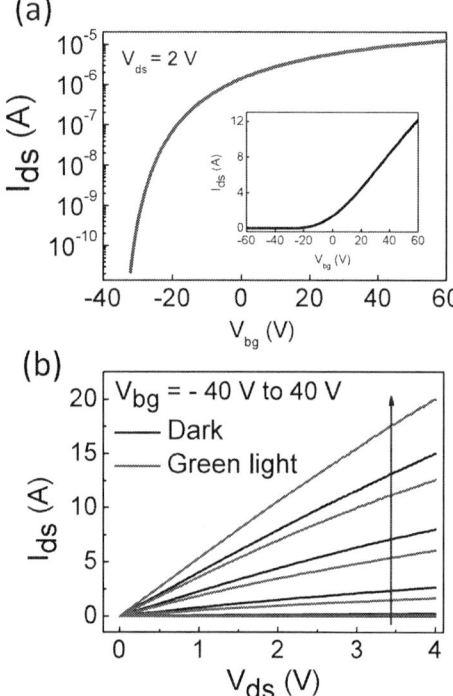

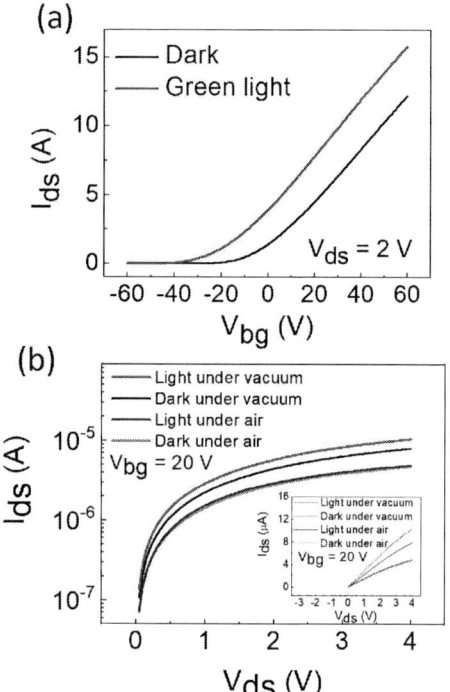

Fig. 2 (a) Transfer characteristic curve (log-scale) of a hBN-ReS₂ phototransistor (dark and vacuum environment). V_{ds} was fixed to 2.0 V. (b) The output characteristics of the device were measured using green light (46.2 mW/cm2) emitted by a green laser.

Fig. 3 (a) Transfer characteristics of the devices under dark and green-light irradiation in vacuum condition. (b) The Ids-Vds characteristics of the hBN-ReS₂ flakes photodetector were shown on a log scale of the y-axis for both dark and green-light (550 nm) conditions. The insert displayed the corresponding curve on a linear scale of the y-axis.

The hBN-ReS₂ flake FET was exposed to green light from a green laser source. The resulting characteristics of the output and transfer devices are displayed in Fig. 2(b) and Fig. 3, respectively. These curves exhibit significant improvement when under green light, indicating the flake's sensitivity to visible light. In addition, the values of μ, V_{th}, and 2D carrier concentration (n_{2d}) under dark and green-light irradiation for the hBN-ReS₂ flakes are extracted from the output characteristics linear region. n_{2d} can be estimated by using equation model I_{ds} at zero V_{bg} as shown in Equation 2, where q is the electron charge.

$$I_{ds} = q n_{2d} W \mu \left(\frac{V_{ds}}{L}\right) \qquad (2)$$

The values of μ, V_{th}, and n_{2d} are all enhanced under green-light irradiation for hBN-ReS₂ flakes compared to hBN-ReS₂ under dark (Table I). The photogenerated carriers contribute to the improved n_{2d} that resulted in the higher negative voltage which needed to deplete them. By applying electrical gating, n_{2d} (the carrier concentration in the 2D material) can be modulated following the principles of the parallel-plate capacitor model. When a positive Vbg (back-gate voltage) is applied to the bottom of the dielectric (SiO₂/hBN), more electrons are induced at the interface between ReS₂ and hBN, forming a conductive channel and leading to an increase in I_{ds}. Conversely, a negative V_{bg} will deplete the electrons at the interface, resulting in a reduction of I_{ds}.

C. Vacuum and air environment effect on the photoresponse

Based on the previous report, they reported that the environment hugely influenced the photoelectrical response properties for the MoS₂ monolayer phototransistors [6]. To further understand this effect for hBN-ReS₂ flake, we measured photosensitive of our device in the vacuum and ambient air environments. In both green-light irradiation and dark conditions, the I_{ds} is higher when measured under vacuum compared to ambient air, as depicted in Figure 3(b). The increase in current is particularly noticeable during green-light irradiation, as illustrated in the inset of Figure 3(b). Besides, under dark condition, there is an apparent threshold voltage shift by 9.2 V (Table I) between the device under vacuum and air environments which is due to charge transfer takes place to the surface defects.

These surface defects might be due to molecules (OH⁻, O2⁻) adsorbents in the air environment (Fig. 1(a)). For the case of a device under green-light irradiation, V_{th} value shifted 23.9 V between the device under vacuum, and air environment, where the surface defects get activation energy from this green-light irradiation, allowing some charge transfer takes place. Besides, when compared the photoelectrical properties between devices under vacuum and air environments, we observed high photosensitive properties with R_λ of 5.5 A/W and EQE of 1240% under broad irradiation of light at $\lambda = 550$ nm, which showed ~205% and ~206% enhancement for R_λ and EQE, respectively. In a vacuum, responsivity is due to bulk trap states in the ReS_2 flake. While in an ambient air environment, responsivity is due to both surface defect and bulk trap states. As we proved from the shifted V_{th} value, the surface adsorbent and oxide layer are present in an air environment. Those surface defect will also contribute to photocurrent by hole trapping.

III. CONCLUSIONS

The research revealed that incorporating green-light irradiation leads to enhanced electrical and optoelectrical properties in the device. This improvement occurs because the green light activates the surface defects, facilitating charge transfer and thereby increasing its photosensitivity. In an air environment, the presence of surface adsorbent molecules renders these surface defects inactive, limiting their response to photo-irradiation.

Table I V_{th}, n_{2d}, and μ of hBN-ReS$_2$

Parameters (hBN-ReS$_2$)	μ (cm^2/Vs)	V_{th} (V)	n_{2d} (10^{11} cm^{-2})	R_λ (A/W)	EQE (%)
Dark under vacuum	1.95	-9.91	7.78	-	-
Green-light under vacuum	2.41	-11.2	12.0	5.5	1240
Dark under ambient	0.91	-19.1	12.8	-	-
Green-light under ambient	1.00	-35.1	22.4	1.8	405

REFERENCES

[1] Hu, Yinhua, et al. "Noble-transition-metal dichalcogenides-emerging two-dimensional materials for sensor applications." *Applied Physics Reviews* 10.3 (2023).

[2] Dutta, Riya, et al. "Optical Enhancement of Indirect Bandgap Two-Dimensional Transition Metal Dichalcogenides for Multi-Functional Optoelectronic Sensors." *Advanced Materials* (2023): 2303272.

[3] Bhatt, Mahesh Datt, Heeju Kim, and Gunn Kim. "Various defects in graphene: a review." *RSC advances* 12.33 (2022): 21520-21547.

[4] Zereshki, Peymon, et al. "Interlayer charge transfer in ReS 2/WS 2 van der Waals heterostructures." *Physical Review B* 99.19 (2019): 195438.

Linear, Efficient and Wideband Emitter Follower Class B Amplifier for Auxiliary Envelope Tracking Supply Modulator

Zubaida Yusoff
Faculty of Engineering
Multimedia University
Cyberjaya, Malaysia
zubaida@mmu.edu.my

Md Mushfiqur Rahman
Faculty of Engineering
Multimedia University
Cyberjaya, Malaysia
1181402135@student.mmu.edu.my

Farid Zubir
Wireless Communication Centre
Universiti Teknologi Malaysia
Johor Bahru, Malaysia
faridzubir@utm.my

Jahariah Sampe
Institute of Microengineering &
Nanoelectronis (IMEN)
Universiti Kebangsaan Malaysia
Bangi, Malaysia
jahariah@ukm.edu.my

Abstract— **In this paper, an efficient and wideband Class B amplifier is designed for the application as a supply modulator of an Envelope Tracking (ET) radio-frequency power amplifier (RFPA). The common collector configuration is employed to drive a low output impedance since the output signal of this amplifier will be used as the bias signal to the drain port of the RFPA. A quasi-complementary design, also referred to as a totem-pole with a pair of n-type transistors, is implemented to build the Class B amplifier. In this design, the transistors are properly biased to avoid crossover distortion and maintain a good level of linearity. The wideband property of the amplifier has been achieved by controlling the values of the coupling capacitors. A good efficiency is recorded by using a step-up center-tapped transformer at the emitter outputs of the transistors. Simulation results show that if the input AC signal level is chosen to be half of the maximum DC voltage level, then a transformer turn ratio of 1:6 will give an efficiency of 87%. Moreover, the bandwidth is recorded to be almost 80 MHz after using a 10 uF coupling capacitor and a 70 uF bypass capacitor. The proposed supply modulator is suitable for the base station applications of the LTE networks.**

Keywords—Class B amplifier, supply modulator, envelope tracking, totem-pole, wideband, efficiency.

I. INTRODUCTION

Envelope tracking (ET) offers a highly efficient power amplifier for the next-generation base station and mobile UE (user equipment) applications. This efficiency enhancement technique can be thought of as a simplified version of the Envelope Elimination and Restoration (EER) method. The performance of an ET power amplifier largely depends on four major parameters: linearity, efficiency, broadband, and power handling capacity.

The design of the supply modulator is an important step that ensures a good trade-off between maximum output power, efficiency, and linearity. Supply modulators can be linear, discrete, or hybrid. Linear modulators are not suitable for modern wireless communication signals with high peak-to-average power ratio (PAPR). Although they are wideband and provide power at higher frequencies, they have low efficiency. On the other hand, discrete modulators are

efficient and provide power to low frequencies, but they are narrowband and expensive. Additionally, they become lossy when transformers are involved and experience high switching stress due to output inversion [2-3].

The issues with linear and discrete modulators can be mitigated by combining them in series or in parallel. Series hybrid modulators are wideband, linear, and highly efficient. However, a major problem is that they use two stages to achieve higher efficiency, which, in turn, requires a sufficiently high switching frequency to operate a multi-stage or multi-phase modulator. In contrast, the parallel hybrid modulator is very wideband, highly efficient, and provides most power in the low-frequency range [4-6].

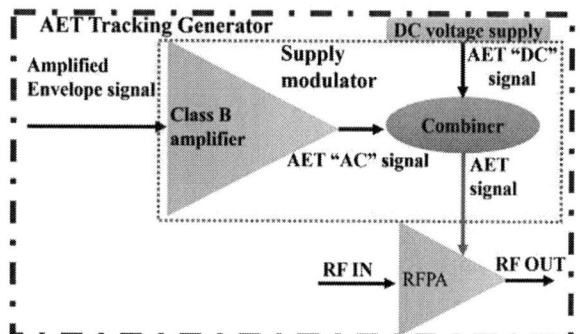

Fig. 1.   Block diagram of AET tracking generator.

Another simpler variant technique of the parallel hybrid topology is the Auxiliary Envelope Tracking (AET), which consists of a linear supply modulator and a broadband RF transformer as a combiner. The basic difference between AET and other hybrid topologies is the mechanism of generating the modulated supply. In AET, the AC and DC supply components are initially separated. The envelope signal is applied directly to the linear modulator, as shown in Fig. 1. The output of the modulator is the AC component, which is then applied to the combiner. Simultaneously, the DC component is supplied to the combiner from an external DC power supply. Finally, the combiner combines them to form the final AET signal. This approach provides easy and cost-effective solutions for ET applications by offering a straightforward circuit. Additionally, it overcomes the challenges and efficiency issues caused by switch-mode power supply modulators [7-9].

The authors would like to acknowledge the FRGS fund FRGS/1/2019/TK04/MMU/02/15 from the Ministry of Higher Education Malaysia (MOHE) for funding this research activity.

In this paper, the focus was on designing a class B amplifier to achieve efficiency more than 70% and has the bandwidth more than 20MHz. This class B amplifier has been designed with an appropriate biasing circuit to maintain the linearity of the output AC signal. An in-house designed RF broadband transformer is used as a combiner which also suppresses the DC offset of the output AC signal. This proposed supply modulator combination of the class B amplifier and the RF broadband transformer contributes in designing a simpler, cost effective and efficient tracking generator while maintaining a good level of linearity rather than conventional ET system.

II. DESIGN OF THE CLASS B AET SUPPLY MODULATOR

The Class B AET supply modulator comprises two major components:

- Class B amplifier

- Combiner

The Class B amplifier is used for producing the AET AC signal from the amplified envelope signal. In this stage, only current amplification is performed. A biasing circuit is utilized to minimize the distortion of the output AC signal. A detailed analysis of the entire circuit of the amplifier is given below:

A. Circuit Analysis

The Class B amplifier is designed using two types of configurations: the push-pull and the totem pole. Both configurations use two transistors to achieve a conduction angle of 180 degrees for each transistor. In the push-pull configuration, the bases of one NPN transistor and one PNP transistor are connected together, ensuring that only one transistor is switched on at a time. Specifically, during the positive half cycle of the AC signal, the NPN transistor is switched on, and during the negative half cycle, the PNP transistor is switched on. However, this configuration can pose a problem when the complementary pair of a transistor is not available.

To address this limitation, the totem pole configuration is employed in this research, where two identical types of transistors are used. A phase splitter, represented by a center-tapped transformer with a 1:1 ratio, generates two out-of-phase AC signals that are applied at the bases of the transistors. As a result, the signals cannot be switched on at the same time. Biasing is applied to the transistor, ensuring it turns on at 0.55 V, but this means that there will be some portion of the AC input signal that will not pass through, leading to minimal cross-over distortion in the output AC signal. The biasing voltage can be controlled by the fixed DC supply through a voltage divider circuit.

In Fig. 2, the Class B circuit without the combiner is shown, and at this stage, the output signal is a pulsating DC voltage waveform. To convert this pulsating DC signal into an output AC signal, another center-tapped transformer is utilized at the output. This transformer effectively transforms the pulsating DC signal into an AC signal.

For simulating the circuit operation, a 5 V peak-to-peak (Vpp) AC signal is applied as the input signal. When this AC signal source is combined with a fixed DC source through a voltage divider circuit, the AC signal will ride on a DC value determined by the voltage divider circuit. The type of

transistor used in this circuit is the NPN transistor 2SCR542P from ROHM Semiconductors. The maximum ratings for this transistor are a collector-emitter voltage of 30 V and a collector current of 5 A. The turn-on voltage of this transistor, also known as the collector-emitter saturation voltage, is 0.4 V.

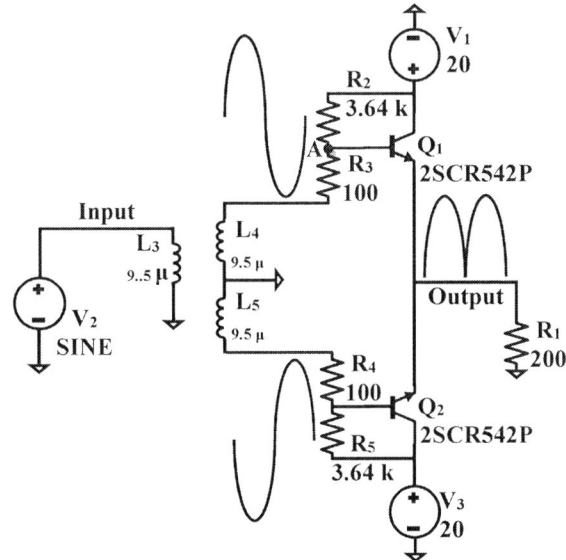

Fig.2. The circuit diagram without the combiner.

Based on the common collector configuration, the input signal is applied to the bases of the transistors, and the output is taken across a load connected at the junctions of their emitters. According to the datasheet, the NPN transistor 2SCR542P requires 0.4 V to turn it on. Without a biasing circuit, any signal below the 0.4 V DC line will be clipped off, resulting in a conduction angle less than 180 degrees. To avoid distortion, a voltage divider circuit is used from a fixed DC supply of 20 volts. However, when the AC signal is combined with this DC biasing circuit, the resulting combined signal drops slightly due to the clipping. To address this issue, the biasing voltage is chosen slightly higher than the transistor's saturation voltage.

The biasing circuit is set to bias the circuit at point A (the input bias voltage at the base) to 0.55 V, with the DC supply voltage V1 set to 20 V. In Fig. 2, R3 is chosen as 100 Ω, and R2 is set to 3.64 kΩ. The same voltage divider circuit is used for the other identical NPN transistor, i.e., R3 = R4 and R4 = R5. With these voltage divider circuits, the input AC signal at the bases will shift up from the 0.53 V DC line (originally set to 0.55 V, but due to simulation adjustment, the DC value is 0.53 V). As a result, the AC signal becomes a 4.86 Vpp signal at point A (Fig.2). Finally, with the bias voltage, the positive peak becomes (4.86 + 0.53) 5.4 V, and the negative peak becomes (-4.86 + 0.53) -4.3 V, as shown in Fig. 3. It is essential to note that the biasing needs to be set closely to 0.55 V because if the biasing voltage chosen is more than 0.55 V, the conduction angle will exceed 180 degrees. Conversely, if it is less than 0.55 V, the conduction angle will be less than 180 degrees.

In Fig. 3, the input signal represents the case of the Q1 transistor. For the Q2 transistor, the input will be the same but 180 degrees phase-shifted. This means that the Q1 and Q2 transistors operate in a push-pull manner, where one is responsible for odd-numbered positive cycles at the output, and the other is responsible for even-numbered positive cycles at the output.

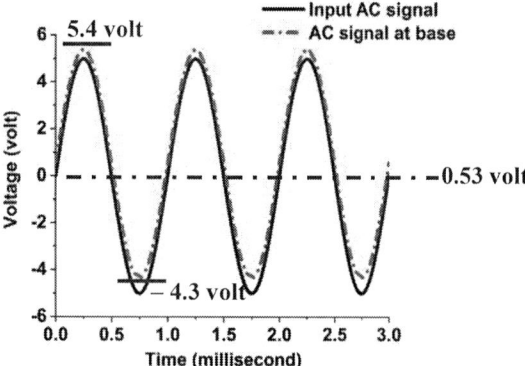

Fig. 3.   The effect of the voltage divider circuit on the input AC signal

Fig. 4 shows the output of the entire circuit, which is the combination of the emitter outputs of the transistors. The push-pull operation ensures that both Q1 and Q2 work in tandem to provide a continuous output signal without any dead zones or distortion.

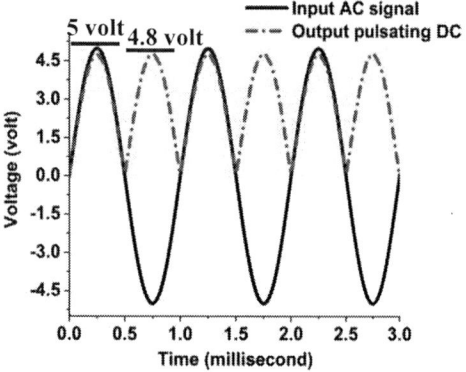

Fig. 4.   The pulsating DC output of the entire circuit

B. Combiner

In the totem pole configuration, the combiner at the output is a center-tapped step-up RF broadband transformer. The center tap is utilized to generate the phase-shifted output of the Q2 transistor. The step-up voltage transformation ratio is employed to reduce the requirements of the tracking signal for the envelope amplifier. However, it should be noted that since the step-up operation affects the amplitude of the output voltage swing, it directly influences the efficiency of the amplifier. The overall circuit of the amplifier with the combiner is depicted in Fig. 5, and the corresponding output signal is shown in Fig. 6.

To improve the linearity of the amplifier, two series capacitors (C1 and C2) are added at the outputs of the first phase shifter transformer. Additionally, these capacitors help control the lower cutoff point of the amplifier bandwidth. Furthermore, two shunt capacitors are added at the emitter outputs, which control the upper cutoff points of the bandwidth.

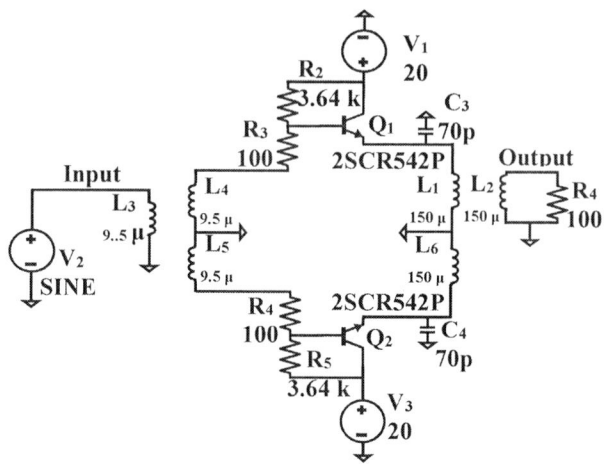

Fig. 5.   The overall circuit of the amplifier with the combiner.

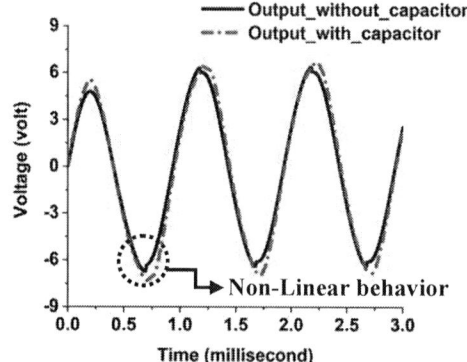

Fig. 6.   The effect of capacitor on the linearity of the final amplifier output

III. RESULTS AND DISCUSSIONS

In this section, the calculation of efficiency is provided, and the relationship between efficiency and the transformer ratio is discussed. The observed bandwidth of the overall simulated circuit is also presented. Additionally, the dependency of bandwidth on coupling and bypass capacitances is discussed.

A. Efficiency

For a Class B amplifier, the theoretical efficiency, η, can be expressed as:

$$\eta = (\pi/4).(V_{output\_max} / V_1) \qquad (1)$$

where V_{output_max} is the peak value of the output swing, and V_1 is the fixed DC value of the DC voltage supply.

In the case of a common collector amplifier in push-pull configuration, the maximum voltage gain is one. Therefore, when V_{output_max} is equal to V_1, the efficiency reaches its maximum value of 78.5%. However, for the totem pole configuration, the value of V_{output_max} is controlled by the turn ratio of the output combiner transformer. This allows the possibility of achieving efficiencies greater than 78.5% because the transformer can step up the voltage swing.

979-8-3503-2369-6/23 $31.00 © 2023 IEEE   30

Nonetheless, it should be noted that if the input AC signal level is low, a very high transformer turn ratio might be necessary to achieve a good efficiency. For an input of 5 Vpp sinusoidal signal, the effect of the transformer ratio on the efficiency can be observed and analyzed as shown in Table I. In the simulation using LTSpice version XVII, various transformer ratios can be tested, and the corresponding efficiencies can be recorded. This analysis will help determine the optimal transformer ratio to achieve the desired efficiency for the specific input signal level.

TABLE I. TURN RATIO VS EFFICIENCY (5 VPP)

Turn ratio	Voutput-max (Volts)	Efficiency (%)
1:2	6.7	26.3
1:6	11.54	45.3
1:10	14.86	58.4
1:14	17.53	68.84
1:18	19.82	77.83
1:22	21.85	85.8

Indeed, a transformer turn ratio of 1:18 is quite high and may lead to challenges such as high heat loss and critical dimension problems in the device. To achieve a good trade-off between efficiency and transformer ratio, selecting the input AC signal level to be half of the maximum DC voltage level can be a practical approach. Suppose an AC signal of 10 Vpp is used. In this case, a lower transformer ratio can be employed to achieve higher efficiency, as shown in the table below:

TABLE II. TURN RATIO VS EFFICIENCY (10 VPP)

Turn ratio	$V_{output-max}$ (Volts)	Efficiency (%)
1:2	12.85	50.46
1:3	15.72	61.73
1:4	18.14	71.23
1:5	20.26	79.56
1:6	22.18	87.1

B. Bandwidth

The 3dB simulated bandwidth of the circuit is observed by conducting an AC analysis in LTSpice. The bandwidth is measured to be almost 80 MHz, ranging from 180 Hz to 80 MHz. A parametric analysis is performed on the input coupling capacitor to observe its effect on the bandwidth. It is found that reducing the value of the capacitor shifts the lower cut-off frequency to a higher frequency region, as shown in Fig. 7. Conversely, increasing the value of the bypass capacitor shifts the upper cut-off frequency to a lower frequency region, as illustrated in Fig. 8.

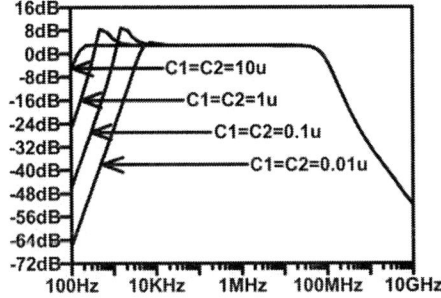

Fig. 7   The effect of coupling capacitors on bandwidth

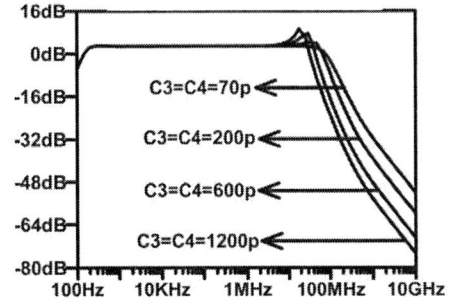

Fig. 8.   The effect of bypass capacitors on bandwidth

IV. CONCLUSIONS

An efficient common collector Class B amplifier is designed for the AET supply modulator application. The totem pole configuration is used, allowing the use of transistors of the same type. A good trade-off can be achieved between the transformer turn ratio and efficiency by selecting the value of the input AC signal to be almost half of the fixed DC voltage. For a 10 Vpp input AC signal, a combiner transformer turn ratio of 1:6 yields an efficiency of about 87%. The simulated bandwidth is recorded to be 80 MHz. The coupling and bypass capacitors play a vital role in improving the linearity of the amplifier and also affect the bandwidth. Overall, the designed amplifier is well-suited for AET supply modulators utilized in cellular base station applications. Its efficiency, bandwidth, and linearity characteristics make it an efficient choice for these applications.

REFERENCES

[1] Z. Wang, Envelope Tracking Power Amplifiers for Wireless Communications, Artech House, 2014.

[2] R. W. Erickson, DC–DC Power Converters, New York: Wiley, 1999.

[3] T. L. Floyd, Electronic Devices, 3rd ed., Upper Saddle River, NJ: Prentice Hall, 1991.

[4] Y. Li, "System and Circuits Investigation of Wideband RF Polar Transmitters Using Envelope Tracking for Mobile WiMAX/WiBRO Applications," Master of Science Thesis, Electrical Engineering, Texas Tech University, Texas, 2009.

[5] T. M. Aitto-oja, "High Efficiency Envelope Tracking Supply Voltage Modulation for High Power Base Station Amplifier Applications," IEEE MTT-S Intl. Microwave Symposium Digest, Anaheim, CA, May 2010, pp. 668–671.

[6] P. M. Asbeck, et al., "High Efficiency WCDMA Envelope Tracking Base-Station Amplifier Implemented with GaAs HVHBTs," Proc. IEEE Compound Semiconductor Integrated Circuit Symposium, Monterey, CA, October 2008, pp. 1–4.

[7] Z. Yusoff, et al., "High Linearity Auxiliary Envelope Tracking (AET) System Using GaN Class-J Power Amplifier," IEEE Power Amplifier Symposium, Arizona, September 2010, pp. 1–2.

[8] Z. Yusoff, et al., "The Benefit of GaN Characteristics over LDMOS for Linearity Improvement Using Drain Modulation in Power Amplifier System," 2011 Workshop on Integrated Nonlinear Microwave and Millimetre4Wave Circuits (INMMIC), April 2011, pp. 1–4.

[9] Z. Yusoff, J. Lees, J. Benedikt, P. J. Tasker and S. C. Cripps, "Linearity Improvement in RF Power Amplifier System Using Integrated Auxiliary Envelope Tracking System," IEEE MTT International Microwave Symposium (IMS) Digest, June 2011, pp. 1–4J. Clerk Maxwell, A Treatise on Electricity and Magnetism, 3rd ed., vol. 2. Oxford: Clarendon, 1892, pp.68–73.

Fabricating SWCNT thin film via Spray coating and Nitric Acid Vapor Treatment

Arulampalam Kunaraj
Dept. of Electrical, Electronic & Systems Engineering
Universiti Kebangsaan Malaysia
Bangi, Selangor, Malaysia
p105752@siswa.ukm.edu.my

Puvaneswaran Chelvanathan
Solar Energy Research Institute (SERI)
Universiti Kebangsaan Malaysia
Bangi, Selangor, Malaysia
cpuvaneswaran@ukm.edu.my

Ahmad AA Bakar
Dept. of Electrical, Electronic & Systems Engineering
Universiti Kebangsaan Malaysia
Bangi, Selangor, Malaysia
ashrif@ukm.edu.my

Avinash A/L Kumaresan
Dept. of Electrical, Electronic & Systems Engineering
Universiti Kebangsaan Malaysia
Bangi, Selangor, Malaysia
a176946@siswa.ukm.edu.my

Iskandar Yahya
Dept. of Electrical, Electronic & Systems Engineering
Universiti Kebangsaan Malaysia
Bangi, Selangor, Malaysia
iskandar.yahya@ukm.edu.my

Abstract— There is an increased call for research aimed at finding viable substitute material for indium tin oxide (ITO) in optoelectronics applications, given the high cost and scarcity of ITO. Carbon nanotubes (CNTs) are among the most promising nanomaterials for creating thin films, not only in optoelectronics but also across various other applications. In comparison with alternative nanomaterials, CNTs offer distinct advantages attributable to their exceptional electronic, optical, and mechanical properties. The quality of thin films composed of single-walled carbon nanotubes (SWCNTs) predominantly hinges on the material itself, the fabrication method employed, and the post-treatment techniques applied. In this research, SWCNT thin films were produced utilizing an automated spray coating technique. These films underwent treatment with nitric acid vapor to enhance their optoelectronic characteristics. Evaluation of the SWCNT thin films encompassed electrical characterization using a 4-point probe, as well as optical characterization via UV spectroscopy and field emission scanning electron microscopy (FESEM). The optimal SWCNT film exhibited a sheet resistance value of 7.23 Ω/sq. and a transmittance percentage of 60.4 at a wavelength of 800 nm. This study thus paves the way for the fabrication of transparent conductors based on SWCNTs that rival the electronic properties of ITO.

Keywords—carbon nanotubes, thin film, spray coating, acid treatment.

I. INTRODUCTION

Recent years have witnessed rapid developments in the miniaturization of electronic devices, driving research towards enhanced thin film production. Thin film utilization is commonplace in engineering and technology, aimed at augmenting electronic performance. A thin film encompasses any film with a thickness ranging from an atom-layer to the millimeter scale. Nanostructured materials are highly sought-after in thin film production due to their lightweight, strength, and flexibility. Carbon nanotubes (CNTs) stand among the finest nanomaterials for optoelectronic thin film fabrication, attributed to their exceptional electrical, optical, and mechanical characteristics [1][2][3].

Material purity, fabrication processes, and post-treatment methods are key determinants of thin film quality [4].

Various methods such as spin coating, dip coating, and vacuum filtering are popular for fabricating CNT thin films. However, challenges persist in achieving uniformity and cost-effectiveness in CNT thin film fabrication. Spray deposition, due to its cost-efficiency, emerges as a promising method for both research and industrial applications [5].

Indium tin oxide (ITO) is a leading thin film material known for its high transmittance (90%) and low sheet resistance (10 Ω/sq) [6]. Given the scarcity of natural indium reserves, alternative materials must be explored. Although CNTs are among the finest ITO alternatives, they still exhibit higher sheet resistance compared to ITO. Consequently, innovative techniques to manipulate CNT electrical properties are urgently required to establish them as potential transparent conductor materials [7]. Post-treatment methods, including acid treatment, have been reported to reduce CNT thin film sheet resistance. Nitric acid, a potent acid, is frequently employed. Gu et al. conducted a study involving different post-treatment solutions such as HNO_3, HCl, and $NaNO_3$. Comparative investigations reveal that HNO_3-treated CNT-TCFs exhibit optimal optoelectronic characteristics, notably transmittance and sheet resistance, owing to the acid's effective surfactant residue removal and subsequent decrease in sheet resistance due to its strong acidity [8].

In this study, we strive to modify the electrical and optical characteristics of SWCNTs thin film by subjecting it to a treatment with nitric acid vapor. This is the primary purpose of our work, and it is accomplished by employing spray coating to create SWCNTs thin film. Using these methods, we were able to create CNT thin films with drastically enhanced electrical and comparatively good optical qualities. Additionally, we demonstrated a rapid and efficient modification of the SWCNT thin film physical morphology, which was validated by field emission electron microscopy methods. In this investigation, the fabricated SWCNTs thin film sheet resistance value is 7.23 Ω/sq., which is a significant improvement over our previous SWCNTs thin film sheet resistance value of 1,330 Ω/sq.[9].

979-8-3503-2369-6/23 $31.00 © 2023 IEEE

II. Experimental Details

A. Material

SWCNTs Nano powder is the primary material that is provided by China's Chengdu Organic Chemicals Co. Ltd., and the purity of this powder is more than 95 %. Sigma-Aldrich provides the surfactant Tx-100 that is used and the primary function is to improve the solution's dispersion(28). The 15.44-molarity Nitric acid was provided by KGA Germany and the posttreatment process is the primary use for this nitric acid.

B. Glass substrate cleaning method

All SWCNT thin-film fabrication experiments were carried out in the open air. The clean glass slide was brushed five times on the front side and five times on the back side and again 5 times in front side. Glass slides were then cleaned with DI water. The glass slides should be placed in the baker, and the sonification process should be carried out using methanol for 10 minutes (min), ethanol for 5 min, and methanol once again for 5 min. After that glass slides clean with DI water and do the sonification with DI water for 10 min. After that, nitrogen gas was used to clean the glass slides, which were then immediately transferred to an 80 °C hotplate for 5 min. Finally, glass slides were subjected to a 10 min ozone treatment [10].

C. Solution preparation method

Tx-100 2 % and 10 mg of SWCNT Nano powder(measured from the Nano balance) add into the 40 ml DI water. This solution was subjected to the homogenizer tip probe ultra-sonification. The whole tip probe was cycled through 6 cycles, each consisting of a 2 min run followed by a 1 min rest. If the tip probe is allowed to run continuously, the CNTs tubular structure may get damaged [11].

D. Automatic spray system Experimental setup.

Fig 1 depicts the experimental setup for SWCNT thin film fabrication.

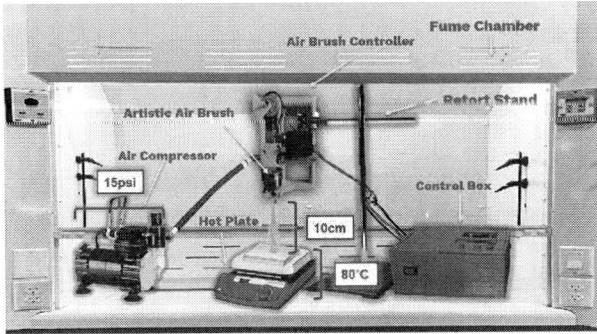

Fig. 1. Experimental setup of Automatic spray system

The air compressor pressure was kept at 15 psi and was connected to the automated airbrush system. The 80 °C temperature was maintained on the hot plate. A 230 V AC provide the electric power supply to the hotplate and air compressor. The main spray parameters in this experiment are 10 cm nozzle distance and 6 spray pluses. Every single spray pulse has a temporal period of one second. The solution for the SWCNTs is sprayed onto the glass, and then immediately thereafter, it is evaporated using a hotplate set to 80 °C. This is done to avoid the SWCNTs from clumping together because of the surface tension [12].

E. Experimental setup of Nitric Acid Treatment

The experimental setup for the nitric acid vapour treatment is shown in Fig 2.

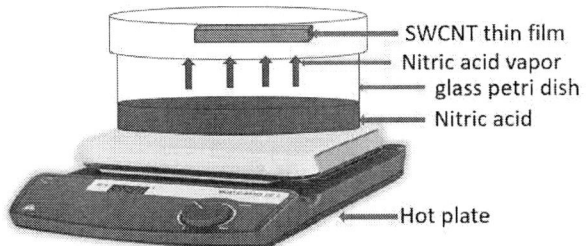

Fig. 2. Experimental setup of Nitric Acid Treatment

In the transparent glass peri dish, half of the available area was taken up by nitric acid. During the vapor treatment process hot plate surface temperature was kept as 85 °C. To treat the SWCNT thin film, 8 molarity nitric acid was used and the duration of acid treatment time is 5 min and 10 min.

F. Measurement and Analysis

To measure the sheet resistance value, 4-point probe was used which is manufactured from Ossila Ltd.UK. To determine the transmission percentage UV-Vis spectroscopy (Perkin Elmer 950) was used. To analyse the SWCNT thin film physical structure, field emission scanning electron microscopy (FESEM) was utilized.

III. Result and Discussion

A. Electrical charateristics.

Fig 3 shows the 5 min nitric acid vapor treatment SWCNT thin film sheet resistance value. Fig 4 shows the 10 min nitric acid vapor treatment SWCNT thin film sheet resistance value

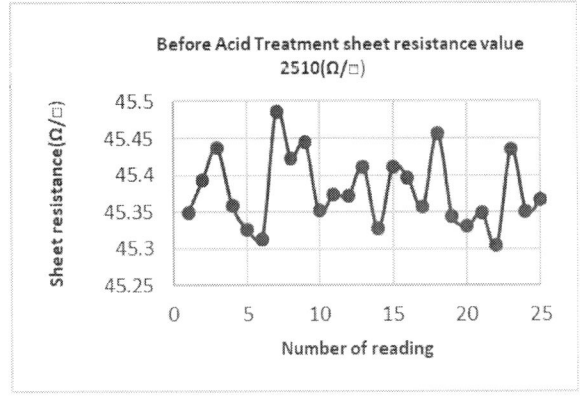

Fig. 3. 5 min nitric acid vapour treatment SWCNT thin film sheet resistance

SWCNT thin film 25 sheet resistance values were measured in each vapor treatment and are shown in Fig 3 and Fig 4 . Before the treatment with nitric acid the SWCNT thin film sheet resistance value is 2510 Ω/sq. The average sheet resistance value for SWCNT thin film throughout the five minutes of acid treatment is 45.38 Ω/sq. SWCNT thin film

979-8-3503-2369-6/23 $31.00 © 2023 IEEE

had an average sheet resistance value of 7.23 Ω/sq. after being subjected to an acid treatment for 10 min. The sheet resistance value suddenly drops after the acid treatment, which is particularly noticeable during the duration of the 10 min. In the research field ITO thin film has the very lower sheet resistance value which is 10 Ω/sq. However, during the 10 min treatment SWCNT thin film shows the lower than the ITO thin film sheet resistance value.

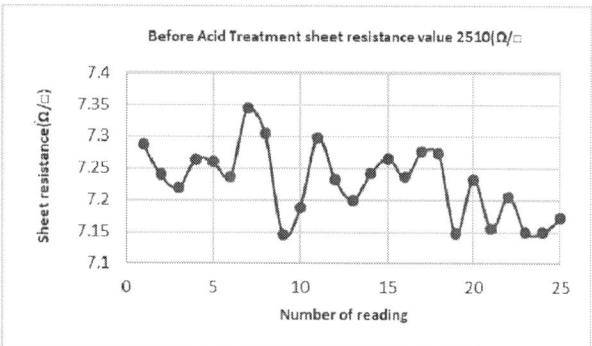

Fig. 4. 10 min nitric acid vapour treatment SWCNT thin film sheet resistance

Before acid treatment, surfactant prevents interaction between CNT tubes; consequently, conductivity is low and sheet resistance is high. During the nitric acid vapor treatment, nitric acid forms n and m ions, which readily permeate the CNT-surfactant bond and increase the likelihood of CNT interconnection. When CNT tubes are interact each other SWCNT thin film conductivity become high and sheet resistance become low [4].

B. UV-Vis-spectrophotometer Transmission spectra Analysis

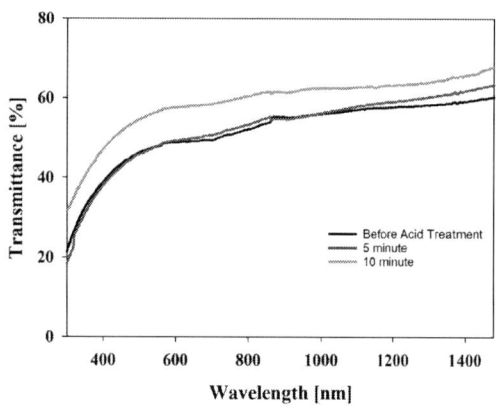

Fig. 5. Transmittance percentage samples from various SWCNT types across a variety of wavelengths.

Fig 5. depicts the UV transmittance curve of SWCNT thin film before it was subjected to acid treatment, as well as after it was subjected to acid treatment for 5 min and 10 min. The UV transmittance curve is able to be analyzed in three wavelength regions, including ultraviolet, visible, and infrared. During a 10 min vapor treatment, SWCNT thin film demonstrates the highest transmittance percentage among the three regions. SWCNT thin film that received a 5 min acid treatment showed a slight increase in transmittance compared

to untreated SWCNT thin film. In the 800nm wavelength, SWCNT thin film transmittance percentage values before acid treatment, 5 min acid treatment, and 10 min acid treatment are 52.1, 53.4 and 60.4, respectively. According to the preceding results, increasing the nitric acid treatment duration has a significant probability of increasing the transmittance percentage. The removal of surfactant percentage and the removal of carbon percentage are the two reasons for increasing the transmittance percentage after acid treatment.

C. FESEM Morphology characteristics

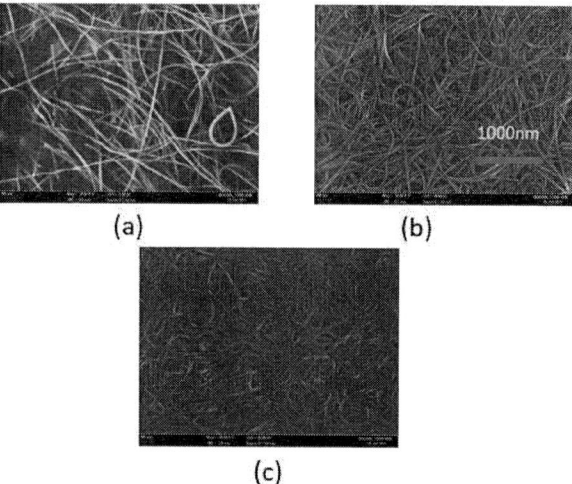

Fig. 6. FESEM images of (30k magnification) of SWCNTs various sample (a) before acid treatment (b) 5 min acid treatment (c) 10 min acid treatment.

Fig 6 displays FESEM images of a SWCNTs thin film prior to acid treatment, after acid treatment for 5 min, and after acid treatment for 10 min. The FESEM images used to observe and analysis the SWCNTs thin film. In SWCNT thin film prior to acid treatment, CNT tubes are plainly visible and fully covered with surfactant. Here surfactant is visible with white color. During the acid treatment time of 5 min, CNTs tube interaction is high, and SWCNT thin film size decreases. In the 5 min treatment time, the percentage of surfactant in SWCNT thin film is also low compared to before acid treatment because the film's white color is diminishing. In a treated SWCNTs thin film, the proportion of surfactant is extremely low after 10 min because, in comparison to other films, the whiteness of the treated film is much less, the interaction between the CNT tubes is quite strong, and the film also becomes thinner. Overall, as the surfactant percentage decreases, there is a higher likelihood that CNT tubes will contact with one another, making the film more conductive.

D. Figure of merit Analysis.

The figure of merit (FOM) is one of the best analyzing methods for describing the efficiency of a thin film for transparent conductivity application. The thin film transparency and sheet resistance value are the primary factors that determine FOM. The equation for the FOM analysis, which may be found in equation 1.

979-8-3503-2369-6/23 $31.00 © 2023 IEEE

$$FOM = \frac{T^{10}}{R_S} \quad \ldots\ldots\ldots (1) \; [13].$$

Where T is the transmittance percentage divided by 100 and R_S is the sheet resistance.

TABLE I.   FOM ANALYSIS CALCULATION OF POST TREATMENT OF SWCNTs THIN FILM

Sample	Rs (Ω/sq.)	T (%) at 550 nm	Figure of Merit (FOM)	Fabrication Method
10 min acid treatment	7.23	56.55	4.63×10^{-4}	Our work
Ref [14]	57	65	2.4×10^{-4}	Spray coating
Ref [15]	59	71	5.5×10^{-4}	Spin coating
Ref [16]	472	85	2.1×10^{-4}	Spray coating
Ref [17]	670	60	9.0×10^{-6}	Spray coating
Ref [18]	100	70	2.8×10^{-4}	Rod coating

Table 1 displays the values for sheet resistance, transmittance, and FOM for various fabrication techniques of SWCNTs thin film. When compared to other methods of manufacturing, our technique's FOM value shows the second highest value; however, when it comes to the spray coating method, our method's FOM value shows the highest value.

IV. CONCLUSION

In this study, the effects of nitric acid treatment at different times on the properties of thin films made of spray-coated SWCNTs were investigated. SWCNTs thin film has an impact on the transmittance value and sheet resistance value during the acid treatment. After acid treatment, SWCNTs thin film demonstrates an enormous reduction in sheet resistance and an increase in transmittance. Finally prepared SWCNTs thin film is good replacement for the ITO ,due to its high conductivity and good transmittance .

V. ACKNOWLEDGEMENT

This research was funded by the Ministry of Higher Education of Malaysia through the Fundamental Research Grant Scheme (FRGS/1/2021/TK0/UKM/02/29) and the Research University Grant (GUP-2022-006) of Universiti Kebangsaan Malaysia (UKM).

VI. REFERENCES.

[1] I. Yahya, M. A. Hassan, N. Nasyifa, M. Maidin, and M. A. Mohamed, "SWCNT Network - FET Device for Human Serum Albumin Detection," 2022.

[2] I. Yahya, L. L. Theng, S. M. Mustaza, H. Abdullah, and N. Amin, "Characterization of transparent conducting carbon nanotube thin films prepared via different methods," Sains Malaysiana, vol. 46, no. 7, pp. 1103–1109, 2017, doi: 10.17576/jsm-2017-4607-13.

[3] I. Yahya, A. Kunaraj, S. M. Mustaza, S. Clowes, and S. R. P. Silva, "Methods for Estimating Composition of Single Walled Carbon Nanotubes Based on Electronic Type," Materials Science Forum, vol. 1055 MSF, pp. 77–86, 2022, doi: 10.4028/p-073fwj.

[4] H. Gao, R. Izquierdo, and V. Van Truong, "Chemical vapor doping of transparent and conductive films of carbon nanotubes," Chemical Physics Letters, vol. 546, pp. 109–114, 2012, doi: 10.1016/j.cplett.2012.07.047.

[5] M. A. Hassan, M. A. Mohamed, A. Kunaraj, and I. Yahya, "Morphology and Electrical Characteristics of Single-Walled Carbon Nanotubes Film Prepared by Air Brush Technique," International Journal of Integrated Engineering, vol. 14, no. 3, pp. 273–279, 2022, doi: 10.30880/ijie.2022.14.03.029.

[6] Y. Zhou and R. Azumi, "Carbon nanotube based transparent conductive films: progress, challenges, and perspectives," Science and Technology of Advanced Materials, vol. 17, no. 1, pp. 493–516, 2016, doi: 10.1080/14686996.2016.1214526.

[7] D. Janas, S. Boncel, A. A. Marek, and K. K. Koziol, "A facile method to tune electronic properties of carbon nanotube films," Materials Letters, vol. 106, pp. 137–140, 2013, doi: 10.1016/j.matlet.2013.04.111.

[8] Z. Z. Gu, S. L. Jia, G. Li, C. Li, Y. Q. Wu, and H. Z. Geng, "Mechanism of surface treatments on carbon nanotube transparent conductive films by three different reagents," RSC Advances, vol. 9, no. 6, pp. 3162–3168, 2019, doi: 10.1039/c8ra09443h.

[9] A. Kunaraj, sinnia kodikama , Iskandar Yahya, Ahmad AA Bakar, and P. Chelvanathan, "Single-walled carbon nanotube (SWCNT) thin film for optoelectronics device prepared by spray coating technique.," IOP Conf. Series: Materials Science and Engineering, pp. 0–9, 2023, doi: 10.1088/1757-899X/1278/1/012007.

[10] [10] E. S. Hossain et al., "Enhancement in structural and optical properties of copper tin sulphide (CTS) thin films via sulphurization process," Materials Science in Semiconductor Processing, vol. 143, 2022, doi: 10.1016/j.mssp.2022.106496.

[11] A. Kunaraj, P. Chelvanathan, A. A. A. Bakar, and I. Yahya, "Single-Walled Carbon Nanotube (SWCNT) thin films via automatic spray coating and nitric acid vapor treatment," Journal of Engineering Research, Jul. 2023, doi: 10.1016/j.jer.2023.07.002.

[12] M. A. Hassan, M. A. Mohamed, H. Abdullah, and I. Yahya, "Effect of channel length on single walled carbon nanotubes thin film characteristics deposited via spray coating technique," Proceedings - 2021 IEEE Regional Symposium on Micro and Nanoelectronics, RSM 2021, no. 2, pp. 169–172, 2021, doi: 10.1109/RSM52397.2021.9511594.

[13] G. Haacke, "New figure of merit for transparent conductors," Journal of Applied Physics, vol. 47, no. 9, pp. 4086–4089, 1976, doi: 10.1063/1.323240.

[14] S. Kim, J. Yim, X. Wang, D. D. C. Bradley, S. Lee, and J. C. DeMello, "Spin-and spray-deposited single-walled carbon-nanotube electrodes for organic solar cells," Advanced Functional Materials, vol. 20, no. 14, pp. 2310 2316, 2010, doi: 10.1002/adfm.200902369.

[15] J. W. Jo, J. W. Jung, J. U. Lee, and W. H. Jo, "Fabrication of highly conductive and transparent thin films from single-walled carbon nanotubes using a new non-ionic surfactant via spin coating," ACS Nano, vol. 4, no. 9, pp. 5382–5388, 2010, doi: 10.1021/nn1009837.

[16] S. Paul and D. W. Kim, "Preparation and characterization of highly conductive transparent films with single-walled carbon nanotubes for flexible display applications," Carbon, vol. 47, no. 10, pp. 2436–2441, 2009, doi: 10.1016/j.carbon.2009.04.045.

[17] Q. Bu, Y. Zhan, F. He, M. Lavorgna, and H. Xia, "Stretchable conductive films based on carbon nanomaterials prepared by spray coating," Journal of Applied Polymer Science, vol. 133, no. 15, pp. 1–8, 2016, doi: 10.1002/app.43243.

[18] B. Dan, G. C. Irvin, and M. Pasquali, "Continuous and Scalable Fabrication of Nanotube Films," ACS Nano, vol. 3, no. 4, pp. 835–843, 2009.

Equivalent Circuit Model and Simulation of 2D Asymmetrical PMUT for Non-Destructive Testing

Darven Raj Ponnuthurai
Institute of Microengineering and Nanoelectronics (IMEN)
National University of Malaysia (UKM)
Bangi, Selangor, Malaysia
p126284@siswa.ukm.edu.my

Anis Amirah Alim
Institute of Microengineering and Nanoelectronics (IMEN)
National University of Malaysia (UKM)
Bangi, Selangor, Malaysia
p103655@siswa.ukm.edu.my

Mohd Farhanulhakim Mohd Razip Wee
Institute of Microengineering and Nanoelectronics (IMEN)
National University of Malaysia (UKM)
Bangi, Selangor, Malaysia
m.farhanulhakim@ukm.edu.my

Rhonira Latif
Institute of Microengineering and Nanoelectronics (IMEN)
National University of Malaysia (UKM)
Bangi, Selangor, Malaysia
rhonira@ukm.edu.my

Abstract— Miniature, low-power transducers and arrays of such transducers that are linked to electronic systems are employed in a variety of ultrasonic applications, such as sensing, actuation, and imaging. One alternative to integrated piezoelectric micromachined ultrasonic transducers (PMUTs) is thin film flexural transducers, which resemble a diaphragm and are usually produced on silicon substrates. First, an equivalent circuit model of the PMUT system has been discussed, followed by an analysis of the thin film piezoelectric transducer using simulation on various mediums. We present measurements of the PMUTs inherent electrical, mechanical, and acoustical characteristics. The presence and absence of solid objects does a significant change in the behavior of the acoustical waves that propagates from the simulated PMUT in different mediums.

Keywords— *Piezoelectric materials, microelectromechanical systems, ultrasonic transducers, equivalent circuit, non-destructive testing*

I. INTRODUCTION

Miniature devices or systems called micro-electromechanical systems (MEMS) combine mechanical parts, sensors, actuators, and electronics on a small scale. They can only be a few micrometres in size and are produced utilising microfabrication processes [1-3]. Compact and effective devices with a variety of capabilities, including signal processing, sensing, and actuation, may be created using MEMS technology. These gadgets have uses in consumer electronics, telecommunications, environmental monitoring, and healthcare. Micro-electromechanical systems (MEMS) are a subset of transducers that include micromachined ultrasonic transducers (MUTs). These elements are integrated onto a chip or substrate, allowing for miniaturization and integration with other MEMS components, such as microfluidic channels, sensors, or signal processing circuitry [4,5].

A micromachined ultrasonic transducer (MUT) is a small and efficient device that converts electrical energy into ultrasonic sound waves and vice versa. It utilizes microfabrication techniques to create a compact structure, often incorporating a thin film membrane material that deforms under electrical stimulation, generating ultrasonic waves [6]. MUTs have diverse applications in fields such as medical imaging, industrial sensing, and distance measurement, offering advantages such as high sensitivity, low power consumption, and the ability to operate at different frequencies.

A typical micromachined ultrasonic transducer (MUT) is built on two guiding principles. When using piezo ceramic or piezo composite materials, the piezoelectric effect has typically used to make thin film, micromachined, and bulk mode piezoelectric ultrasonic transducers. The second technique uses a capacitive micro-machined ultrasonic transducer (CMUT) made by electrostatically deflecting a flexible membrane electrode [7].

PMUT is a type of transducer that utilizes the piezoelectric effect and microfabrication techniques to generate and detect ultrasonic waves. PMUTs consist of tiny vibrating structures made of piezoelectric material, which deform when an electrical signal is applied, producing ultrasonic waves. These transducers are compact, efficient, and capable of operating at a wide range of frequencies. PMUTs find applications in various fields, including medical imaging, industrial sensing, and consumer electronics [8]. The performance of the PMUT mainly depends on the optimization of various parameters to achieve its goal and to enhance its capabilities [9]. By refining design parameters such as electrical characteristics and damping factors the transducer's sensitivity, bandwidth, and efficiency can be significantly improved. Simulation tools, like the Mason equivalent circuit model, aid in iteratively fine-tuning the design. Experimentation and validation refine the optimized design, leading to a PMUT with heightened sensitivity, wider bandwidth, and superior overall performance for diverse ultrasonic applications, ranging from medical imaging to industrial sensing, non-destructive testing and so on. The primary focus of this study is to optimize the effects of electrical and mechanical impedance in enhancing the performance of PMUT using Mason equivalent circuit.

II. METHODOLOGY

A. Equivalen Circuit Model

Despite advancements in PMUT research, lower effective electromechanical coupling and bandwidth in manufactured devices cannot be improved with predictive modelling and optimisation. The definitions of the Mason equivalent circuit's combined mechanical, electrical and acoustical impedance characteristics are a feature of more thorough techniques that improve model possibilities. Analytical models should concentrate on the Mason model equivalent circuit for practicality since it clearly specifies the needs for bandwidth, transmit and receive sensitivity, and power amplification circuits.

Using Mason equivalent circuit model, a direct method is used to increase the performance of the 2D PMUT. The

---

This research is funded under the grant GUP-2022-070 from the National University of Malaysia.

979-8-3503-2369-6/23 $31.00 © 2023 IEEE

properties of this medium play a critical role in determining the sensitivity and bandwidth of the transducer. The equivalent circuit includes components for electrical impedance, mechanical impedance, and acoustic impedance, which are all influenced by the properties of the medium [10]. The mechanical impedance of the transducer is related to the mechanical domain properties of the medium, and it affects the sensitivity of the transducer. Specifically, the mechanical impedance affects the amount of mechanical energy that is transferred between the transducer and the medium, and therefore influences the sensitivity of the transducer to changes in the medium properties.

$$Z_m = \frac{M_s s^2 + B_m s + K_m}{s^2 + (B_m + M_m R_{m2})s + K_m} \quad (1)$$

Equation (1) shows mechanical impedance of the system using second-order transfer function where M_m, B_m, K_m, R_m and s represents the mass, the damping coefficient, stiffness, resistance of the system and complex frequency variable [10]. The numerator of the equation represents the force acting on the system due to its mass, damping, and stiffness. It can be seen as the mechanical input to the system. The denominator of the equation represents the mechanical response of the system. It characterizes how the system reacts to the applied force.

$$Z_e = \frac{(\frac{1}{C_e s})}{(\frac{1}{C_e s}) + R_e + L_e} \quad (2)$$

The electrical impedance of the transducer is related to the electrical domain properties of the medium, and it affects the impedance performance of the transducer. Equation (2) shows the electrical impedance of the system using second-order transfer function where C_e, R_e, L_e and s represents the capacitance, resistance, inductance of the system and complex frequency variable [10]. The numerator of the equation represents the electrical current flowing through the system due to its capacitance. It can be seen as the electrical input to the system. The denominator of the equation represents the electrical response of the system. It characterizes how the system reacts to the applied electrical current.

The acoustic impedance of the medium is related to the acoustic domain properties of the medium, and it affects the sensitivity and impedance performance of the transducer. The acoustical impedance is not explicitly represented as a second-order transfer function for this model. Instead, the acoustical impedance is typically incorporated into the overall mechanical impedance of the transducer. The acoustical impedance is typically captured through the acoustic load impedance element in the model.

B. Simulation Study

The software COMSOL Multiphysics version 5.5. was used to simulate the asymmetrical 2D PMUT analysis. The Eigen frequencies are simulated using modal analysis in this programme and the software solves the coupled structural-electrostatic equations to simulate the behaviour of the piezoelectric transducer. A 2D component design is used and the piezoelectric material Barium Titanate (BaTiO₃) was chosen due to its processing flexibility, biocompatibility, and good mechanical and piezoelectric properties [11].

As shown in Fig. 1, the circular diaphragm is set to a fixed constraint at all sides and the boundary is set as water and air medium. The inclusion of a perfectly matched layer (PML) at the boundary of the medium can be used to minimize reflection and artificially simulate an infinite domain. The PML layer acts as an absorber for acoustic waves, reducing the reflection of waves at the boundaries of the computational domain. It helps provide a more accurate representation of the PMUT's response and its interaction with the medium under investigation. Each side of the piezoelectric transducer has been fixed its own ground, terminal, voltage and resistor. For this simulation study, acoustic structure boundary and piezoelectric effect were chosen to understand the effect caused by the piezoelectric material on the output signal and the propagation of the waves.

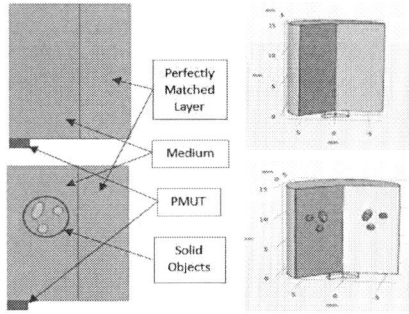

Fig. 1. The 2D asymmetrical and 3D view of the PMUT model with the presence and absence of objects in air and water mediums.

For the analysis, the total acoustic pressure inside the mediums with solid objects and without it is studied as well as the terminal voltage obtained. From the time dependent studies, the displacement field in z-axis is also obtained. A few parameters have been changed such as the presence and absence of solid objects and the different medium used. The list of parameters used in the simulation model is listed in the Table 1. Optimization of parameters needed to be done in order to simulate a transducer that has higher efficiency and sensitivity to detect the solid object.

TABLE 1. Parameters applied in the simulations

Parameters	Values
Thickness of piezoelectric membrane	0.5 mm
Width of piezoelectric membrane	2 mm
Initial operating frequency	100 kHz
Density of water	997 kg/m³
Density of air	1.225 kg/m³
Young's Modulus of BaTiO₃	67 GPa [11]
Piezoelectric constant of BaTiO₃, d₃₃	160-350 pC/N [11]
Relative permittivity of BaTiO₃, ε'	1680 [11]
Acoustic velocity of BaTiO₃	1790 m/s [11]

III. RESULTS

From the data obtained from the simulation there was 2 signal spikes that can be clearly seen from Fig. 2. This indicates the initial transmitted signal frequency at 100 kHz that has the highest amplitude and the amplitude of the received signal that is smaller than the transmitted signal.

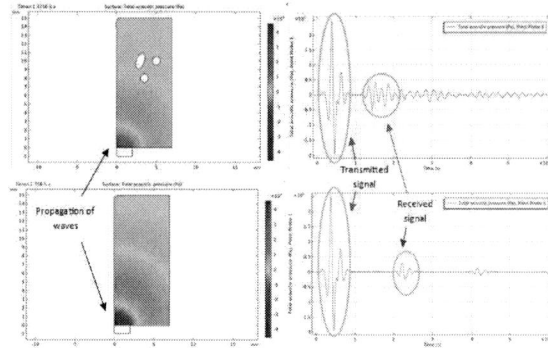

Fig. 2. The total acoustic pressure within the water medium with the presence and absence of solid object.

From the observed data, when there are no solid objects in the medium, the acoustic waves can propagate more freely and travel at a higher speed. This is because the absence of solid objects reduces scattering, absorption, and other interactions that may slow down the wave propagation. On the other hand, when solid objects are present in the medium, they can act as obstacles or scatterers for the acoustic waves. These objects can cause wavefront distortion, reflections, and diffraction, which may alter the path and speed of the acoustic waves. Consequently, the received signal arrives at the PMUT faster due to the direct or reflected paths provided by the solid objects. This observation highlights the importance of considering the influence of objects or obstacles in the medium when analysing and interpreting the received signals from PMUTs in practical applications such as imaging, sensing, or non-destructive testing.

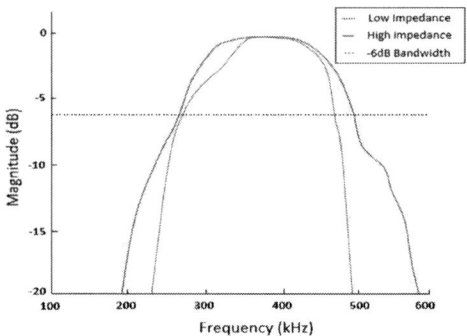

Fig. 3. The bandwidth of the PMUT in water medium by lowering and increasing its impedance

The data from the COMSOL simulation was used to perform Fast Fourier Transform (FFT) analysis to calculate the bandwidth as shown in Fig. 3. Optimizing parameters to achieve low impedance configuration, through careful tuning of parameters such as capacitance, inductance, and mechanical properties such as damping, fosters efficient energy transfer between mechanical and electrical domains. This results in a broader bandwidth, enabling the PMUT to effectively respond to a wider range of frequencies. In contrast, a high impedance configuration can restrict energy exchange and lead to resonance at specific frequencies, potentially narrowing the transducer's bandwidth. From the optimization of the Mason equivalent circuit a bandwidth of 64% was calculated from the initial model that has 58% bandwidth.

The terminal voltage and total acoustic pressure data from the COMSOL simulation have been used to estimate the amplitude sensitivity. The amplitude sensitivity in Fig. 4 shows observable difference in peaks of the amplitude sensitivity between the air and water medium. This can be attributed to the contrasting acoustic properties of the two mediums. In general, air has a lower density and speed of sound compared to water, resulting in a smaller acoustic impedance. This impedance mismatch between the PMUT and air allows for a more efficient transfer of energy, leading to a higher amplitude sensitivity in air. On the other hand, water, with its higher density and slower speed of sound, poses a greater impedance, resulting in a reduced amplitude sensitivity.

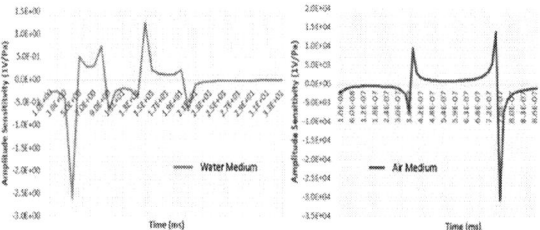

Fig.4. The amplitude sensitivity of the PMUT with the presence of solid objects in both air and water medium.

IV. CONCLUSION

Based on the simulation result obtained, utilizing an equivalent circuit model to optimize the impedance of PMUTs holds immense potential for expanding their bandwidth and enhancing their performance in diverse applications such as non-destructive testing and medical imaging. By strategically tuning the electrical and mechanical parameters within the circuit, including capacitance, inductance, stiffness, damping, the impedance can be effectively lowered, resulting in a broader frequency response. This improved impedance matching enables PMUTs to efficiently transmit and receive signals across a wider spectrum, thereby enabling more accurate and versatile ultrasonic sensing capabilities. This advancement holds significant promise for driving advancements in various fields reliant on high-performance ultrasonic technology.

ACKNOWLEDGMENT

This research was funded by Universiti Kebangsaan Malaysia, grant number GUP-2022-070.

REFERENCES

[1] Aqil, M. M., Azam, M. A., Aziz, M. F. & Latif. R. "Deposition and Characterization of Molybdenum Thin Film Using Direct Current Magnetron and Atomic Force Microscopy", Journal of Nanotechnology, vol. 2017, Article ID 4862087,10 pages, 2017. https://doi.org/10.1155/2017/4862087.

[2] Zawawi, S.A.; Hamzah, A.A.; Majlis, B.Y.; Mohd-Yasin, F. A "Review of MEMS Capacitive Microphones". Micromachines 2020, 11, 484. https://doi.org/10.3390/mi11050484

[3] Latif, R.; Noor, M.M.; Yunas, J.; Hamzah, A.A. "Mechanical Energy Sensing and Harvesting in Micromachined Polymer-Based Piezoelectric Transducers for Fully Implanted Hearing Systems: A Review". Polymers 2021, 13,2276.https://doi.org/10.3390/polym1314 2276

[4] Latif, R. B. Y. Majlis, & R. Cheung. "MEMS design and modelling based on resonant gate transistor for cochlear biomimetical

application". Microsystem Technologies, 2016. 23(7), 2329–2342. doi:10.1007/s00542-016-2937-9.

[5] Sun, S.; Wang, J., Zhang, M. *et al.* "MEMS ultrasonic transducers for safe, low-power and portable eye-blinking monitoring". *Microsyst Nanoeng* **8**, 63, 2022. https://doi.org/10.1038/s41378-022-00396-w.

[6] Tuan Yaakub, T.N.; Yunas, J.; Latif, R.; Hamzah, A.A.; Razip Wee, M.F.M.; Yeop Majlis, B. Surface Modification of Electroosmotic Silicon Microchannel Using Thermal Dry Oxidation. *Micromachines* 2018, *9*,222.https://doi.org/10.3390/mi905 0222.

[7] Wang, J *et al.*, "An Ultra-Low Power, Small Size and High Precision Indoor Localization System Based on MEMS Ultrasonic Transducer Chips," in *IEEE Transactions on Ultrasonics, Ferroelectrics, and Frequency Control*, vol. 69, no. 4, pp. 1469-1477, April 2022, doi: 10.1109/TUFFC.2022.3148314.

[8] T. -H. Hsu, A. A. Zope, M. -H. Li and S. -S. Li, "A Compact Monolithic CMUT Receiver Front-End in a TiN-C CMOS-MEMS

Platform," *2020 IEEE International Ultrasonics Symposium (IUS)*, Las Vegas, NV, USA, 2020, pp. 1-4, doi: 10.1109/IUS46767.2020.9251733.

[9] He, L.-M.; Xu, W.-J.; Wang, Y.; Zhou, J.; Ren, J.-Y. Sensitivity—Bandwidth Optimization of PMUT with Acoustical Matching Using Finite Element Method. *Sensors* 2022, *22*, 2307. https://doi.org/10.3390/s22062307

[10] S. Akhbari, F. Sammoura, & L. Lin, "Equivalent circuit models for large arrays of curved and flat piezoelectric micromachined ultrasonic transducers," IEEE Trans. Ultrason., Ferroelectr., Freq. Control, vol. 63, no. 3, pp. 432–447, Mar. 2016.

[11] Hu, S.; Luo, C.; Li, P. *et al.* Effect of sintered temperature on structural and piezoelectric properties of barium titanate ceramic prepared by nano-scale precursors. *J Mater Sci: Mater Electron* 28, 9322–9327 (2017). https://doi.org/10.1007/s10854-017-6670-7.

Effect of biasing under illumination on GaAsBi/GaAs multiple quantum wells for solar cell performance

Faezah Harun
Electronics Technology Section
British Malaysian Institute, Universiti
Kuala Lumpur
Gombak, Malaysia
faezah@unikl.edu.my

Robert D. Richards
Department of Electronic and Electrical
Engineering University of Sheffield
Sheffield, UNITED KINGDOM
r.richards@sheffield.ac.uk

John P.R David
Department of Electronic and Electrical
Engineering University of Sheffield
Sheffield, UNITED KINGDOM
j.p.david@sheffield.ac.uk

Abstract— **Gallium arsenide bismide (GaAsBi) is a potential candidate to replace InGaAs as the middle-junction structure in a multiple junction solar cell due to its lower level of lattice strain in the structure. The level of strain in the lattice structure is challenging as it can manipulate the total current collected in a solar cell, affecting the total output power. In this study, a GaAsBi multiple quantum wells with GaAs barriers (GaAsBi/GaAs) p-i-n structure was grown by using MBE machine and fabricated. A current-voltage (I-V) under illumination measurements were conducted to study the effect of biasing with the presence of light. This result is compared with a strained-balance InGaAs/GaAsP device. From the photocurrent output, it is shown that GaAsBi can achieve a longer cut-off wavelength, around 1053nm compared to InGaAs/GaAsP, with only 930 nm cut-off wavelength. The result also shows that the device is extracting low carriers from the photon sweeps at zero bias. A small amount of reverse bias is needed to allow carrier enhancement and increase the number of carriers collected by the device. As the forward bias is applied, photocurrent value drops as the forward dark current dominates the total current output. In conclusion, GaAsBi/GaAs MQW can be a competitive alternative to InGaAs in achieving a 1eV material system for photovoltaic, especially when the growth and structural component is optimized, improving its strain level and dark current density.**

Keywords— bismuth-containing material, III-V semiconductor, multijunction photovoltaics, 1eV bandgap, current-voltage

I. INTRODUCTION

The incorporation of bismuth into III-V semiconductor material systems allows a large reduction in band gap value. Bismuth can reduce the bandgap about 80 meV/%, much larger compared with indium (16%) or antimony (21%).[1] This creates a lot of interest in different technology application especially in telecommunication, spintronics and solar cell.[2-4] In solar cell or photovoltaic application, currently InGaP/InGaAs/Ge layers has been proven to reach 40.7% efficiency rate under research and development phase.

However, InGaAs can impose lattice-mismatch during the growth and causes dislocation, therefore introducing strain to the lattice structure. In order to achieve higher theoretical efficiency available for near-lattice matched top junction, 1 eV material is needed. Researchers

are looking into a potential III-V semiconductor material combination where the band gap can be adjusted to 1eV since there is no known binary semiconductor material with this bandgap value. Therefore, it is of interest that bismuth can be a competitive material to replace InGaAs in junction's structure. [5]

According to study, introducing just 6% of bismuth into GaAs will allow the bandgap to reach 1 eV.[6] The rate of change of bandgap per unit strain on a GaAs based ternaries is 750meV/% strain with bismuth, compared to 200meV/% strain with indium. The rate of change suggests that bismuth imposed less strain level when it is grown on GaAs, only 0.7% compared to indium on GaAs, which is 1.9% for the same bandgap value.[7]

Several studies have been done to successfully incorporate bismuth into GaAs. A higher bismuth incorporation percentage can be achieved by using a low temperature during the growth process, typically below 400°C.[1,8] However, the material quality is compromised due to the formation of growth defects and localized density of states above the valence band.[9] The surface roughness on the grown sample is obvious. It can also lead to defect-assisted recombination to occur and decreases the minority carrier lifetime. In turns, the dark current density will increase. The increase in dark current is undesired because it causes performance losses and impact negatively to the electrical performance of a solar cell. This is the electrical parameter that need to be well-adjusted in order to obtain higher efficiency, without compromising the quality of the material.

In this paper, the opto-electrical characterizations of strained-GaAsBi MQW device is studied and compared with strained-balanced material system. The material system will be noted as (*well/barrier*) devices for the rest of the work e.g GaAsBi/GaAs is GaAsBi wells grown in between GaAs barriers. Illuminated current-voltage measurements were done to study the material's behavior under the presence of photon or light, mimicking the Sun. A p-i-n layer structure is used in this study because it is a simple structure that can be used to predict a semiconductor material's behavior. This is useful in identifying the efficiency and electrical behavior of GaAsBi when it is

979-8-3503-2369-6/23 $31.00 © 2023 IEEE

actually working as one of the junctions in a multiple junction solar cell structure.

II. METHODOLOGY

A. Device structure

A GaAsBi/GaAs multiple quantum wells (MQW) p-i-n diode was by using a molecular beam epitaxy (MBE) machine. This device is a strained GaAsBi based device and it has 20 quantum wells of GaAsBi alloy. Each well's thickness is fixed at 8nm of 3% of bismuth while the GaAs barrier thickness is set at ~22nm each. The device under test will be called as QW20 and its layer structure is as shown in Fig. 1, where n in the structure means number (of wells). Previous work done on this p-i-n diode, including the growth specification, fabrication procedures and other characterizations has been extensively discussed. [10-12]

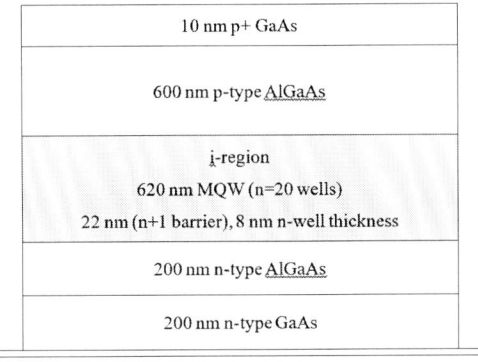

| 10 nm p+ GaAs |
| 600 nm p-type AlGaAs |
| i-region
620 nm MQW (n=20 wells)
22 nm (n+1 barrier), 8 nm n-well thickness |
| 200 nm n-type AlGaAs |
| 200 nm n-type GaAs |
| GaAs n+ substrate |

Fig. 1: QW20 device structure.

The experimental work for QW20 is compared with a strain-balanced InGaAs/GaAsP MQW from Quantasol Ltd. The layer structure is as shown in Fig. 2. The material device is chosen because it has nearly zero strain percentage property, which is the ideal condition for a lattice-matched photovoltaic application. The Quantasol's InGaAs/GaAsP will be stated as QT1879 for the rest of this work.

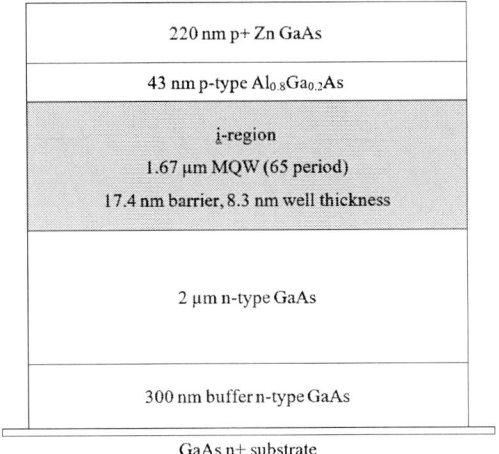

| 220 nm p+ Zn GaAs |
| 43 nm p-type Al$_{0.8}$Ga$_{0.2}$As |
| i-region
1.67 µm MQW (65 period)
17.4 nm barrier, 8.3 nm well thickness |
| 2 µm n-type GaAs |
| 300 nm buffer n-type GaAs |
| GaAs n+ substrate |

Fig. 2: QT1879 device structure.

B. Experimental method

A set of photocurrent measurements as function of wavelengths were taken at zero bias, forward biases and reverse biases voltage by using a monochromator and source-measure unit (SMU). The current created due to different wavelengths are collected and quantified by using a phase-sensitive lock-in detection method. A GaAs filter was used to filter out light absorption from the barrier structure in the devices. The wavelength for GaAs and Al$_{0.8}$Ga$_{0.2}$As barriers is 873nm and 699nm, respectively. This procedure will allow only current collection due to the carrier excitation from the quantum wells only, thus improving the accuracy of the current measured.

Firstly, the photocurrent measurements taken at zero bias to study the photocurrent created due to the incident lights with presence of built-in voltage only. Then, a low reverse bias is applied to ensure a full depletion of i-region width. This process is continued until no increase in current collected when reverse bias is increased. Thirdly, the effect of forward dark current of the device on the total current is observed while photocurrent measurements are taken. This is done by applying forward biases during the photocurrent measurement takes place.

III. RESULTS AND DISCUSSION

A. Photocurrent results

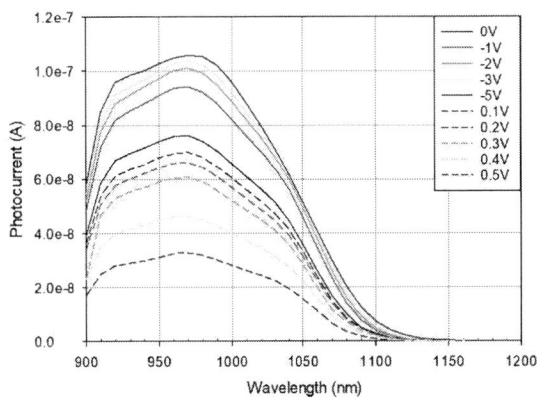

Fig. 3: Photocurrent values with respect to wavelength at different biases.

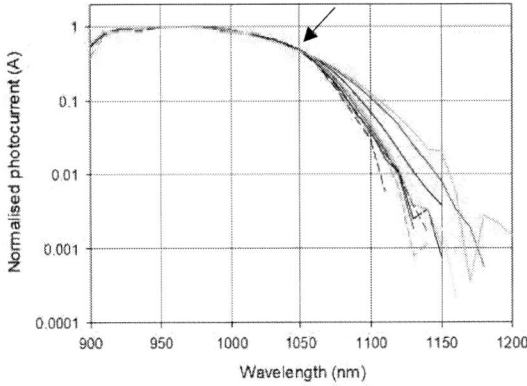

Fig. 4: Photocurrent values with respect to wavelength when the data is normalized to unity.

Fig. 3 shows the photocurrent data for QW20 taken at 0 V, reverse bias and slight forward bias, taken at room temperature. From 0 V to reverse bias applied, the photocurrent collected increases up to 40%. The photocurrent collected by QW20 reaches saturation when the reverse voltage is -5 V. This shows that at zero bias, the carrier extractions due to built-in voltage by the material is incomplete and there are carriers trapped in the wells. The result also suggest that the carrier trapped is actually holes in valence band, as bismuth-based material incorporation is normally caused by a raising of the valence band energy. The manipulation of the data from Fig. 3 can be represented in Fig. 4. Y-axis is set at log scale for a better data representation.

From the figure, after about 1050 nm, the photocurrent value starts roll-off. The pointed wavelength, where the clear pinch off is seen before the photocurrent disperse for different reverse bias is at 1053 nm, and it is considered as the cut-off wavelength of QW20. The cut-off wavelength or also known as absorption band edge agrees with the room temperature PL peak emission reported in our previous work, with ±10 nm difference. At forward bias, the roll-off of the photocurrent after the band edge are the same except at 0.5 V where the roll-off has a slight deviates from the rest.

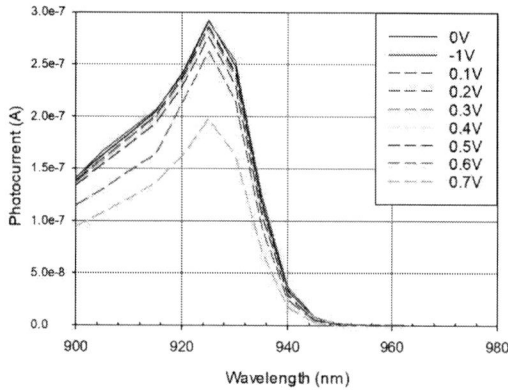

Fig. 5: Photocurrent measurement of QT1879 at different biases with respect to wavelength.

Zero bias, forward and reverse-biased photocurrent measurements were also taken on QT1879 for comparison, as shown in Fig. 5. The photocurrent value remains unchanged as both forward and reverse bias is applied, and only showing reduction in current when the forward voltage is slightly higher, at around 0.5V. Since this device is a strain-balanced material, it is expected to see little to no change to its photocurrent as the biases are manipulated. Although QT1879 has a really good electrical properties and stable photocurrent collection with respect to biases, its cut-off wavelength is limited to 930nm absorption, or a bandgap value of 1.33 eV minumum. If the bandgap for InGaAs/GaAsP is stretched beyond this value to achieve a 1 eV material, it may lose its strain-balance properties.

B. Comparison on Effect of Biasing

Fig. 6 shows the percentage data of photocurrent collected for QW20 and QT1879 at their photocurrent spectrum peak, 980nm and 925nm, respectively. Both values at zero bias are set as 100% and other values follow accordingly.

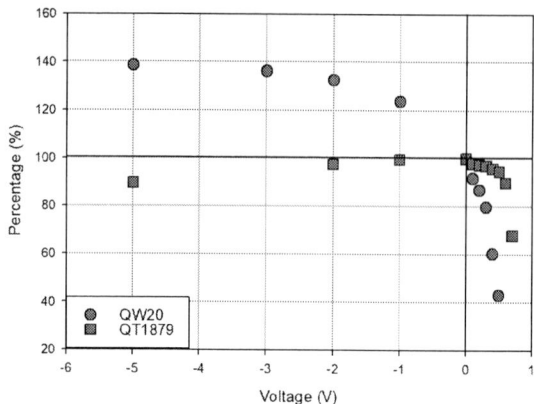

Fig. 6: Percentage change for photocurrent values relative to zero bias.

By comparing the value obtained in Fig. 6 with depletion width calculation (from C-V measurement, not shown here), it is confirmed that photocurrent increases with negative bias is due to escaped carriers trapped, instead from the un-depleted i-region of the device. In addition to that, slight forward bias is causing the the photocurrent magnitudes to be dropped dramatically, up to ~60% reduction when the biasing goes to 0.5V. The dark current in QW20 at forward bias is significant compared to the photocurrent created in the i-region, therefore the total current value drops. To mitigate this issue, the photocurrent measurements will need to use a very high resistor value in order to filter out the dark current. This is critical as the photocurrent is the important element in identifying the ideal power output or known as "fill factor".

On the contrary, QT1879 shows a very small change to the percentage change after biasing. It can be conclude that carrier extraction is collected completely at zero bias. The reverse bias applied do not improve the value of photocurrent collected because there is no carrier trapped in the well. Besides that, the drop in photocurrent value in forward bias is also negligible and can only be seen after 0.5 V forward bias applied, compared to QW20 which is seen as soon as small forward bias is applied. This is because, QT1879 has a lower dark current, about two magnitude lower compared to QW20. Therefore, it allows constant photocurrents collection at forward bias.

IV. CONCLUSIONS

In conclusion, the study shows that GaAsBi/GaAs MQW has the potential to reduce the bandgap from 1.42eV to 1.17eV (1053nm band edge) by having 3 percent of bismuth incorporation. The biasing applied during wavelength sweep shows that photocurrent collected can changed at different potential difference. While reverse bias allows complete carrier extraction and improving the photocurrent collected, forward bias reduce photocurrent collection due to dark current domination at forward bias.

Overall, a better-quality bismuth-based material with higher efficiency can be obtained, if the carrier collection component and source of dark current is identified and studied. It is also important to figure out the optimum growth conditions of bismuth-based material, where low growth temperature can still be imposed for bismuth incorporation, and at the same time the quality of the material is not compromised.

REFERENCES

[1] S. Tixier *et al.*, "Molecular Beam Epitaxy Growth of GaAs1-xBix,", *Applied Physics Letters,* Article vol. 82, no. 14, pp. 2245-2247, Apr 2003, doi: 10.1063/1.1565499

[2] Sweeney, S. J. *et al.*, The potential role of bismide alloys in future photonic devices. In 13th *International Conference on Transparent Optical Networks*, (IEEE) (2011).

[3] Marko, I. P. & Sweeney, S. J., Progress toward III–V bismide alloys for near-and midinfrared laser diodes. *IEEE J. Sel. Top. Quantum Electron.* **23**(6), 1–12 (2017).

[4] Thomas, T. *et al.* Requirements for a GaAsBi 1 eV sub-cell in a GaAs-based multi-junction solar cell. *Semicond. Sci. Technol.* **30**(9), 094010 (2015)

[5] Richards, R. D., *et al.* GaAsBi: An alternative to InGaAs based multiple quantum well photovoltaics. In *43rd Photovoltaic Specialists Conference (PVSC)*, 1135–1137 (2016)

[6] T. Thomas *et al.*, "Requirements for a GaAsBi 1eV Sub-cell in a GaAs-based Multijunction Solar Cell," *Semiconductor Science and Technology,* Article vol. 30, no. 9, p. 6, Sep 2015, Art no. 094010, doi: 10.1088/0268-1242/30/9/094010

[7] J. F. Geisz *et al.*, "40.8% Efficient Inverted Triple-Junction Solar Cell with Two Independently Metamorphic Junctions," , *Applied Physics Letters,* Article vol. 93, no. 12, p. 3, Sep 2008, Art no. 123505, doi: 10.1063/1.2988497

[8] Young, E. C. *et al.* Bismuth incorporation in GaAs1−xBix grown by molecular beam epitaxy with in-situ light scattering. *Phys. Status Solidi C* **4**(5), 1707–1710 (2007).

[9] F. Harun *et al.*, "Effect of bismuth flux on the optical and morphological properties of GaAsBi grown by Molecular Beam Epitaxy," *2022 IEEE 8th International Conference on Smart Instrumentation, Measurement and Applications (ICSIMA)*, pp. 127-131, doi: 10.1109/ICSIMA55652.2022.9929153.

[10] R. D. Richards, "Molecular beam epitaxy growth and characterisation of GaAsBi for photovoltaic applications," ed, September 2014.

[11] F.Harun, "Characterisation of GaAsBi/GaAs Multiple Quantum Wells for Photovoltaic Applications", ed, February 2020

[12] F.Harun *et al.*, "Opto-Electronic Characterisation of GaAsBi/GaAs Multiple Quantum Wells for Photovoltaic Applications." Solid State Phenomena, vol. 343, Trans Tech Publications, Ltd., 30 May 2023, pp. 99–104. doi:10.4028/p-j0b6ku.

979-8-3503-2369-6/23 $31.00 © 2023 IEEE

The effects of particle sizes of neodymium iron boron microstructure on the magnetic characteristics

Siti Aisyah Ishak
Malaysia-Japan International Institute of Technology (MJIIT)
Universiti Teknologi Malaysia (UTM)
Kuala Lumpur, Malaysia
saishak@graduate.utm.my

Jumril Yunas
Institute of Microengineering and Nanoelectronics (IMEN)
Universiti Kebangsaan Malaysia (UKM)
Bangi, Malaysia
jumrilyunas@ukm.edu.my

Abdul Manaf Hashim
Malaysia-Japan International Institute of Technology (MJIIT)
Universiti Teknologi Malaysia (UTM)
Kuala Lumpur, Malaysia
abdmanaf@utm.my

Abstract—In this work, the magnetic properties of neodymium iron boron ($Nd_2Fe_{14}B$) powder with different particles sizes are studied using a vibrating sample magnetometer (VSM). The results show that as the particles size decrease, the saturation magnetization (B_s) and remanent magnetization (B_r) decrease while coercivity (H_c) increases to its maximum and subsequently declines towards zero when a particle's critical radius is reached. In particular, it has been observed when the particle size reduces to the range of 1.0 to 2.0 μm, the B_s and B_r of the particles significantly drop to 120.37 Am^2/kg and 84.38 Am^2/kg, respectively and its H_c increases to 0.69 T. Further decrease in size in the mass powder, however causing the H_c to decline to 0.08 T and the B_s and B_r reduce to 99.55 Am^2/kg and 32.54 Am^2/kg, respectively as the particle's critical radius is reached. Despite having almost identical average particle sizes, the presence of a notable amount of smaller particles in a particular powder causes the magnetic properties value to shift significantly. The magnetic behavior suggested a superparamagnetic behavior of the smallest NdFeB particles studied.

Keywords—magnetic particles, ferromagnetic, NdFeB

I. INTRODUCTION

Magnetic particles in micro and nano sizes have emerged as crucial material for a wide range of applications such as robotics, biomedical, healthcare, data storage and microelectromechanical systems (MEMS) [1-3] due to the availability of large surface area for interactions and effective dispersion, providing enhanced reactivity and binding capacity. Additionally, their magnetic properties allow for easy and remote manipulation and actuation using external magnetic fields [2], enabling efficient recovery and reusability. Furthermore, magnetic microparticles also offer excellent stability, biocompatibility, and minimal toxicity, making them ideal for biomedical applications [1-2].

Anisotropic neodymium iron boron (NdFeB) magnetic particles possess excellent permanent magnetic properties. Few factors such as their shape, crystal structure, defects, surface chemistry of the particles, density, volume, grain alignment and size distribution, can influence or affect their magnetic properties [2-4]. Despite numerous benefits, magnetic particles also come with certain restrictions. Agglomeration of particles can occur, particularly in fabrication process [5], hindering their dispersion and maximum capacity and reactivity. Additionally, the

manufacturing of uniform-sized magnetic microparticles with precise control over their shape and size distribution is vital and can be challenging [4-5]. For reliable performance in applications and constant magnetic characteristics, achieving monodispersity is crucial [2] In this study, the effects of particle sizes of NdFeB powder on the magnetic characteristics are investigated.

II. EXPERIMENTAL

The NdFeB alloy powder were acquired from ShenZhen HongMin Magnet, China and reduced in size using ball-milling process. By varying the milling time, two other different particles sizes were prepared. The as-received NdFeB powder and the two other different sizes were labelled as NdFeB A, NdFeB B and NdFeB C, respectively. The milling time of NdFeB B is shorter than NdFeB C which means that the average particle size of NdFeB B should be larger than NdFeB C but smaller than NdFeB A. The average particle size, D for each sample was calculated by fitting the distribution of particle size histogram to the log-normal distribution function:

$$f(D) = \left(\frac{1}{\sqrt{2\pi}\sigma_D} \right) \exp\left[-\frac{\ln^2 D/D_0}{2\sigma^2} \right]$$

(1)

where D_0 is the median diameter and σ is the standard deviation [3], [6-7]. The magnetic particles morphology was observed through field emission scanning electron microscopy system (FE-SEM, JSM-7500 F, JEOL). To determine the magnetic properties of these particles powder, a vibrating sample magnetometer (VSM, 7404, Lake Shore) test was conducted. The measurement was carried out at room temperature over 1.5 T applied magnetic field.

III. RESULTS AND DISCUSSION

Fig. 1(a-c) show the morphology of NdFeB particles. The as-received powder NdFeB A (Fig. 1(a)) shows a mixture various shapes of coarse particles including angular and irregular block shapes. Fractured surface and pointed edge can be seen, showing brittle behaviour of NdFeB alloy. When the milling process is applied, both size and shape of the particles change significantly. Compared to NdFeB A, NdFeB B and C show relatively 100 times smaller particle size. The particles size of powder NdFeB A shows a wide range from 25 μm to

979-8-3503-2369-6/23 $31.00 © 2023 IEEE

500 μm, indicates a polydisperse distribution. Fig. 1(b) shows the morphology of the NdFe B powder which is milled with a shorter time than C, showing a mixture of fine and coarse particles ranging from 0.2 μm to 1.0 μm and agglomerated onto larger particles (see inset in Fig. 1(b)). As shown in Fig. 1(c), the majority of the powder particles show improved uniformity of the particle sizes and less agglomeration or clustering. It is speculated that the agglomeration or clustering occurs due to the absence or removal of oxide coating on the particle surfaces by the milling performed in the vacuum [8].

Typical fitting to particle size distribution histogram for all magnetic powder is illustrated in Fig. 1(d-f). The estimated average particle sizes for A, B and C are 55.88 μm, 0.52 μm, and 0.69 μm, respectively and their standard deviation values

are 9.42 μm, 1.20 μm and 1.15 μm. respectively. Although the average particle sizes are comparatively the same for NdFeB B and NdFeB C but -NdFeB C as portrayed in Fig. 1(f), seems to exhibit sizes tightly clustered around the 0.2 μm to 0.5 μm. Meanwhile, NdFeB B as shown in Fig. 1(e) exhibits a notable number of particles with sizes around 0.5 μm to 1.0 μm.

The result of VSM measurements that show magnetization vs. applied magnetic field is illustrated in Fig. 2. This graph demonstrates distinct magnetic behaviour for the three magnet particles sizes studied. The disparity in the magnetic characteristics occurred because of the size variation. The derived saturation magnetization (B_s), remanent magnetization (B_r) and coercivity (H_c) were listed in Table 1. True to form, the hard magnetic powder A with the largest

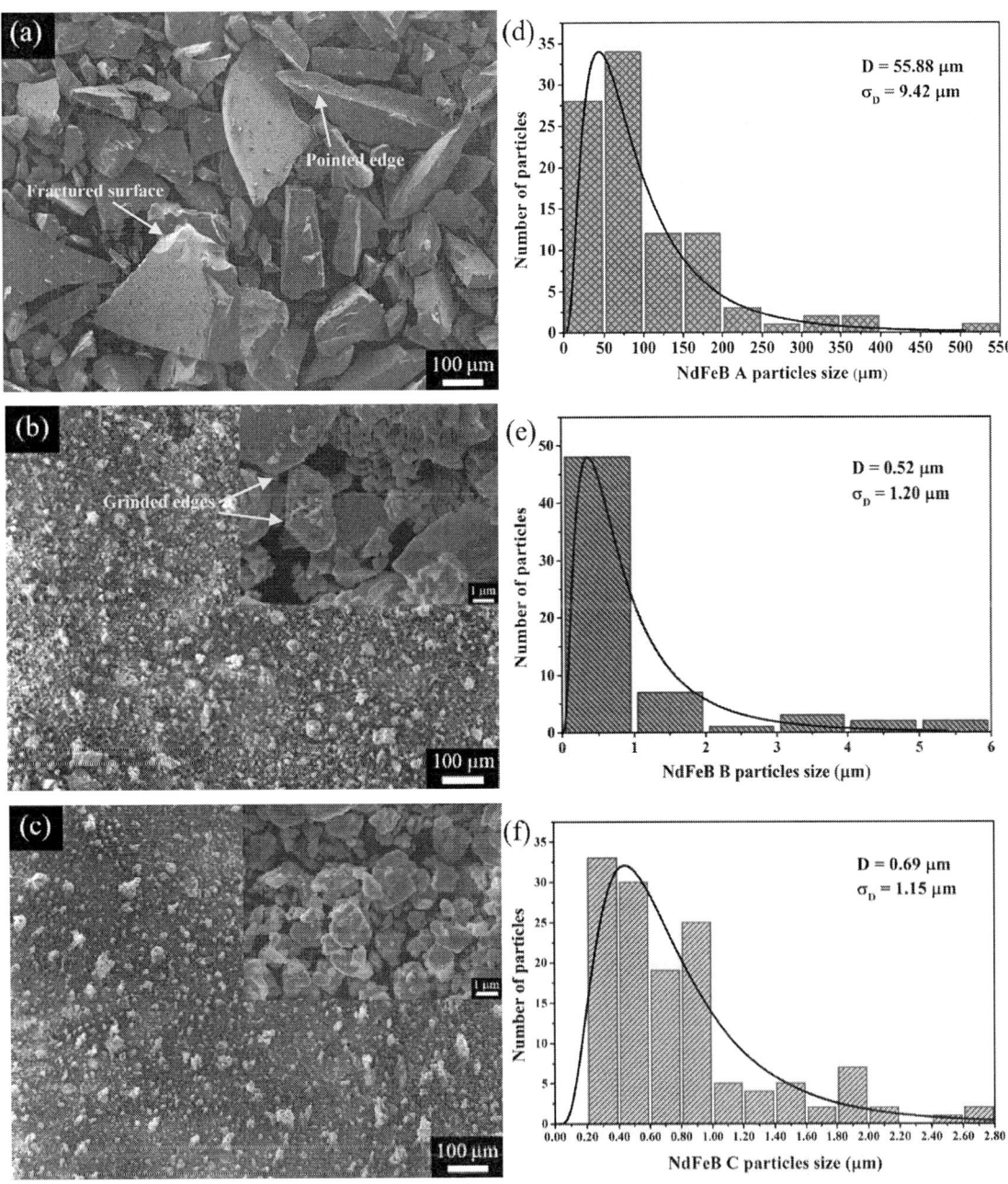

Fig. 1. The results of (a) NdFeB A, (b) NdFeB B, and (c) NdFeB C FE-SEM analysis;
(d) NdFeB A, (e) NdFeB B, and (f) NdFeB C particles size distribution fitted with a log-normal distribution function.

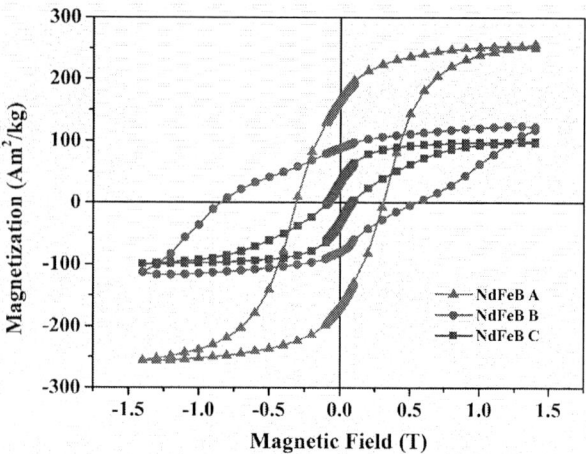

Fig. 2. Room temperature magnetization curves of NdFeB A, B and C samples measured.

dimension (see Fig. 1(a)) displays the highest B_s and B_r as shown in Table 1. The B_s and B_r were observed to be decreasing with the decrease of particle sizes [4-5]. With a decrease in particle size, the average magnetic domain size shrinks, hence, increase the surface to volume proportion of atoms. Inevitably, the increase of surface effects such as canted spin and dead layer at the particle surface due to crystal defects, core spin, disorganized surface spin, broken exchange bonds, dislocation, and lattice strain, causing a decrease in B_s [3], [6].

On the other hand, the values of H_c as in Table 1 show otherwise. H_c, which correlates with the width of the curve, is majorly size-dependent when dealing with minuscule particles [2], [9]. H_c of the as-received magnetic powder NdFeB A, increases as compared to other samples with smaller size, is in accordance with other studies [4-5]. As the powder being crushed into smaller particles, their domain wall energy will increase [9] making it more difficult for the domains to move. Consequently, it causes the H_c of the particles to increase [2]. However, as the particles reach their critical diameter, the H_c (usually represented in Oe) will start decreasing to zero and can be described as reaching a superparamagnetic behaviour [2], [7], [9]. This behavior can be seen from NdFeB C powder hysteresis curve. It is noted that the mean particle size of powder NdFeB C is slightly larger than powder NdFeB B, however the majority submicron particles in powder C are believed to cause such drastic changes in the magnetic properties. The depletion in H_c for powder NdFeB C indicates the formation of single domain particles in the sample [2][7]. When the magnetic field is reduced and completely removed, the magnetization drops significantly approaching zero [7], before increasing back to a significant value when an opposite magnitude of magnetic field is being applied. Among these three group size particles, NdFeB C having the lowest B_r and H_c.

The hard ferromagnetic character of the powder is demonstrated by the finding of wide magnetic hysteresis loops for all studied magnet powder. It is worth noting, every time the dipoles of magnetic particles change in direction with the applied magnetic field, there is heat generated and lost to the environment. This is resulted from magnetic anisotropy, which serves as an energy barrier to stop moments from switching away from the magnetic easy axis. This is represented by the area of the hysteresis loop. Smaller loops show less hysteresis loss and vice versa. The smaller the particles anisotropy, which is influenced by magnetic domain structure, the bigger energy losses [3], [6]. This can be seen in Fig. 2 where the area of loop of powder NdFeB A is smaller than of NdFeB B. Nonetheless, the single domain particles in powder NdFeB C generated much less heat due to rapid fluctuation of the magnetic moment.

IV. CONCLUSION

Three different sizes of microstructured magnetic NdFeB particles have been used to investigate their effects on the magnetic properties. The results show that by altering the physical characteristics of magnet particles, it can modify the magnetic properties of NdFeB particles. It has been observed that the value of B_s and B_r are directly proportional to the sizes of NdFeB particles. Whereas H_c ascends to a maximum as particle size decreases, and then weakens towards zero upon reaching the particles' critical radius.

ACKNOWLEDGMENT

This work was financially supported by the Universiti Teknologi Malaysia through UTM-FR research grant Vot: 22H16.

REFERENCES

[1] V. Iacovacci et al., "Polydimethylsiloxane films doped with NdFeB powder: magnetic characterization and potential applications in biomedical engineering and microrobotics," Biomed Microdevices, vol. 17, pp. 1–2, 112, December 2015.

[2] A. Akbarzadeh, M. Samiei, and S. Davaran, "Magnetic nanoparticles: preparation, physical properties, and applications in biomedicine," Nanoscale Res Lett, vol. 7, pp. 1–5, February 2012.

[3] H. Jalili, B. Aslibeiki, A. G. Varzaneh, and V. A. Chernenko, "The effect of magneto-crystalline anisotropy on the properties of hard and soft magnetic ferrite nanoparticles," Beilstein J. Nanotechnol., vol. 10, pp. 1351–1352, 1353–1354, July 2019.

[4] S. Namkung, D. H. Kim., and T. S. Jang, "Effect of particle size distribution on the microstructure and magnetic properties of sintered NdFeB magnets," Rev. Adv. Mater. Sci., vol. 28, pp. 187–188, July 2011.

[5] D. Shin et al., "Investigation of the magnetic properties and fracture behavior of Nd-Fe-B alloy powders during high-energy ball milling," Mater. Res. Express, vol. 7, pp. 1, 7, September 2020.

[6] S. K. Paswan et al., "Optimization of structure-property relationships in nickel ferrite nanoparticles annealed at different temperature," J. Phys. Chem. Solid, vol. 151, pp. 12–13, 17–19, January 2021.

[7] S. Savliwala et al., "Magnetic nanoparticles," in Nanoparticles for Biomedical Applications, E. J. Chung, L. Leon and C. Rinaldi, Eds. Gainesville, FL, USA: Elsevier, 2020, ch. 13, pp. 198–200.

[8] T. Miko, F. Kristaly, K. Bohacs, M. Sveda, A. Sycheva, and D. Janovszky, "The effect of process control agents and milling atmosphere on the structural changes of $Ti_{50}Cu_{27,5}Ni_{10}Zr_{10}Co_{2,5}$ master alloy during short time milling," IOP Conf. Ser.: Mater. Sci. Eng., vol. 426, p. 2, 2018 [11th Hungarian Conf. Mater. Sc, 2017.

[9] B. Gogoi and U. Das, "Magnetization dynamics of iron oxide super paramagnetic nanoparticles above blocking temperature," Mater. Today: Proc., vol. 65, pp. 2636–2637, May 2022.

TABLE 1   MAGNETIC PROPERTIES OF NDFEB POWDER

Powder	Magnetic properties		
	B_s (Am²/kg)	H_c (T)	B_r (Am²/kg)
NdFeB A	256.37	0.31	167.78
NdFeB B	120.37	0.69	84.38
NdFeB C	99.55	0.08	32.54

Evaluation of Cross-Contamination Risk during CMOS Devices Fabrication in an Industrial Silicon Wafer Processing

Amir Zulkefli
Quality Reliability Engineering,
Infineon Technologies (Kulim)
Sdn. Bhd.,
Kulim, Malaysia.
MohdAmir.Zulkefli@infineon.c
om

Ismail Umar
Quality Reliability Engineering,
Infineon Technologies (Kulim)
Sdn. Bhd.,
Kulim, Malaysia.
Ismail.Umar@infineon.com

Vanita Manaoogaran
Process Integration Engineering,
Infineon Technologies (Kulim)
Sdn. Bhd.,
Kulim, Malaysia.
Vanita.Manaoogaran@infineon.
com

Wan Hidayatulhusna
Production WET Process
Engineering,
Infineon Technologies (Kulim)
Sdn. Bhd.,
Kulim, Malaysia.
WanHidayatulhusna.WMohama
dRani@infineon.com

Oh Guan Kai
Process Integration Engineering,
Infineon Technologies (Kulim)
Sdn. Bhd.,
Kulim, Malaysia.
GuanKai.Oh@infineon.com

Deyline Samail
Process Integration Engineering,
Infineon Technologies (Kulim)
Sdn. Bhd.,
Kulim, Malaysia.
Deyline.Samail@infineon.com

Izzuddin Iskandar
Quality Reliability Engineering,
Infineon Technologies (Kulim)
Sdn. Bhd.,
Kulim, Malaysia.
Izzuddin.Iskandar@infineon.co
m

Abstract—In the last decade, large-scale production of semiconductor device technologies and applications have been fabricated in a state-of-the-art fabrication facility. For instance, CMOS smart power device, discrete and Integrated Circuit (IC) products have been mass production in the fabrication foundry. However, these device applications have been processed in the same fabrication line which composed of ion implantation, wet chemical etching, lithography, and others. For instance, ion implantation process could lead to contamination from Arsenic element from implant residue which leads to early electrical breakdown of the device. In order to avoid cross contamination, the mixed fabrication process line of CMOS smart power device and other IC products is taken care and additional containment process is introduced before the high temperature step. The following study demonstrates cross contamination effect analysis on CMOS smart power device and the containment action in order to prevent cross contamination in the semiconductor mass production line.

Keywords— *Cross contamination risk, CMOS smart power device, Fabrication facility, Containment action*

I. INTRODUCTION

Smart power device of CMOS-DMOS chip based on integrated circuit on silicon materials are achieving new benchmarking in smart power devices application in harsh environment (Figure 1) [1, 2]. Depending on the targeted applications, several integration routes of fabricating smart power device have been considered to achieve higher device performance as well as from an economical point of view. Besides, smart power device fabrication processes involve sharing expensive CMOS state-of-the-art processes with other device technology applications, which could lead to cross contamination especially around wet chemical treatments, ion implantation and others [3]. In this manner, cross-contamination risk must be evaluated in case of tool sharing or process sharing between each technology to avoid

any possible cross contamination, which can cause device performance and reliability degradation. In this paper, we propose to evaluate cross-contamination risk from smart power technology perspective when wet chemicals steps are performed separately from other technology like IC Bi-CMOS products.

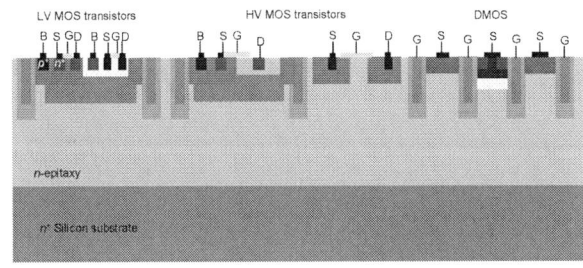

Fig. 1. Schematic diagram of smart power device of CMOS-DMOS chip on Si substrate.

II. EXPERIMENT AND RESULT

In this investigation, eight discipline problem solving (8D) methodology [4] (Figure 2) is used to investigate the electrical characteristic variation namely abnormal QBD distribution (blue line: before containment action where the lots were run near to end of chemical life while magenta line refers to lots running at early of chemical life) for CMOS smart power device as shown in Figure 3a, 3b, 3c, and 3d for DMOS, NMOS, PMOS, and Gate oxide capacitor (CGOXP), respectively. The hypothesis established in 4D state verified that cross contamination could originated from other technology during CMOS smart power device fabrication processes that undergoes silicon cleaning steps before sacrificial oxide oxidation. Based on the affected wet chemical traceability suggested that during the cleaning process, all affected CMOS smart power devices are batched together with other IC products in wet bath after implantation of Arsenic and Phosphorus, which could be come from cross contamination from this IC product implanted surface.

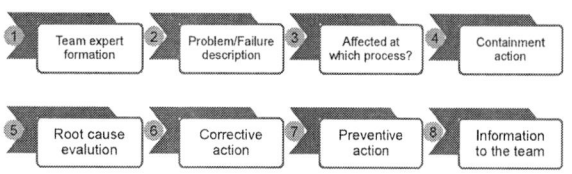

Fig. 2. 8D problem solving methodology for investigating the issue of electrical characteristic variation: abnormal QBD distribution of CMOS smart power devices.

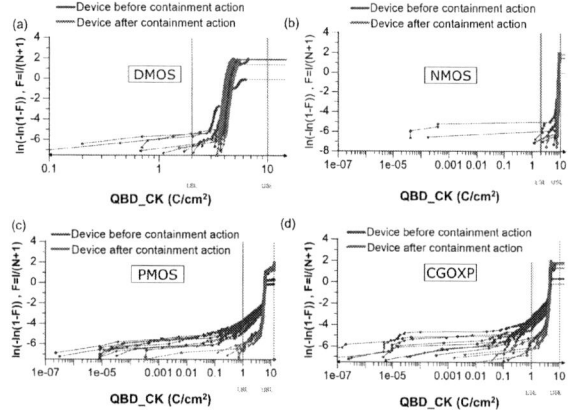

Fig. 3. Graph of probability, F=(I-0.5)/N vs. QBD_CK value for (a) DMOS, (b) NMOS, (c) PMOS, and (d) CGOXP. Blue line and magenta line refer to devices processed before and after containment action plan (normal QBD distribution), respectively.

To verify the hypotheses, two design of experiment (DOE) [5] were performed (Figure 4). DOE#1 is performed to re-produce the failure where suspected contaminating IC product were batched together with CMOS smart power device in the same wet batch. Then, DOE#2 is performed to identify the implant element that could cause the contamination in the wet batch where Bi-CMOS were implanted with Arsenic and Phosphorus separately in different wafers in order to segregate the contamination source. After that, these wafers were batched together in wet process into two groups (group#1: Bi-CMOS wafers implanted Arsenic+power device wafers; group#2: Bi-CMOS wafers implanted Phosphorus+power device wafers).

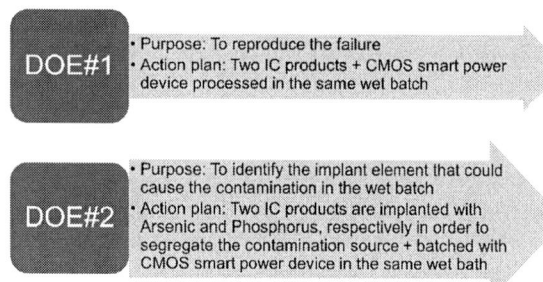

Fig. 4. Design of experiments were performed to verify the hypotheses of possible cross contamination from other IC products fabrication processes.

Figure 5 shows DOE#1 of current density measurement on the CMOS smart power devices resulted in an abnormal QBD distribution, which is out of lower spec. limit while Figure 6 shows DOE#2 of current density measurement on CMOS smart power devices batched in the same wet bath with IC products with implanted Arsenic (magenta line) shows extrinsic and tailing characteristics in QBD distribution while implanted phosphorus (blue line) shows normal QBD distribution, which suggested that contamination is originated from implanted Arsenic. As a containment action to eliminate the contamination, wet cleaning of IC products after implanted Arsenic and Phosphorus batching is removed from silicon cleaning steps of CMOS smart power devices before sacrificial oxide oxidation process wet bath, whereby this cross contamination is fully eliminated and will never happen in the production line (as shown in magenta line: Figure 3 showing normal QBD distribution, which is within the spec. limit after the containment action plan).

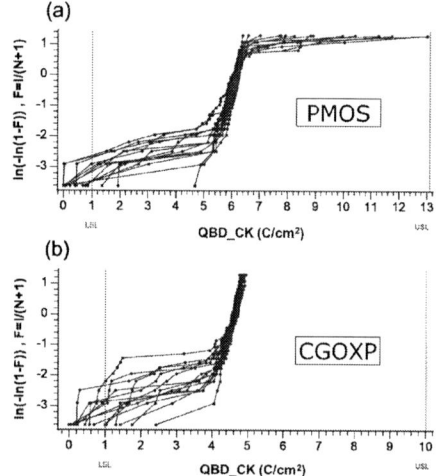

Fig. 5. DOE#1 of current density measurement on the CMOS smart power devices for (a) PMOS and (b) CGOXP.

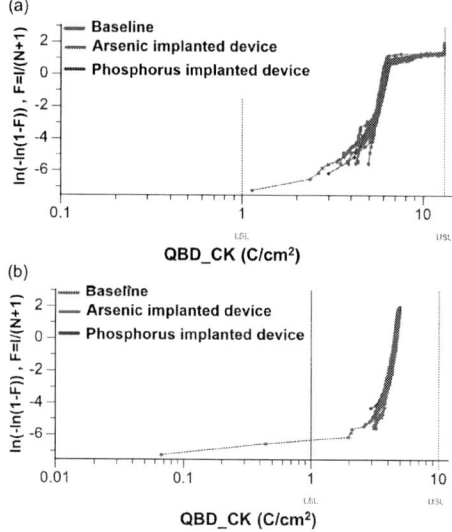

Fig. 6. DOE#2 of current density measurement on the CMOS smart power devices for (a) PMOS and (b) CGOXP. Magenta line and blue line refer to Arsenic and Phosphorus implanted device, respectively.

III. CONCLUSION

Based on the cross-contamination risk evaluation study, it is verified that the possible cross contamination of CMOS smart power device originated from other IC products of Arsenic implanted surface, which degraded the gate oxide quality of CMOS smart power device, causing electrical characteristic variation. As a permanent containment action for eliminating this cross contamination, additional step of reducing the temperature process at the affected wet chemical cleaning is introduced before processing at the high temperature fabrication process, thus, this cross contamination will never happen at the mass production line. Nevertheless, more efforts are still needed to further understand the details of these cross-contamination issues. Properties like elemental analysis of XRD for contaminated devices and contaminated elemental concentration of Arsenic/Phosphorus on the affected devices are to be considered for the future cross-contamination risk evaluation analysis.

ACKNOWLEDGMENT

We would like to express sincere thanks to Zool Helmi Hamzah for supporting the wet chemical process and Vijay Anand Ramadass for the fruitful discussion.

REFERENCES

[1] X. Du, J. Zhang, G. Li, Y. Yi, C. Qian, and R. Du, Thermal Reliability of Power Semiconductor Device in the Renewable Energy System. Springer Nature (2022) 9.

[2] G. Croce, Extended Abstracts of the 2022 IEEE International Memory Workshop (2022) pp. 1-4.

[3] A. Gupta et al., Advanced Etch Technology and Process Integration for Nanopatterning XI (2022) 12056.

[4] Y. T. Prasetyo et al., Extended Abstract of the 2021 IEEE 8th International Conference on Industrial Engineering and Applications (2022) pp. 360-367.

[5] M. Joshi, Design of experiments (DOE) in the semiconductor industry. IEEE/SEMI Advanced Semiconductor Manufacturing Conference and Workshop (1991) pp. 17-24.

NBTI Defects Characterization Using Energy Profiling Simulation Technique

Hanim Hussin
Integrated Microelectronic
System and Applications,
School of Electrical
Engineering,
College of Engineering,
Universiti Teknologi MARA,
40450, Shah Alam, Selangor
*hanimh@uitm.edu.my

Sharifah Fatmadiana Wan
Muhamad Hatta
Faculty of Engineering,
University of Malaya, 50603,
Kuala Lumpur
sh_fatmadiana@um.edu.my

Norhayati Soin
Faculty of Engineering,
University of Malaya, 50603,
Kuala Lumpur
norhayatisoin@um.edu.my

Yasmin Abdul Wahab
Nanotechnology & Catalysis
Research Centre,
University of
Malaya,
50603, Kuala Lumpur
yasminaw@um.edu.my

Maizan Muhamad
Integrated Microelectronic
System and Applications,
School of Electrical
Engineering,
College of Engineering,
Universiti Teknologi MARA,
40450, Shah Alam, Selangor
maizan@uitm.edu.my

Nurul Ezaila Alias
Faculty of Electrical
Engineering,
Universiti Teknologi Malaysia,
81310 Johor Bahru
ezaila@fke.utm.my

Abstract— **A numerical simulation framework to simulate the positive charges based on location of energy levels is conducted in this work. This framework utilizes an energy profiling approach, where the behavior of hole traps under stress conditions is studied. In this process, a recovery voltage is applied to facilitate charged hole traps release. The recovery voltage is incrementally increased in the positive direction, to move the Fermi level thus creating the desired energy spectrum. Experimental data suggests that the defect charges responsible for this phenomenon arise from various sources and simulation results able to probe these types of charges.**

Keywords— NBTI, p-MOSFET, positively charged defect, energy profiling, reliability component.

I. INTRODUCTION

The ongoing reduction in transistor size to below 20nm necessitates increased focus on device reliability [1]–[3]. One of the device reliability issues is NBTI which studied to have degradation mechanisms vary depending on the experimental method employed by prominent researchers in this field [4]–[6]. Each mechanism introduces distinct defects that contribute to the degradation process. Degradation lead to a Δvth of p-MOSFET, hence reduced the device performance [7][8]. HfO-based devices that contains more defect sites as compared to conventional devices give serious problem due to NBTI [9][10].

In the conventional NBTI experimental method, the mechanism is due to the created dangling bonds resultant from Si-H bonds breaking [1]. Another finding which is the deep level hole trap (DLHT) effect become positive charge, described in [11] exhibits a time exponent of 0.3. Apart from DLHT, another type of charge known as anti-neutralization positive charges (ANPC), which resides beyond the silicon E_c, is identified in [5]. The energy profiling method described in [5] highlight that the ANPC, CPC and AHT are distributed in different energy spectrum. Energy levels assigned to E'center/Pb H complex – interface trap, bulk defect E'center, and precursor as described by [12] is different as compared to in [5]. These energy levels is seen to be associated as highlighted by [13], that positively charged N_{it} are located beneath the E_i. Consequently, it is necessary to conduct a simulation-based study to assess the degradation caused by charges located beneath, between, and above the energy band gap of silicon.

II. SIMULATION FRAMEWORK USING ENERGY PROFILING APPROACH

Each trap are located at different energy levels as illustrated in [5]. In this work, the Two-stage NBTI model is employed [12]. . This model highlights that the contribution toward the NBTI degradation is associated with E'center/Pb H complex – interface trap, bulk defect E'center, and precursor. A comparison of newly defined energy levels presented in this study with those found in the literature can be seen in Table 1. To validate the accuracy of the NBTI simulation, various stress biases were applied. The trap density is extracted as well as the surface potential. Then, during the stress and un-stressed period, the resulting trap concentration, hence the vth were assessed. The pmosfet was subjected to a stress period of 10000s at 125°C, which aligns with the experimental conditions described in most previous works [5].

The new energy level exhibited discharge attributes that were comparable to those observed in experimental studies [5]. The extensive range of $V_{discharge}$ variations allowed for sweeping E_f across a wide spectrum. ranging from below the E_v, to above E_c. To ensure the complete discharge of all positive charges within the dielectric below E_f, a discharge duration of 3000s was employed at a specific $V_{discharge}$ value [5].

TABLE I   TYPE OF CHARGES AND THE LOCATIONS

	Defect charges	Energy Levels	Literature	
			Defect charges	Ref.
Energy profiling method	Precursor	Beneath E_v	AHT	[5]
			Precursor	[14]
	E'center	Within bandgap	CPC	[5]
			Switching Oxide	[12][15]
	E'center/ Pb H Complex	Above E_c	ANPC	[5]
			DLHT	[11]
Default [15]	Precursor	Beneath E_v	[16]	
	E'center	Fermi level – valence level = 0.3ev	[16][17]	
	E'center/ Pb H Complex	Fermi level – valence level = 0.5 eV	[8][9]	

The study utilized p-type devices that featured high-k as the dielectrics together with the SiO_2, and these devices were simulated using the 32-nm CMOS process established by the foundry. as referenced in [18].

III. RESULTS AND DISCUSSION

The discharging characteristics resulting from the original energy spectrum are shown in Fig. 1 (a). The discharging characteristics for the new energy level are shown in Fig. 1 (b). Both figures clearly demonstrate distinct discharging behaviors. In Fig. 1 (a). the Δvth during the settling progression saturates, indicating no further discharging effect beyond 3000s of discharge time. Conversely, as depicted by Fig. 1 (b), the Δvth almost completely recovers, approaching full recovery around 3000s. This disparity can be justified by the new energy level, by looking at the E' center/Pb H complex location. They are placed over the conduction band. The resultant mechanism is non-recoverable characteristics which examined by [3]. The non-recoverable degradation defect is named ANPC. Due to their energy level above E_c, these positively charged species are not easily neutralized by the application of positive bias, as electrons find it challenging to reach them [3]. In

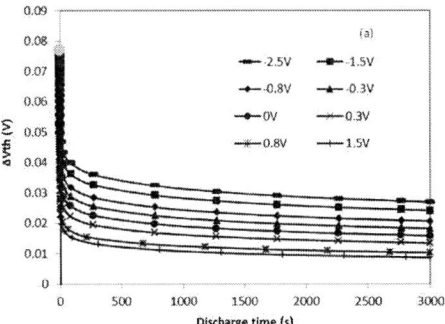

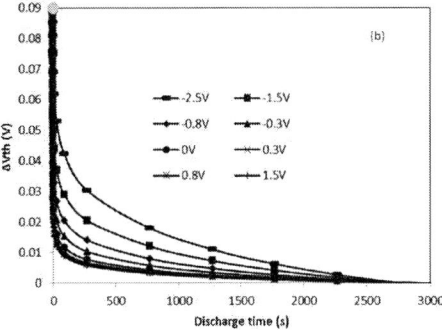

Fig. 1. (a) New energy location (b) original energy location.

contrast, the original energy location situates the Pb center differently. The location is below the energy mid-band gap, resulting in poor recoverability. However, the Pb center is able to recover for further discharge time. It is observed that full recovery near 3000s of discharge time.

Fig. 2 (a), (b), and (c) illustrate three different conditions of energy band diagrams. The shape of the energy band reflect the location of the fermi level as the wide range of stress voltage applied. At each specific biasing voltage, the extracted values of fermi level and valence level were used. The difference between fermi level and valence level, denoted as E_f - E_v, was then utilized to outline the overall charges obtained, as depicted in Fig. 2. Additional information regarding the overall charges is presented to provide a comprehensive explanation. Fig. 3 (a) and (b) display the total positive charges' contribution to the degradation concerning $V_{discharge}$ and E_f - E_v, respectively. Considering this work applies the two-stage NBTI concept, the Δvth, as provided in [12], is expressed as follows.

$$\Delta V_{th}(t) = -\frac{\Delta Q_{ox}(t) + \Delta Q_{it}(t)}{C_{ox}} \qquad (1)$$

The sum of ΔQ_{ox} and ΔQ_{it} is represented in Fig. 3. The distribution of each defect contributes to N_{eff}, and their energy levels have been extensively explained before. In accordance with [12], N_{ox} is the collective count of oxide traps from overall charges, excluding state 1 (the precursor state) is,

$$N_{ox}(t) = 1- < f_1(t) > \qquad (2)$$

In this study, the likelihood that source of degradation

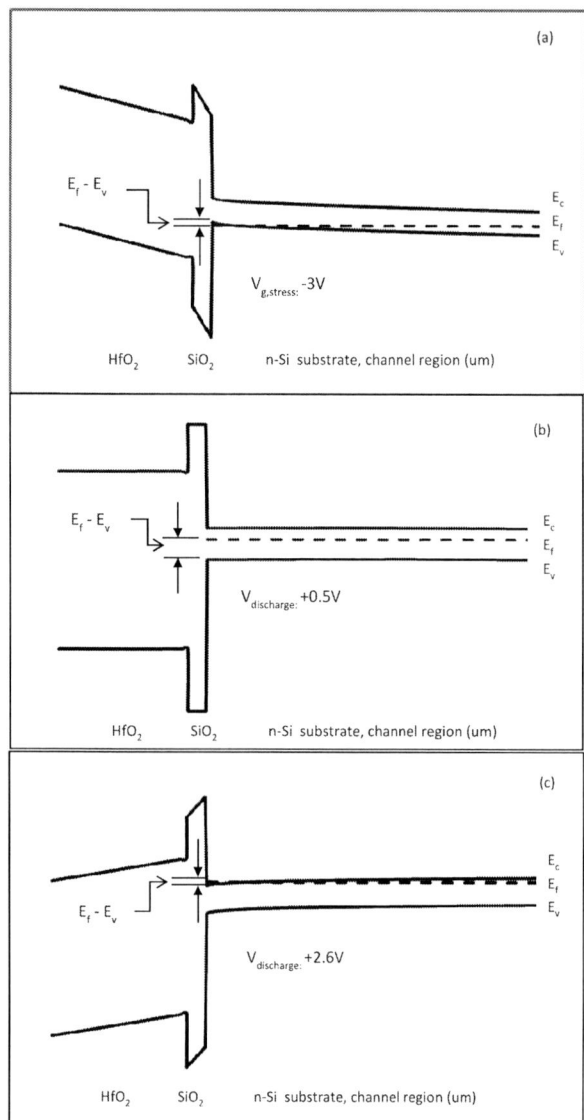

Fig. 2. The three different conditions of energy band diagrams under wide range of biasing levels

occurring at condition 1 is denoted as <f₁(t)>. The total Δvth in this research is determined considering that the precursor energy level is defined as being lower than E_v, hence

$$\Delta V_{th} = -\frac{qN_{eff}}{C_{ox}} \quad (3)$$

The N_{eff} is the total positive charges. The energy levels associated with these positive charges are located inside and above the bandgap. The component of this N_{eff} is denoted as AHT, CPC, and ANPC according to references [5]. The next section will provide validation for this proposition. The validity of this concept is confirmed through data obtained in [5] which demonstrates that positively charged defects located in various energy spectrum. By graphing the N_{eff}, in relation to various levels of biasing discharge voltage and E_f - E_v, further investigation into the different energy spectrum for each positive charge able to be investigated.

The relationship between the overall positive charges and

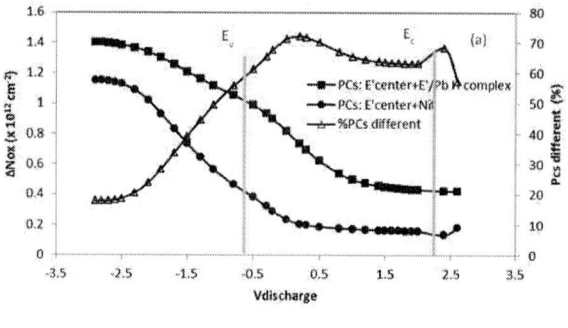

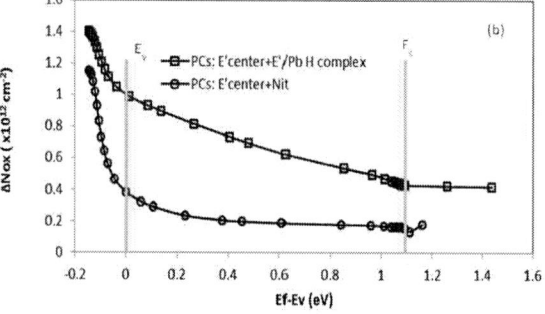

Fig. 3. Total ΔN_{ox} versus (a) wide range of discharge bias (b) the extracted E_f - E_v

the energy difference between the E_c and E_v, as depicted in Fig. 3 (b) for the redefined energy levels, exhibits a similar pattern to that observed in reference [5]. However, it is important to note that the default energy level used in this study differs from the one used in the reference. The spreading of the N_{it} is nearly consistent between E_c and E_v [5][19].

In this work, the new energy levels for N_{eff} components are properly investigated. Each defect that significantly contributed to NBTI degradation was accounted, that was not observed as in [1], [10]. Earlier studies presents two different characteristics of components that can be recover and cannot recover. The resultant Δvth solely based on the recover and non recoverable components. Furthermore, the non recoverable component specifically denoted by the N_{it} that is located below the energy mid-gap. Fig. 4 (a) and (b) show the ΔD_{ox} for original and new defined energy levels .

In Fig. 4 (a), the trap density associated with the default energy level was found to be concentrated solely below E_i, spanning the energy difference between E_f and E_v, which was defined as 0.5 eV. Consequently, the original energy spectrum, that exclusively considered interface traps and oxide traps with energy levels below E_i, did not contribute to the presence of positive charges in between the E_c and E_v as well as above the E_c. that is described in [5]. There is another inconsistency in the original energy level. The ΔD_{ox} is missing at 0 to -0.3, E_f - E_v. The discrepancy that arises is associated with energy location that is assigned to the precursor. The energy location is at the -0.3 to -0.5. Thus, trap density corresponding to E_f - E_v values between 0 to -0.3 were not considered in the simulation. To address these issues, a new energy level, depicted in Fig. 4 (b), was proposed, providing an energy distribution for the trap density.

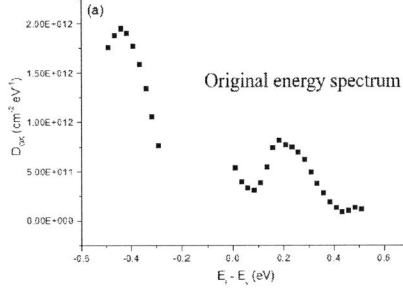

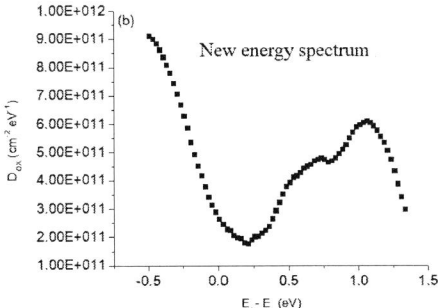

Fig. 4. The ΔD_{ox} for (a) original energy spectrum and (b) new energy spectrum

More recently, CV measurements can probe the source of defects. It is found that the spreading of energy for each trap is within the E_c and E_v. The hydrogen responses are associated with the positive charge buildup [20]. All defects, between dielectric and semiconductor and within the dielectric, collectively contribute to the NBTI hence reduced the device performance.

IV. Conclusion

The examination of energy profiling technique highlights the NBTI degradation mechanism is contributed significantly by the different generated defects that properly distributed across wide energy spectrum. By adjusting the energy level of the positive charges individually, it becomes possible to investigate the AHT, CPC, and ANPC across the entire energy spectrum, which is located below E_v, within bandgap and above E_c respectively.

Acknowledgments

College of Engineering, UiTM and HIR Grant (UM/MOHE HIRGA D000019-16001).

References

[1] M. N. Hamzah, N. E. Alias, and M. L. Peng Tan, "Negative Bias Temperature Instability Analysis of a 15 nm p-channel Junctionless Fin Field Effect Transistor (p-JLFinFET)," in *2022 IEEE 20th Student Conference on Research and Development, SCOReD 2022*, Institute of Electrical and Electronics Engineers Inc., 2022, pp. 73–76.

[2] S. Kim *et al.*, "Reliability Assessment of 3nm GAA Logic Technology Featuring Multi-Bridge-Channel FETs," in *IEEE International Reliability Physics Symposium Proceedings*, Institute of Electrical and Electronics Engineers Inc., 2023.

[3] K. Tselios *et al.*, "Revealing the Impact of Gate Area Scaling on Charge Trapping employing SiO2 Transistors," *IEEE Trans. Device Mater. Reliab.*, 2023.

[4] C. Bogner, C. Schlunder, M. Waltl, H. Reisinger, and T. Grasser, "Modeling of NBTI Induced Threshold Voltage Shift Based on

Activation Energy Maps Under Consideration of Variability," in *IEEE International Reliability Physics Symposium Proceedings*, Institute of Electrical and Electronics Engineers Inc., 2023.

[5] S. W. M. Hatta *et al.*, "Energy distribution of positive charges in gate dielectric: Probing technique and impacts of different defects," *IEEE Trans. Electron Devices*, vol. 60, no. 5, pp. 1745–1753, 2013,

[6] R. Gao *et al.*, "NBTI-Generated Defects in Nanoscaled Devices: Fast Characterization Methodology and Modeling," *IEEE Trans. Electron Devices*, vol. 64, no. 10, pp. 4011–4017, 2017,

[7] P. Kumar, K. Koley, S. S. A. Askari, A. Maurya, and S. Kumar, "Assessment of Negative Bias Temperature Instability Due to Interface and Oxide Trapped Charges in Gate-All-Around TFET Devices," *IEEE Trans. Nanotechnol.*, vol. 22, pp. 157–165, 2023,

[8] K. Tselios *et al.*, "Impact of Single Defects on NBTI and PBTI Recovery in SiO2Transistors," in *IEEE International Integrated Reliability Workshop Final Report*, Institute of Electrical and Electronics Engineers Inc., 2022.

[9] W. J. Sung *et al.*, "Investigation on NBTI Control Techniques of HKMG Transistors for Low-power DRAM applications," in *IEEE International Reliability Physics Symposium Proceedings*, Institute of Electrical and Electronics Engineers Inc., 2023.

[10] B. Ye *et al.*, "NBTI Mitigation by Optimized HKMG Thermal Processing in a FinFET Technology," *IEEE Trans. Electron Devices*, vol. 69, no. 3, pp. 905–909, 2022.

[11] D. S. Ang, S. C. S. Lai, G. A. Du, Z. Q. Teo, T. J. J. Ho, and Y. Z. Hu, "Effect of hole-trap distribution on the power-law time exponent of NBTI," *IEEE Electron Device Lett.*, vol. 30, no. 7, pp. 751–753, 2009.

[12] T. Grasser, B. Kaczer, W. Goes, T. Aichinger, P. Hehenberger, and M. Nelhiebel, "A Two-Stage Model for Negative Bias Temperature Instability," in *IEEE Proc.*, I E E E, 2009.

[13] Dieter K. Schroder, "Negative bias temperature instability: What do we understand?," *Microelectron. Reliab.*, vol. 47, no. 6, 2007.

[14] W. Goes, T. Grasser, M. Karner, B. Kaczer, and • * Christian, "A Model for Switching Traps in Amorphous Oxides," in *IEEE Proc.*, 2009.

[15] Z. Q. Teo, D. S. Ang, and C. M. Ng, "Non-hydrogen-transport characteristics of dynamic negative-bias temperature instability," *IEEE Electron Device Lett.*, vol. 31, no. 4, pp. 269–271, Apr. 2010.

[16] S. Gupta, B. Jose, K. Joshi, A. Jain, M. Ashraful Alam, and S. Mahapatra, "A Comprehensive and Critical Re-assesment of 2-Stage Energy Level NBTI Model," in *IEEE Proc.*, 2012.

[17] Synopsis, "Simulation of NBTI Degradation with Two-Stage Model," 2011.

[18] H. Hussin, N. Soin, M. F. Bukhori, S. Wan Muhamad Hatta, and Y. Abdul Wahab, "Effects of gate stack structural and process defectivity on high- k dielectric dependence of nbti reliability in 32 nm technology node PMOSFETs," *Sci. World J.*, vol. 2014, 2014.

[19] H. Hussin, N. Soin, S. Wan Muhamad Hatta, and M. F. Bukhori, "Characterization of NBTI-induced positive charges in 16 nm FinFET," *Proc. 2015 IEEE Int. Conf. Electron Devices Solid-State Circuits, EDSSC 2015*, pp. 365–368, 2015.

[20] F. Palumbo, M. Klebanov, G. Monreal, and S. Chetlur, "Physical origin of the permanent components of the positive charge buildup resulting from NBTI/PBTI stress in nMOS/pMOS transistors," *Proc. Int. Symp. Phys. Fail. Anal. Integr. Circuits, IPFA*, vol. 2022-July, pp. 1–5, 2022,.

Surface Morphology of Fabricated TiO₂-Graphene Thin Film by Spin-Coating Technique for pH Sensing Electrode Application

Anis Nabilah Daud
School of Electrical Engineering
College of Engineering,
Universiti Teknologi MARA
13500 Permatang Pauh,
Malaysia
anismd98@gmail.com

Aina Syakirah Mohd Masri
School of Electrical Engineering
College of Engineering,
Universiti Teknologi MARA
40450 Shah Alam, Malaysia
syakirahh21@gmail.com

Nur Syahirah Kamarozaman
School of Electrical Engineering
College of Engineering,
Universiti Teknologi MARA
40450 Shah Alam, Malaysia
nursyahirahk@uitm.edu.my

Muhammad Alhadi Zulkefle
School of Electrical Engineering
College of Engineering,
Universiti Teknologi MARA
40450 Shah Alam, Malaysia
alhadizulkefle@gmail.com

Zurita Zulkifli
School of Electrical Engineering
College of Engineering,
Universiti Teknologi MARA
40450 Shah Alam, Malaysia
zurita101@uitm.edu.my

Sukreen Hana Herman
School of Electrical Engineering
College of Engineering,
Universiti Teknologi MARA
40450 Shah Alam, Malaysia
hana1617@uitm.edu.my

Abstract—**This paper presented the fabrication of TiO₂-graphene hybrid thin film using spin-coating technique. The surface morphology and materials concentration has been studied using Field Emission Scanning Electron Microscopy (FESEM) and Surface Enhancement Raman Spectroscopy (SERS), respectively. Three parameters were varied for the study which are number of graphene solution drop on TiO2 thin film, spin coating speed and the dispersion time. From the FESEM images, the surface morphology resulted smooth surface for the sample of 3 drops of graphene at 3000 rpm for one-minute dispersion time. The peak of graphene (G) on the samples can be observed from the Raman spectra at 1600nm-1. The longer dispersion time of spin-coating showed the lower peak of G. The best sample was used as the sensing electrode for pH sensor by connecting the electrode with the Extended Gate Field Effect Transistor (EGFET) circuit. The TiO₂-graphene sensing electrode was immersed in pH buffers to evaluate the sensing performance by calculating the linearity and sensitivity. The result shows high linearity and promising sensitivity value of 0.935 and 38 mV/pH, respectively.**

Keywords— *TiO₂-Graphene; Spin Coating; Sensing Electrode; Surface Morphology, Raman Spectra*

I. INTRODUCTION

Titanium Dioxide (TiO₂) nanoparticles have its own characteristics such as low toxicity, high stability and easy to prepare. One of the most widely used and promising photocatalytic materials is TiO₂ [1]. This is because of the high photocatalytic activity, low toxicity and high stability make the TiO₂ is good for environmental cleaning, hydrogen production and dye sensitized solar cells [2]. Spin coating is one of the methods that commonly used to deposit on flat substrates [3]. It consists small amount of coating material that are being placed in the centre of the substrate and the material is spins at the controlled speeds until it is completely covered by centrifugal force [4]. The coating thickness decreases as the rotational speed increases and vice versa [5]. Besides the rotational speed, the time taken also plays their role to give the best deposition on the sample [6]. The solvents used on the sample are usually non-polar volatile chemicals that can evaporate quickly. The effectiveness of this technique seems mostly from the ability to control certain process variables such as rotation speed and time or the mechanism and method of deposition.

The graphene plays an important role in the field of nanotechnology and nanoscience from photocatalysis to sensor applications [7-8]. Graphene possesses exceptional mechanical strength and stability. Incorporating graphene into the TiO₂ electrode can improve its structural reliability and durability, making it more resistant to mechanical stress and electrode degradation [9]. Surface morphology is important for a sensing membrane, rougher or smooth of the surface may give different sensing output. The inconsistency of fabrication of the sensing electrode might affect the output reading and may give inaccurate result. Thus, the study of surface morphology at different parameters in spin-coating technique for pH application is significant. The sensing performance was based on the linearity and sensitivity of the electrode which can be calculated using the Nernst equation as shown in equation (1).

$$E(T) = Eo(T) - 2.303\, RT/nF\, pH \qquad (1)$$

Where,

$E(T)$ = Measured potential *mV* at temperature *T (Kelvin)*
$Eo(T)$ = Constant, standard potential mV at temperature *T (Kelvin)*
2.303 = Factor to convert ln to log
R = Molar gas constant (8.3144 *J mol⁻¹ K⁻¹*)
F = Faraday constant 96485 *C mol⁻¹*
T = Temperature *K (kelvin)*

The sensing electrode was connected to the EGFET circuit to measure the dc output characteristic. The advantage of using EGFET is the electrode connected to the gate of the EGFET can be changed without having to change the whole circuit system. The research objective is to study the surface morphology of fabricated TiO$_2$-Graphene thin film by spin-coating potential for sensing electrode application. The method used to fabricate was spin coating method. Glass and indium tin oxide (ITO) were used as the substrates as the TiO$_2$-based sensing electrode

II. METHODOLOGY

ITO substrates was cut in 1cm x 2cm by using the diamond cutter followed by substrate cleaning with methanol and sonicated the substrates in the Hwashin Powersonic 405 ultrasonic bath at the temperature of 50°C for 10 minutes. The process continued by using deionized water for the same temperature and duration

A. TiO$_2$ Solution Preparation

The chemicals used were Titanium (IV) isopropoxide (TTIP), Absolute Ethanol, Glacial Acetic Acid (GAA), deionized water (DI water) and Trion-X 100. Next, the Acetic Acid was added to stabilize the chemical reaction and to avoid the precipitation of the solution [10]. Then, the Absolute Ethanol was used as a solvent and the Trixon-X was used as the surfactant. Two beakers were prepared in which beaker A contained 1.5 ml of Titanium (IV) isopropoxide (TTIP) with 23 ml of Absolute Ethanol and 2.5 ml of GAA and beaker B was fill with 23 ml of Absolute Ethanol and 1 drop of Triton-X 100 and 0.2 ml of deionized water [11]. Both solutions in the Beaker A and Beaker B were being stir for one hour separately at room temperature by using magnetic stirrer. Then both solutions were mixed and stirred for another one hour to get well stirred solution.

B. Maintaining the Integrity of the Specifications

The sample preparation was conducted using spin coating technique by varying the speed of the rotation from 1000 rpm to 4000 rpm, the dispersion time from 60s to 150s. The ITO glass was placed on the spin coating holder. The prepared solution was dropped 10 times on the ITO glass using the pipet within 10 seconds.

C. TiO$_2$-Graphene Sample Preparation

The commercial liquid Graphene was a mixture of water and Graphene nanoparticles (0.4–0.5 wt% Graphene) with opaque in color used as the source of graphene in the experiment. The chemicals used for TiO2 were Titanium (IV) isopropoxide (TTIP), Absolute Ethanol, Glacial Acetic Acid (GAA), deionized water (DI water) and Trion-X 100. Acetic Acid was added to stabilize the chemical reaction and to avoid the precipitation of the solution. The Absolute Ethanol was used as a solvent and the Trixon-X also was used as the surfactant. Fig.1 shows the configuration of TiO2-graphene thin film prepared by spin coating technique. The function of the Kapton tape is to differentiate the original ITO glass with the material deposition area. The parameter for graphene was varied from 1 drop to 4 drops while the speed of spin coat was varied from 1000 – 4000 rpm. The spin coat time also was varied from 60 – 150s. All samples were dried at 100°C for 10 minutes in the furnace.

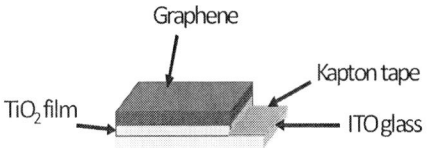

Fig. 1. Configuration of TiO$_2$-Graphene Thin Film

D. Characterization

The prepared TiO$_2$ and TiO$_2$-graphene thin film were characterized for the surface morphology using Field-Emission Scanning Electron Microscopy (FESEM) and the concentration of material based on the peak of Raman Spectra by using Surface Enhancement Raman Spectroscopy (SERS). Finally, the chosen sample with best characteristic as sensing electrode was selected to be characterized for the sensing electrode performance using Semiconductor Device Analyzer (SDA) Keysight B1500A.

III. RESULT AND DISCUSSION

The findings were divided into three parts. The results from Field-Emission Scanning Electron Microscopy (FESEM) were discussed first followed by results from Surface Enhancement Raman Spectroscopy (SERS) and lastly the result obtained from sensing measurement. The result from B1500A was plotted to find the linearity and sensitivity.

A. Surface Morphology

Fig. 2 (b-e) shows the FESEM images for different drops of graphene on the TiO2 thin film while Fig.2 (a) shows the pristine TiO$_2$ thin film surface morphology. The spin coater speed was fixed at 3000 rpm and dispersion time taken was 60 seconds. Based on the surface morphology, Fig. 2 (d) exhibited uniform surface morphology in the most of the surface area compared to the other samples.

A smoother surface of a sensing electrode is generally preferred over a rougher surface for to ensures better contact and a more uniform interface between the sensing electrode and the sample solution. This improves the electrochemical response and reduces variations in the sensing performance [12]. In contrast, a rough surface may introduce irregularities and gaps, leading to uneven contact and less reliable measurements.

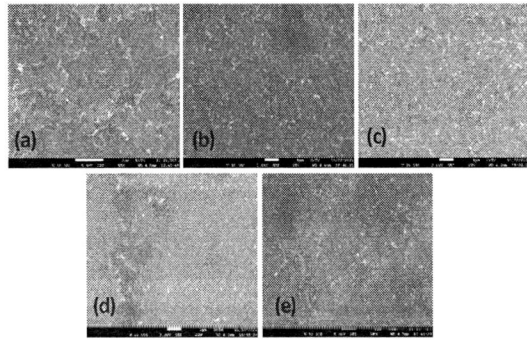

Fig. 2. Field-Emission Scanning Electron Microscopy (FESEM) images of (a) TiO$_2$ thin film, thin film with graphene drop; (b) 1 drop, (c) 2 drop, (d) 3 drop and (e) 4 drop deposited using spin coat at 3000 rpm for 60s

979-8-3503-2369-6/23 $31.00 © 2023 IEEE

To further investigate the surface morphology, the rotation of spin coater and dispersion time was varied. When these two parameters were varied the drop of graphene was fixed for 3 drops. As depicted in Fig. 3 (a) the speed of 1000 rpm showed the non-uniform surface morphology. Fig. 3 (b – d) shows the surface morphology of TiO$_2$-graphene thin film for the rotation speed of 2000 rpm, 3000 rpm and 4000 rpm, respectively. Based on the observation on the FESEM images, the thin films deposited at 3000 rpm for 60 s showed the uniform surface morphology structure.

Fig. 4 (a – d) shows the TiO$_2$-graphene thin film surface morphology at different dispersion time while fixed the rotation speed at 3000 rpm and 3 drop of graphene. From the results, 60s (Fig. 4a) and 150s (Fig. 4d) were observed showed the most uniform surface morphology compared with 90s and 120s.

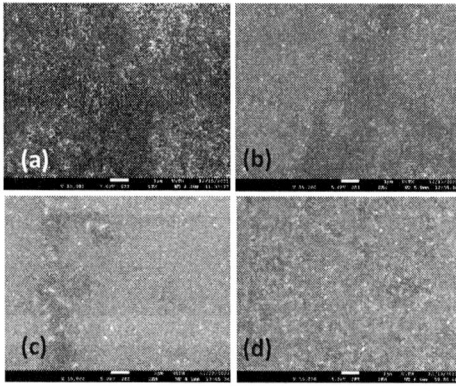

Fig. 3. FESEM images of TiO$_2$ –graphene film at different rotation speed (a) 1000 rpm, (b) 2000 rpm, (c) 3000 rpm and (d) 4000 rpm

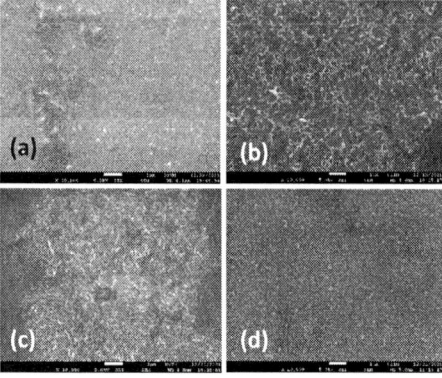

Fig. 4. FESEM images of TiO$_2$-graphen thin film at different dispersion time (a) 60s, (b) 90s, (c) 120s and (d) 150s

B. Surface Enhancement Raman Spectroscopy (SERS)

SERS was used for the structural analysis of materials deposited on the thin films. Fig. 5 shows the Raman spectra for different drops of graphene solution. It shows that, the graphene peak can be observed at Raman shift of 1600 cm^{-1} for all samples. The intensity of the D band between 1300-1370 cm^{-1} refers to a higher oxidation degree corresponds to greater disorder and consequently a more intense D band. The higher peak of D band is due to higher oxidation of the sample. The higher drop of graphene on the TiO$_2$ sample shows the D band slightly shifted towards the lower wavenumber. This condition was believed due to the rougher surface morphology on surface of thin film with the higher drop of graphene.

Raman spectra for different rotation of spin coat is shown in Fig. 6. The surface morphology may affect the shifted peak of Raman D band. Fig. 7 shows the Raman spectra for different dispersion time. The longer the dispersion time, the more uniform the surface hence shifted Raman spectra can be seen for the longer dispersion time. The G peaks are very weak at 120s and 150s samples due to the dispersed solution of graphene during the spin coating process.

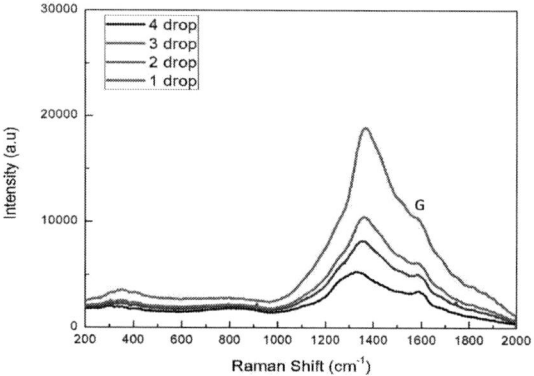

Fig. 5. Raman spectra for different graphene drop on TiO$_2$ thin film

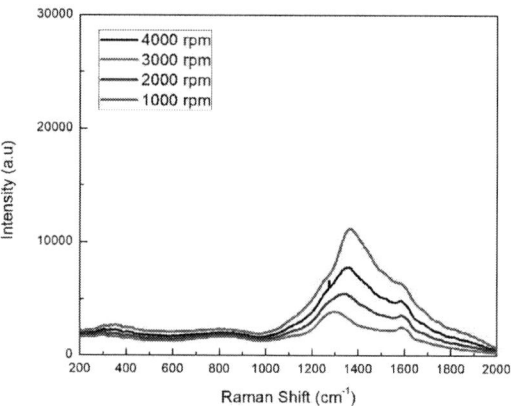

Fig. 6. Raman Spectra for different spin coat speed

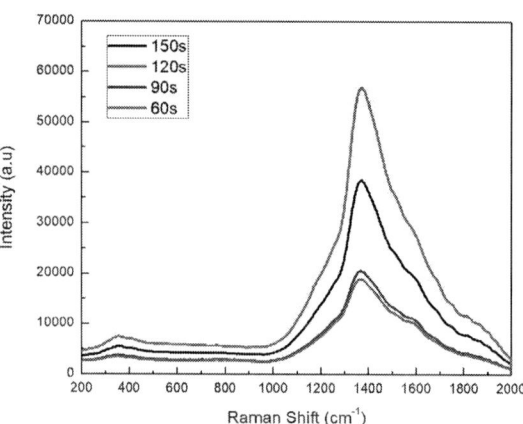

Fig. 7. Raman Spectra for different dispersion time

C. TiO₂-Graphene Film as pH Sensing Electrode

Fig. 8 shows the output voltage curve of pH measurement. The sample choose for pH characerization was sample deposited at 3000 rpm with 3 drops of graphene for 60s. The TiO₂-graphene thin film was employed as the sensing electrode and immersed in pH buffer solution. Based on the output voltage at different pH value, the linearity of the measurement obtained at 0.935 and the sensitivity of the electrode was 38 mV/pH. The sensitivity value is quite low compared to the commercial value of pH but still can consider as the promising pH sensor with further improvement.

The low sensitivity of TiO₂-graphene electrode was observed due to the peel-off layer when immersed several times in the pH buffer solution. This condition could be improved in the next experimental work by introducing hydrophobic polymer into the TiO₂-graphene solution.

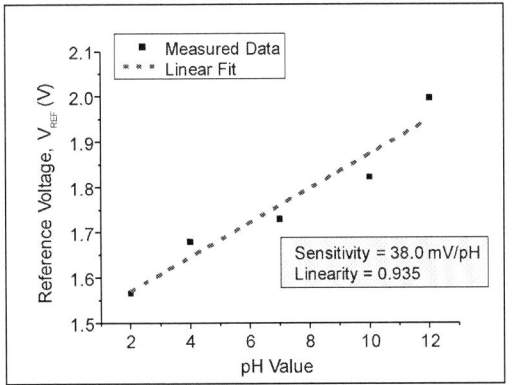

Fig. 8. Output voltage versus pH value of TiO₂-graphene sensing electrode

IV. CONCLUSION

As the conclusion, the surface morphology of fabricated TiO₂-Graphene thin film by spin coating technique varied with the drop of graphene, spin speed and dispersion time. The different surface roughness affects the vibrational molecules on the thin film and thus affect the sensing performance. Based on the experiment, the optimized parameter was 3 drop of graphene at 3000 rpm for 60s. The sensing electrode can further improve by incorporated the solution with any hydrophobic polymer to improve the peel-off issue. Yet, the sample is still considered as the promising pH sensing electrode based on the value of linearity and the sensitivity obtained from the measurement.

ACKNOWLEDGMENT

The authors would like to acknowledge NET Laboratory UiTM Shah Alam and College of Engineering, Universiti Teknologi MARA, Shah Alam for the financial support.

REFERENCES

[1] Jyoti V. Patil, Sawanta S. Mail, Jasmin S. Shaikh, Akhilesh P. Patil, Pramod S. Patil, "Influence of reduced graphene oxide-TiO₂ composite nanofibers in organic indoline DN350 based dye sensitized solar cells" Synthetic Metals, Volume 256, 2019.

[2] Kazutoshi Sekiguchi, Ken-ichi Katsumata, Hiroyo Segawa, Takayuki Nakanishi, Atsuo Yasumori, "Effects of particle size, concentration and pore size on the loading density of silica nanoparticle monolayer arrays on anodic aluminum oxide substrates prepared by the spin-coating method", Materials Chemistry and Physics, Volume 277, 2022.

[3] Petrucci E, Orsini M, Porcelli F, De Santis S, Sotgiu G. "Effect of Spin Coating Parameters on the Electrochemical Properties of Ruthenium Oxide Thin Films", Electrochem. 2(1) 2021 pp:83-94

[4] Paulina Aulema-Pullupaxi, Lenys Fernandez, Alexis Debut, Cristian P. Santacruz, William Villacis, Carola Fierro, Patricio J. Espinoza-Montero, "Photoelectrocatalytic degration of glyphosate on titanium dioxide by so-gel/spin-coating on boron doped diamond (TiO2/BDD) as a photoanode" Chemosphere, Vol. 278, 2021, 130488

[5] Kyesmen Pannan I., Nombona Nolwazi, Diale Mmantsae, "Effects of Film Thickness and Coating Techniques on the Photoelectrochemical Behaviour of Hematite Thin Films", Frontiers in Energy Research, Vol.9, 2021

[6] Chapman N, Chapman M, Euler WB. "Evolution of Surface Morphology of Spin-Coated Poly(Methyl Methacrylate) Thin Films", Polymers, 13(13), 2021, 2184

[7] Liu J, Bao S, Wang X. "Applications of Graphene-Based Materials in Sensors: A Review", Micromachines (Basel). 13(2) 2022, 184.

[8] Neupane, S., Subedi, V., Thapa, K.K. et al. "An alternative pH sensor: graphene oxide-based electrochemical sensor",Emergent Mater. 5, 2022, pp: 509–517.

[9] Nurdin, M., Maulidiyah, M., Watoni, A.H. et al. Nanocomposite design of graphene modified TiO₂ for electrochemical sensing in phenol detection. Korean J. Chem. Eng. 39, 2022, pp 209–215.

[10] Muhammad Syahin Firdaus Aziz Zamri, Norzahir Sapawe, "Performance studies of electrobiosynthesis of titanium dioxide nanoparticles (TiO₂) for phenol degradation" Materials Today: Proceedings, Vol 5, 2018, pp: 21797-21801

[11] A.S. AlShammari, M.M. Halim, F.K. Yam, N.H. Mohd Kaus, "Synthesis of Titanium Dioxide (TiO2)/Reduced Graphene Oxide (rGO) thin film composite by spray pyrolysis technique and its physical properties" Materials Science in Semiconductor Processing, Vol. 116, 2020, 105140.

[12] Tae-Yeol Jeon, Seung-Ho Yu, Sung J. Yoo, Hee-Young Park, Sang-Kyung Kim, "Electrochemical determination of the degree of atomic surface roughness in Pt–Ni alloy nanocatalysts for oxygen reduction reaction", Carbon Energy, Vol.3, 2021, pp:375-383.

Graphene-Based Hybrid Sensor for the Detection of Cancer Cells Using K-SPR Technology

P Susthitha Menon
Institute of Microengineering
and Nanoelectronics (IMEN),
Universiti Kebangsaan Malaysia
UKM Bangi, Malaysia
susi@ukm.edu.my

Nur Shahirah Shaári
Institute of Microengineering
and Nanoelectronics (IMEN),
Universiti Kebangsaan Malaysia
UKM Bangi, Malaysia
shahirahshaari99@gmail.com

Vatsala Pithaih
Institute of Microengineering
and Nanoelectronics (IMEN),
Universiti Kebangsaan Malaysia
UKM Bangi, Malaysia
vatsalaesther1998@gmail.com

Siti Nasuha Mustaffa
Institute of Microengineering
and Nanoelectronics (IMEN),
Universiti Kebangsaan Malaysia
UKM Bangi, Malaysia
nasuhamustaffa@gmail.com

Affa Rozana Abdul Rashid
Faculty of Science and
Technology, USIM, Bandar
Baharu Nilai, 71800 Nilai,
Negeri Sembilan, Malaysia.
affarozana@usim.edu.my

Vikneswary Ravi Kumar
Department of Obstetrics and
Gynaecology, Faculty of
Medicine, UKM, Kuala Lumpur
56000, Malaysia
vickee10925@gmail.com

Nor Haslinda Abd Aziz
Department of Obstetrics and
Gynaecology, Faculty of
Medicine, UKM, Kuala Lumpur
56000, Malaysia
norhaslinda.abdaziz@ppukm.uk
m.edu.my

Nirmala Kampan
Department of Obstetrics and
Gynaecology, Faculty of
Medicine, UKM, Kuala Lumpur
56000, Malaysia
nirmala@ppukm.ukm.edu.my

Abstract— **The Malaysian National Cancer Registry (MNCR) 2012-2016 report has found that more than 90% of cancer cases are confirmed at stages 3 or 4 with a 5-year survival rate ranging from 3% - 32% depending on the type of cancer. Hence, there is an urgent need for the development of sensitive point-of-care technologies (POCT) including biosensors that are able to detect cancer at its early stages. In this work, a Kretschmann-based surface plasmon resonance sensor (K-SPR) using graphene (Gr) and graphene oxide (GO) on chromium (Cr)/gold (Au) was analyzed for the detection of two types of cancer cell, rat kidney epithelial cell (RKEC) and HeLa cervical cancer cell (HC). Numerical simulations using Finite Difference Time Domain (FDTD) method for the Cr/Au/Gr(GO) thin films have been optimized to achieve the best performance of the sensors in terms of its sensitivity (S) and quality factor (Q). The results show that all 4 sensors are capable of detecting the RKEC with highest S value of 188.41°/RIU at 670 nm wavelength and best Q of 57.41 RIU at 785 nm wavelength. HeLa cervical cancer cells were detected with highest S value of 219.80 °/RIU at 670 nm wavelength and best Q of 47.94 RIU at 785 nm wavelength.**

Keywords—biosensor, surface plasmon resonance (SPR), Kretschmann, cancer, FDTD, graphene

I. INTRODUCTION

Cancer is the second most common disease in the world, with an increasing fatality rate in recent years. Due to the limitations of cancer diagnosis and treatment, the patient's survival rate is unknown. Early detection of cancer is critical for effective treatment. However, due to the lack of distinguishable physical symptoms and sensitive screening techniques, early detection of cancer is quite challenging [1]. A biomarker-based cancer diagnosis could help doctors detect cancer earlier and treat it more effectively. Biosensors are important in the identification of biomarkers because they are simple to use, portable, and can-do real-time analysis.

Refractive index (RI) of a live cell has recently drawn a lot of interest as an alluring sign of cell abnormalities. The

This work was funded by Universiti Kebangsaan Malaysia (UKM) under the grant *Geran Ganjaran Penerbitan* (GP-K017739) and *Dana Impak Perdana (DIP)* grant with code of DIP-2021-016

prospective application in cell physiology and pathology has been taken into consideration thanks to the cell RI due to optical interaction of light field with cellular organelles, which is instructive for assessing chemical composition within cellular structures [2]. A good criterion for the quantitative diagnosis of cell malignancy is thought to be that cancer cells have RIs that are substantially greater than normal cells, which is widely recognised in the field of cancer biology [3]. The surface plasmon resonance (SPR) technology has established a new tool for real-time and label-free detection of refractive index (RI) sensors. Its use is expected to rise as technology becomes more widely available. SPR-based optical sensor have shown to have a lot of potential in biological, chemical, and biomedical applications. SPR technology can be used to measure the analyte concentration in a complicated sample, as well as its selectivity, affinity, and kinetics of biomolecular interaction. SPR is an evanescent wave that occurs when light interacts with a metal layer under the condition of total internal reflection. The Kretschmann configuration is widely utilised for SPR biosensing applications and is based on angular interrogation mode. Fig. 1 shows a metal film put at the interface between a medium having a high refractive index (prism) and a medium with a low refractive index (air or solution of interest).

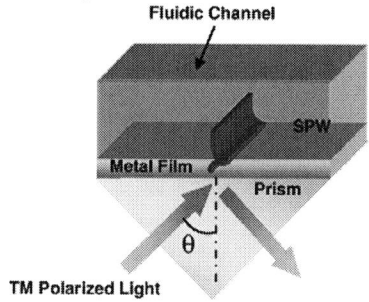

Fig. 1. SPR setup for Kretschmann configuration.

Detector can detect reflection of light as it passes through the prism and hits the metal layer. This yields SPR spectra, which plot reflectance as a function of source incidence angle. The resonance angle at which the least reflectance is attained is provided by the resonance spectra. Refractive index varies in that region when analyte binding takes place close to a metal surface. As a result of the alteration, the resonance angle shifts, and the SPR spectrum shifts as well.

One atom thick graphite is organised in a honeycomb lattice to form the two-dimensional substance known as graphene (Gr). When graphene is utilised as a detecting layer for the SPR sensor, its sensitivity rises significantly [3]. Multiple characteristics may be seen in it, including great optical transparency, low reluctance, excellent carrier mobility, and tunability [4]. In many optical systems, including K-SPR sensors, graphene formed excellent quantum efficiency for the interaction of light and matter. The sensing capability of a gold (Au) SPR sensor with graphene coating may be improved for ultrasensitive biological and chemical detection [5].

In this study, a numerical modelling of graphene coated SPR biosensor is proposed for detecting cancer cell specifically rat kidney epithelial cell line (RKEC) and HeLa cell, which is cancer cell for cervical cancer based on the RI values obtained from previous work [6]. SPR sensor is designed using chromium (Cr), gold (Au) and graphene (Gr) and graphene oxides (GO) to detect the cancer cell. Due to the distinctive properties of graphene, a graphene monolayer is used as an efficient light-absorbing medium between the gold film and sensing dielectric. Using numerical solution software, a comparable finite difference time domain (FDTD) analysis was performed to determine the behaviour of the SPR signal in the sensor and to evaluate the performance parameters. The acquired results are compared to simulated data from numerical analysis in wavelength of 670 nm and 785 nm with or without the Gr and GO layers.

II. METHODOLOGY

A Finite Difference Time-Domain (FDTD) simulator was used to numerically simulate the graphene-coated gold SPR sensor to assess its effectiveness for the detection of rat kidney epithelial cell line (RKEC) and HeLa cell (HC). There are five elements that make up the arrangement which are prism BK7, 0.5 nm chromium (Cr) adhesion layer, 50 nm gold (Au) that are placed in hybrid with graphene and graphene oxide (GO) layer as the outermost layer of the thin film as shown in Fig. 2. To guarantee that the theoretical and simulation results in perfect agreement, all simulation data were checked. L is the number of graphene layers, and the thickness of graphene is given as 0.34 nm [5]. To find the resonance angle at which the SPR mode was excited, a parameter sweeps of the source incidence angle from 36° to 88° was done. A plane wave source was configured as either a Bloch or periodic type at the required wavelengths.

This analysis uses a two-dimensional (2D) FDTD simulation. Simulations were performed using FDTD Solutions software. All materials refractive index (real and imaginary) values were entered into the FDTD software plus additional fitting accuracy control. The sensor was exposed to a source of plane waves with a wavelength of 670 nm and 785 nm.

The completely fitted layer profile, which was positioned at a sharp angle, served as the boundary condition for calculating the complex propagation constant of the modes. There were fixed monitors for time, transmission, and reflectance. The ultimate results were obtained using mesh sizes of 0.3 nm.

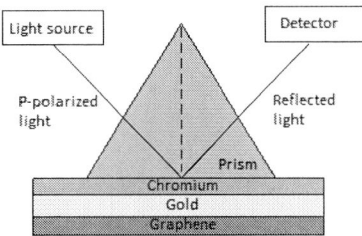

Fig. 2. Kretschmann configuration of graphene-Au based SPR sensor.

Air functioned as the simulation's background (n=1.00). To obtain the resonance angle, angular interrogation was performed using parameter sweep over a variety of source angles. 70 simulations at angles ranging from 36 to 88 degrees were executed using the parameter sweep. To get reflection for every simulation, a theoretical calculation-based script was executed. As soon as the parameter sweep is over, the script graphs the findings against theoretical to verify every simulation. Finally, sensing capabilities of Cr/Au/Gr/GO sensors were examined, compared, and evaluated using rat kidney epithelial cell line (RKEC) and HeLa cell (HC) as simulation backgrounds. RKEC and HC's wavelength-dependent refractive indices which was used as a reference medium were obtained from [6].

Performance of the SPR sensor is influenced by sensitivity (S), full-width-half-maximum (FWHM), and quality factor (Q). The ratio of resonance angle shift ($\Delta\theta_{res}$) for a specific change in refractive index (Δn) in refractive index units (RIU) is used to calculate the sensitivity (S), which is written as shown in (1):

$$S = \Delta\theta_{res}/\Delta n \qquad (1)$$

The FWHM of the SPR spectra was obtained by calculating the full width at half maximum of reflectance drop ($\Delta\theta_{0.5}$). Finally, the quality factor which is defined as the ratio of sensitivity (S) to the FWHM, and their unit is RIU^{-1} and given by (2):

$$Q=S/FWHM \qquad (2)$$

III. RESULTS AND DISCUSSION

Our previous work has shown that the comparison of experimental and FDTD simulation results of 50 nm Au-graphene on BK7 at 670 nm and 785 nm wavelength in air which demonstrates their excellent correlation with only less than 5% change in reflectivity [7]. Further analysis is conducted on the numerical investigation of the SPR sensor in two alternative parameter configurations which are presence of a single and double layer of graphene sheet, graphene oxide, and types of analytes, while maintaining all other parameters constant. The metal thickness will be quantitatively optimised for maximum reflectance and FWHM in all combinations. The largest energy transfer from the incident light to the SP wave, which results in the maximum enhancement of the SPR field in resonance condition, is ensured by the highest value of reflection spectra (equal to the lowest value of the reflection dip).

The mean refractive index of healthy rat kidney epithelial cell lines (RKEC) is 1.353 ± 0.008 and for cancer cell is 1.371 ± 0.014, which is much higher than healthy one and similar to those of other cancer cell studies (full field). The simulation for minimum and maximum refractive index for both healthy and cancer cell was calculated and obtained for RKEC. Fig. 3 illustrates the reflection spectra with incident angle for bare gold, one layer of graphene, two-layer graphene and graphene oxide for maximum refractive index of RKEC. The dotted line represents the cancer cell and the straight line represent the healthy cell. The results compared the healthy and cancer cell for RKEC at both 670 nm (black line with square-shape symbol) and 785 nm (red line with circle-shape symbol) wavelength. Larger wavelengths result in producing narrower resonance peaks and enable accurate estimation of modulation angles, therefore SPR curves at 785 nm are narrower than those of SPR sensors at 670 nm [8].

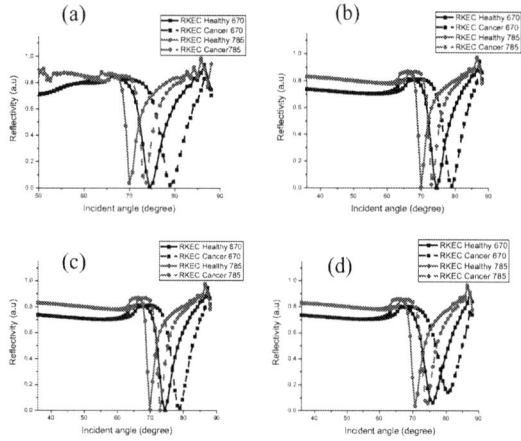

Fig. 3. SPR curves for simulation detecting healthy and cancer cell of RKEC at 670 nm (black line) and 785 nm (red line) wavelength for (a) Au, bare gold (b) Au-Gr, one layer graphene (c) Au-Gr2, two-layer graphene and (d) Au-GO, graphene oxide.

The data for simulation detecting healthy and cancer cell of RKEC at 670 nm and 785 nm wavelength consists of four performance parameters namely, minimum reflectance (R_{min}), FWHM, sensitivity (S) and quality factor (Q). The two parameters which are sensitivity and quality factor are calculated by using (1) and (2) respectively. It can be observed that the R_{min} and FWHM for cancer cell are higher compared to healthy cell. Additionally, the value of FWHM dropped following the addition of a double layer of graphene because, in accordance with (2), the quality factor of the suggested SPR sensor has been increased significantly.

HeLa cell, a cervical cancer cell is also used in this research paper. By using their respective refractive index for healthy (1.368) and cancer cell (1.392), the simulation was executed for varying sensors at both 670 nm and 785 nm

wavelength. Fig. 4 demonstrates SPR curves comparing the healthy and cancer cell (HC) for cervical cancer at varying sensors: bare gold, one-layer graphene, two layers of graphene and graphene oxide. The dotted line represents the cancer cell and the straight line represent the healthy cell at 670 nm (black line with square-shape symbol) and 785 nm (red line with circle-shape symbol) wavelength. Result shows that all sensors manage to detect the HeLa cell and gives the best reflection spectra at both wavelengths. It is observed that the SPR curves at 785 nm are somewhat thinner than those of sensors at 670 nm due to longer wavelengths produce narrower resonance peaks and permit accurate measurement of modulation angles [8]. Similar with RKEC, the R_{min} and FWHM for cancer cell are higher compared to the healthy cell. The value of R_{min} and FWHM shows a rise after adding one-layer and two-layer of graphene.

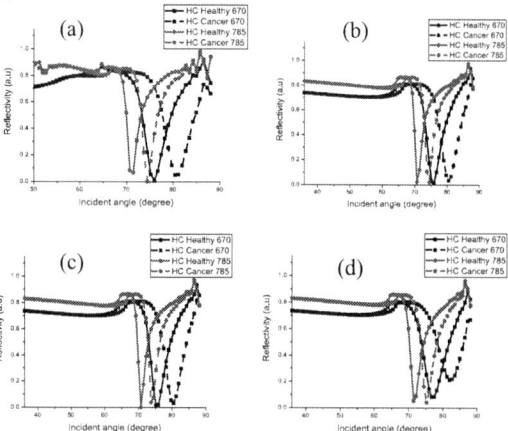

Fig. 4. SPR curves for simulation detecting healthy and cancer cell of HeLa Cervical cell (HC) at 670 nm (black line) and 785 nm (red line) wavelength for (a) Au, bare gold (b) Au-Gr, one layer graphene (c) Au-Gr2, two-layer graphene and (d) Au-GO, graphene oxide.

The sensitivity and quality factor were calculated for the addition of graphene and graphene oxide layers on top of metal layers using (1) and (2). Based on Table I, the trend shows a slight increase which is also the highest sensitivity for RKEC with value of $188.41°/RIU$ for sensor with graphene oxide at 670 nm wavelength. This was due to its specific hydrophilic affinity, electrostatic, hydrogen bonding and π-π stacking interactions that affect the adsorption capacity of biological molecule immobilization [9]. Moreover, due to the constructive interference from various layers that is brought on by radiation fields where the thickness of the thin flakes is considerably lower than the light wavelength, the sensitivity for sensors at 785 nm reduced drastically compared to 670 nm [10]. The sensitivity value at 785 nm dropped when adding graphene layers and increased with sensor with graphene oxide. Their values also are much lower than 670 nm because the higher the wavelength of incident light source, the lower the sensitivity. On the contrary, the quality factor values show a descending trend and gives the highest value of 33.72 RIU and 57.31 RIU for 670 nm and 785 nm, respectively.

TABLE I.   SENSITIVITY AND QUALITY FACTOR OF RAT KIDNEY EPITHELIAL CELL (RKEC) FOR VARYING SENSORS AT 670 NM AND 785 NM WAVELENGTH

Sensor Type	Sensitivity (°/RIU)		Q-factor (RIU)	
	670	785	670	785
Bare gold (Cr/Au)	188.40	157.00	33.72	57.31
1 layer Graphene (Cr/Au/Gr1)	188.40	125.60	31.55	40.75
2-layer Graphene (Cr/Au/Gr2)	188.40	125.60	32.60	40.75
Graphene oxide (Cr/Au/GO)	188.41	157.00	25.08	37.04

Next, for HeLa cell, the sensitivity increases after adding two layers of graphene and graphene oxide on the metal layers as shown in Table II. This gives the highest value of sensitivity for both cell with value of 219.80 °/RIU at 670 nm wavelength. While there is inconsistent trend at 785 nm where the value rise after adding one layer of graphene but dropped after two-layer graphene. Then, the value rises again for sensor using graphene oxide at value of 157.0 °/RIU. For the quality factor, 670 nm wavelength gives the highest value of 35.65 RIU at sensors with two layers of graphene and 47.94 RIU at one layer graphene sensor at 785 nm wavelength. To summarize, two-layer graphene and graphene oxide sensor detecting HeLa cell shows the highest and better sensitivity of all sensors compared to rat kidney epithelial cell with value of 219.80 °/RIU at 670 nm wavelength. While bare gold sensors for detecting RKEC gives us the highest quality factor of 57.31 RIU at 785 nm wavelength. The indifference in sensitivity could be due to the very thin layers used for the simulations [11]-[12].

TABLE II.   SENSITIVITY AND QUALITY FACTOR OF HeLa CELL FOR VARYING SENSORS AT 670 NM AND 785 NM WAVELENGTH.

Sensor Type	Sensitivity (°/RIU)		Q-factor (RIU)	
	670	785	670	785
Bare gold (Cr/Au)	188.40	125.60	32.60	36.76
1 layer Graphene (Cr/Au/Gr1)	188.40	157.00	31.55	47.94
2-layer Graphene (Cr/Au/Gr2)	219.80	125.60	35.65	38.35
Graphene oxide (Cr/Au/GO)	219.80	157.00	27.83	35.43

CONCLUSION

In conclusion, FDTD simulation was successfully used to examine Cr/Au on BK7 with addition of graphene layers and graphene oxide layers for detecting healthy and cancer cell of rat kidney epithelial cell and HeLa cervical cell. The analytical data acquired by theoretical calculation and the FDTD simulation data match each other completely. The simulation results give us highest sensitivity of 219.80 °/RIU

for detecting HeLa cervical cancer cell for Cr/Au sensor with two-layer graphene and Cr/Au/GO sensor at 670 nm wavelength. The capability of graphene to absorb is what caused the greater sensitivity which proves that using graphene can be more efficient than using standard SPR biosensors. Finally, this simulation study can discriminate between healthy and cancer cells, and because of the increased sensitivity, it has a lot of potential for use in early stages of cancer diagnosis in the future using point-of-care-technologies (POCT).

REFERENCES

[1] G. A. Atallah, N. H. A. Aziz, C. K. Teik, N. Kampan, and M. N. Shafiee, "New predictive biomarkers for ovarian cancer," Diagnostics, vol. 11, pp 465, 2021.

[2] W. J. Choi, D. I. Jeon, S. -G. Ahn, J. -H. Yoon, S. Kim, and B. H. Lee, "Full-field optical coherence microscopy for identifying live cancer cells by quantitative measurement of refractive index distribution," Optics Express, vol. 18, pp. 23285, 2010.

[3] H. Xu, J. Guo, L. Wu, X. Dai, X., and Y. Xiang, "Ultrasensitive biosensors based on long-range surface plasmon polariton and dielectric waveguide modes," Photonics Research, vol. 4, pp. 262-266, 2016.

[4] L. Wu, J. Guo, X. Dai, X., Y. Xiang, and D. Fan, "Sensitivity Enhanced by MoS2-Graphene Hybrid Structure in Guided-Wave Surface Plasmon Resonance Biosensor," Plasmonics, vol. 13, pp. 281-285, 2018.

[5] P. S. Menon, N. A. Jamil, G. S. Mei, A. R. Md Zain, D. Hewak, C. C. Huang, M. A. Mohamed, ... N. Bhat, "CVD-grown Graphene-on-Au characterization and sensing using Kretschmann-based SPR," IEEE Journal of the Electron Devices Society, vol. 8, pp. 1227-1235, 2021.

[6] M. A. Jabin, K. Ahmed, M. J. Rana, B. K. Paul, M. Islam, D. Vigneswaran, and M. S. Uddin, "Surface Plasmon Resonance Based Titanium Coated Biosensor for Cancer Cell Detection," IEEE Photonics Journal, vol. 11, pp. 345, 2019.

[7] F. A. Said, P. S. Menon, V. Rajendran, S. Shaari, and B. Y. Majlis, "Investigation of graphene-on-metal substrates for SPR-based sensor using finite-difference time domain," IET Nanobiotechnology, vol. 11, pp. 981-986, 2017.

[8] P. S. Menon, B. Mulyanti, N. A. Jamil, C. Wulandari, H. S. Nugroho, ... D. D. Berhanuddin, "Refractive index and sensing of glucose molarities determined using Au-Cr K-SPR at 670/785 nm wavelength," Sains Malaysiana, vol. 48, pp. 1259-1265, 2019

[9] N. F. Chiu, C. T. Kuo, T. L. Lin, C. C. Chang, and C. Y. Chen, "Ultra-high sensitivity of the non-immunological affinity of graphene oxide-peptide-based surface plasmon resonance biosensors to detect human chorionic gonadotropin," Biosensors and Bioelectronics, vol. 94, pp. 351-357, 2017.

[10] P. S. Menon, N. A. Jamil, G. S. Mei, A. R. M. Zain, D. W. Hewak, C. -C. Huang, M. A. Mohamed, B. Y. Majlis, R. K. Mishra, S. Raghavan, N. Bhat, "Multilayer CVD-Graphene and MoS ethanol sensing and characterization using Kretschmann-based SPR," IEEE Journal of the Electron Devices Society, vol. 8, pp. 1227-1235, 2020.

[11] F. A. Said, P. S. Menon, S. Shaari, and B. Y. Majlis, B. Y., "FDTD Analysis on Geometrical Parameters of Bimetallic Localized Surface Plasmon Resonance-Based Sensor and Detection of Alcohol in Water," Int. J. Simul. Syst. Sci. Technol., vol. 16, pp. 6.1-6.5, 2015.

[12] P. S. Menon, K. Loganathan, N. A. Jamil, N. R. Mohamad, C. F. Dee, M. F. M. R. Wee, M. A. Mohamed, H. Soleimani, B. Y. Majlis, A. A. Hamzah, "Plasmonic biosensing of kidney wastes using carbon-based derivatives," Proceedings of the 2022 IEEE International Flexible Electronics Technology Conference (IFETC), pp. 1-2, 2022.

Enhancing Industrial Machine Monitoring with IoT: A Wireless Solution

Maizatul Zolkapli*
School of Electrical Engineering
College of Engineering
Universiti Teknologi MARA
Shah Alam, Selangor, Malaysia
maizatul544@uitm.edu.my
*Corresponding Author

Yusof bin Johan
Flexitech Sdn. Bhd. (F18B)
Lot 124 & 126 Jalan Lapan
Kompleks Olak Lempit
Banting, Selangor, Malaysia
yusof_johan@flexiss.com

Ahmad Sabirin Zoolfakar
School of Electrical Engineering
College of Engineering
Universiti Teknologi MARA
Shah Alam, Selangor, Malaysia
ahmad074@uitm.edu.my

Rozina Abdul Rani
School of Mechanical
Engineering
College of Engineering
Universiti Teknologi MARA
Shah Alam, Selangor, Malaysia
rozina7370@uitm.edu.my

Abstract— This paper describes the development of an experimental internet of things (IoT) platform which monitors vibrations and temperature of a computer numerical controller (CNC) milling machine. Processed data is sent to the web server NodeRED via Wi- Fi. The aim of the project is to design and utilize the platform which will be able to monitor machine condition for temperature inside the drilling area, sliding door and vibrations of the machines during drilling. The project is conducted by using a microcontroller which is connected to the sensors placed on a CNC milling machine. A microcomputer served as a Message Queuing Telemetry Transport (MQTT) broker to connect with the NodeRED application via the internet. Condition of the machine can be viewed by using a mobile application or by accessing the NodeRED dashboard. When an irregularity of vibration is detected by the sensors, an email notification will be sent to the user thus giving an early warning for immediate intervention.

Keywords— Internet of Things (IoT), Computer numerical controller (CNC), milling machine, drilling

I. INTRODUCTION

The Internet of Things (IoT) is a network that facilitates the interaction between real-world objects through a unique identifier (ID) [1]. The utilization of IoT is experiencing rapid growth on an annual basis [1]. It encompasses a network of physical objects or things that consist of electronic software, sensors, and network connectivity, enabling these objects to gather and exchange data. Additionally, IoT has the potential to enhance productivity, management, and overall throughput [2, 3].

The advent of emerging technologies has led to significant advancements in modern wireless telecommunications, capturing considerable attention and holding the potential to bring numerous benefits to various applications. Conventional machine maintenance typically involves analyzing and troubleshooting, resulting in downtime [4, 5]. However, by implementing a condition monitoring system on a specific machine, production can proceed smoothly without the need to halt the machine for extended repair periods. This is because potential problems can be detected in advance, preventing their occurrence [6].

The aim of this project was to create a design and implementation of an Internet of Things (IoT) system that monitors the vibration of various drill pins in a CNC milling machine. The system utilizes wireless communication techniques to establish a flexible and long-range connection between the industrial environment and the user. The system offers several benefits, including continuous monitoring of industrial applications and prevention of operations that exceed their limits, which could result in damage to the machine or breakage of the drill pins, leading to extended periods of downtime.

II. METHODOLOGY

A. Machine Monitoring System

Fig. 1.   The Computer Numerical Control (CNC) milling machine.

The project comprised of two main components: sensor development and communication network development. Once the circuit and experimental setup are completed, the project will undergo testing by installing different sensors in the CNC milling machine, as depicted in Figure1.

During the operation of the CNC milling machine, a high-powered drill is utilized to cut materials based on the layout plan and drawings provided by the production department. This process generates significant vibration due to the friction between the material and the drill head pin. If the pin is weak or the material is excessively tough, it can result in excessive vibration, potentially causing damage to both the machine turret and the material itself.

To address this issue, a vibration sensor is positioned on the drill turret, enabling easy detection of vibrations. The collected data is then transmitted to a microcontroller, specifically the NodeMCU, which can wirelessly transmit the information via Wi-Fi to the control center or the primary computer in the production department. In case any irregularities are identified, such as vibrations exceeding a value rate of 1500, a notification is sent to alert the user.

Additionally, for safety purposes, temperature and door sensors are incorporated into the system. This is crucial as cutting through hard materials can sometimes lead to hazardous situations like fire, caused by sparks and flying debris resulting from friction between the drill pin and the material. If the temperature inside the machine exceeds 70°C, a notification is triggered to notify the user of the potential danger.

B. Communication Network

This system operates by detecting temperature, humidity, vibration, and a door switch on a Computer Numerical Controller (CNC) milling machine, commonly used in the industry for milling and grinding materials to create finished products. As illustrated in Figure 2, the sensors within the CNC milling machine chamber transmit their output to the NodeMCU. Equipped with an ESP8266 Wi-Fi chip, the NodeMCU connects to the cloud network through the Message Queuing Telemetry Transport (MQTT) network and utilizes the Node-RED online service to display values and data from the sensors placed within the CNC machine.

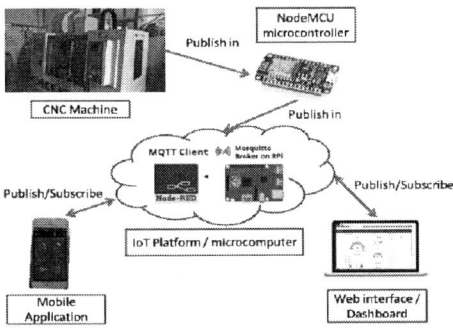

Fig. 2. The IoT network for the monitoring system.

To monitor the live data, the user can access an application on their smartphone or computer that is connected to the same router and shares the same IP address. Node-RED serves as a programming tool for connecting hardware devices, APIs, and online services as part of the Internet of Things (IoT). Its built-in functional nodes simplify the process of creating flows and connecting them together without requiring complex programming. Each node serves a specific purpose and is triggered by data, which can either be generated within the node itself or received from a previous node in the flow.

Figure 3 shows the web service interface, with each sensor having its own nodes that are readable by the Node-RED dashboard through the MQTT node (colored in purple). The blue nodes represent charts and graphs that display the sensor values within the dashboard, as depicted in Figure 4. The yellow node is a function node responsible for initiating email notifications when a high sensor value is detected.

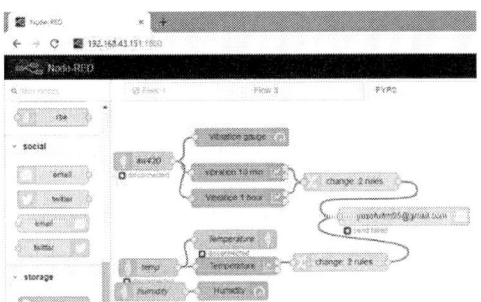

Fig. 3. The IoT network for the monitoring system.

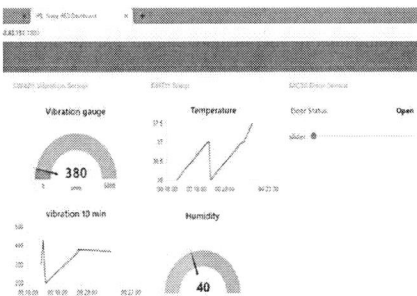

Fig. 4. The Node-RED dashboard.

Both the NodeMCU and a mobile phone, acting as MQTT clients, have the capability to publish and/or subscribe to messages simultaneously with specific topics to the MQTT broker . Each message contains a topic that the broker uses for filtering purposes. When publishing a message, the payload, along with the topic, is included. On the other hand, when subscribing, a list of topic names to which the subscriber is connected is provided. These messages can originate from various sources, such as resource-constrained devices like sensors, backend servers, or graphical clients, all of which are capable of communicating using MQTT [7] as shown in Figure 5.

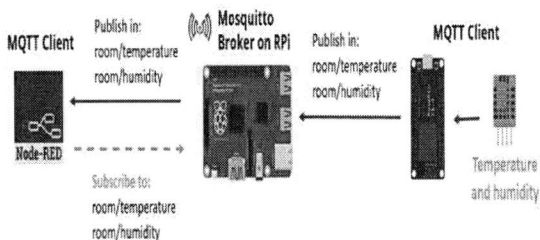

Fig. 5. The MQTT network with the NodeMCU and temperature sensor.

C. Experimental Setup

The setup of the project is visually represented in Figure 6. Each sensor in the setup operates based on its unique working principle, tailored to receive specific input data. The DHT22 sensor employs a single-wire serial interface (SPI) to calculate humidity (in %) and temperature (in Celsius). It utilizes a negative temperature coefficient (NTC) for temperature measurement. Activation of the sensor requires a voltage range of 3-5.5V.

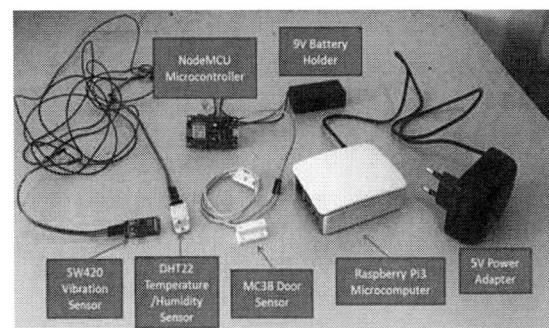

Fig. 6. The monitoring system incorporates various components for its operation.

The vibration sensor detects movement by triggering the piezo coil within the sensor. This module includes an adjustable potentiometer, a vibration sensor, and an LM393 comparator chip. The digital output of the module can be adjusted based on the level of vibration. When triggered, the module produces a logic level high (VCC) and turns on an LED. When no movement is detected, it outputs a low (GND) signal.

The door sensor consists of a reed switch that completes the circuit when a magnet is brought near. In the absence of a magnetic field, the circuit remains open (normally open). The sensor is connected to a signal pin and requires only a few volts to operate. The NodeMCU serves as the microcontroller responsible for reading the data from the sensors connected to its input pins. It can be powered using a 9V battery or a USB cable.

Figure 7 depicts the utilization of plastic wedges to demonstrate the varying effects of different drill types on the material during the drilling process. These wedges are positioned on a tri-axis board, which can be controlled either through a pendant controller or by manual manipulation. Before being mounted to the turret, the drill bit is inserted into a collet chuck.

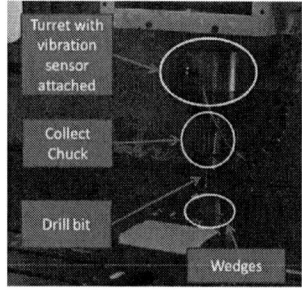

Fig. 7. The standard D_2Q_9 lattice for velocity and temperature mode.

Two distinct types of drill bits for boring machines are employed in the tests. The first type features a sharp and clean lip with a point angle of 118°, while the second type is a used and unpolished drill bit characterized by a dented and rough lip with a point angle of 60°. All tests are conducted at a consistent speed of 500 rpm. Figure 8(a) and (b) illustrate the faulty and good drill bits, respectively. The good drill bit showcases a sharp lip edge and a smoothly relieved area, indicating it is in good condition. Conversely, the faulty drill bit exhibits a rough and dented edge surrounding the lip.

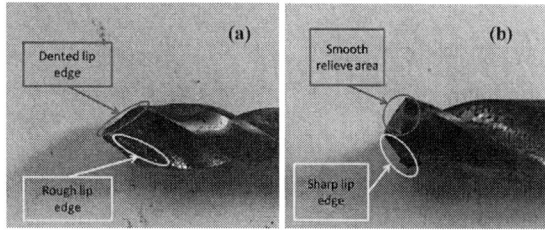

Fig. 8. (a) The faulty drill bit, (b) The good drill bit .

III. RESULTS AND DISCUSSION

TABLE I.   CONDITION OF GOOD DRILL BIT VS. FAULTY DRILL BIT

Item	Good drill bit	Faulty drill bit
Diameter (mm)	4.60	4.70
Length (mm)	90.6	74.5
Point angle	118°	60°
Type	Carbon steel	Solid carbide

Table 1 presents a comparison of the condition of the good and faulty drill bits. Table 2 and Table 3 display the measurements of the diameter and depth of the holes generated by the good and faulty drill bits, respectively. The data reveals that the faulty drill bit exhibits a significant margin of error, as the diameter width of the holes falls outside the acceptable range of +/-0.05mm. The widest deviation of -0.08mm is observed during the 5th test. On the other hand, the good drill bit performs well, as it consistently creates holes within the acceptable range of +/-0.05mm. Although the 3rd test shows a measurement of +0.03mm, it is still deemed acceptable.

TABLE II.   GOOD DRILL BIT

Number of Test	1	2	3	4	5
Diameter (4.60 mm)	-0.01	-0.02	+0.03	+0.01	0.00
Depth (5.0 mm)	+0.01	-0.01	+0.01	+0.03	+0.02

TABLE III.   FAULTY DRILL BIT

Number of Test	1	2	3	4	5
Diameter (4.70 mm)	-0.05	-0.05	-0.06	-0.03	-0.08
Depth (5.0 mm)	+0.01	+0.05	+0.0	+0.01	+0.10

A faulty and unsharpened drill pin can lead to undesirable side effects during the drilling process. These effects include an uneven and deformed material surface due to the excessive feed applied by the drill pin, resulting in the spurring of excess material and leaving traces around the opening area. Figure 9(a) illustrates the traces of rough penetration and scratch marks on the side wall of the hole caused by such a faulty drill pin.

Furthermore, the diameter and penetration depth are also impacted by the condition of the drill pin. In contrast, a sharpened and well-conditioned drill pin, as depicted in Figure 9(b), performs admirably by leaving minimal to no traces of excess material on both the surface and inside of the drilling area. The sidewall of the hole appears clean and smooth when using a properly conditioned drill pin.

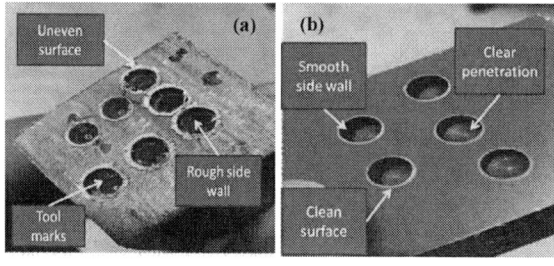

Fig. 9. (a) Rough penetration and scratch marks are evident when using faulty drill bit; (b) When utilizing a good drill bit, there are no visible traces of excess material on both the surface and inside of the drilling area.

The vibration sensors, positioned above the drill pin or turret, capture the vibrations generated during the drilling process. These vibrations are transmitted to the user via the MQTT network, and the results can be accessed through the Node-RED dashboard. Figure 10 illustrates the vigorous movements produced by the faulty drill pin as it penetrates the material, with the maximum amplitude value indicating the need for drill sharpening. On the other hand, Figure 11 showcases the steady vibrations exhibited by the good drill, with a slight increase during the 3rd test.

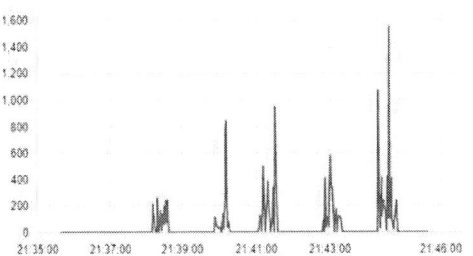

Fig. 10. The vibration graph representing the performance of the faulty drill bit.

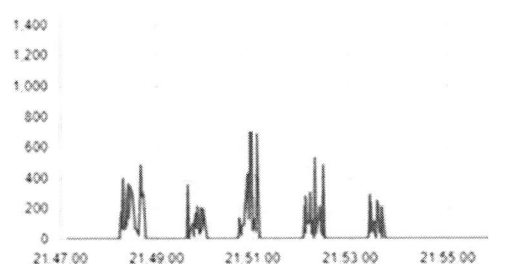

Fig. 11. The vibration graph representing the performance of the good drill bit.

The Node-RED platform has the capability to exhibit real-time data with a refresh rate as fast as 1 second, although the actual display speed may vary depending on the internet connection. While there is a slight delay of around 2 to 3 seconds in presenting the results, they still provide significant data to demonstrate the distinction in vibrations between the drill bits. Notably, the faulty drill recorded the highest vibration level of 1550 units during the 5th test, whereas the good drill reached a maximum of 700 units during the 3rd test.

The obtained results indicate that the faulty drill bit recorded the highest value of 1550. Additionally, the system is designed to activate the email notification function if any value surpasses the rate of 1500. This function is responsible for sending an email to the user through Gmail, as illustrated in Figure 12.

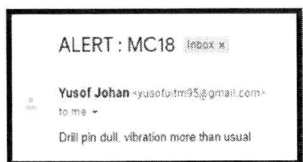

Fig. 12. The Node-RED platform generates and sends the email to the user.

The MQTT network enables users to access live data through a mobile application on any connected device. To view the data, the device must be connected to the same internet protocol address. Figure 24 depicts the live viewing output obtained using the MQTT Dash application.

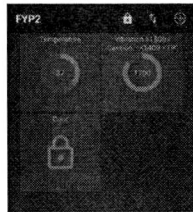

Fig. 13. The output of the MQTT Dash mobile application is displayed.

IV. CONCLUSION

In summary, a monitoring system based on IoT has been created to track the vibration and temperature of the drill pin in a CNC milling machine. The system effectively notifies users when the drill bit begins to exhibit faults. The developed system offers continuous monitoring of industrial applications, helping to prevent operations from exceeding their limits and causing damage to the machine or breakage of the drill pin, ultimately minimizing downtime.

ACKNOWLEDGMENT

The authors wish to thank En. Jaa'far Bin Othman,the owner of Micro Precision Machining Sdn Bhd for allowing us to use the CNC milling machine and School of Electrical Engineering, College of Engineering UiTM for the facilities and administrative support.

REFERENCES

[1] Y. Liu, C. Chi, Y. Zhang and T. Tang, "Identification and Resolution for Industrial Internet: Architecture and Key Technology," in IEEE Internet of Things Journal, vol. 9, no. 18, pp. 16780-16794, 15 Sept.15, 2022.

[2] K. H. Tantawi, I. Fidan and A. Tantawy, "Status of Smart Manufacturing in the United States," 2019 SoutheastCon, Huntsville, AL, USA, 2019, pp. 1-3.0.

[3] B. -Y. Ooi, W. -K. Lee, M. J. W. Shubert, Y. -W. Ooi, C. -Y. Chin and W. -H. Woo, "A Flexible and Reliable Internet-of-Things Solution for Real-Time Production Tracking With High Performance and Secure Communication," in IEEE Transactions on Industry Applications, vol. 59, no. 3, pp. 3121-3132, May-June 2023.

[4] P. Patel, M. I. Ali and A. Sheth, "From Raw Data to Smart Manufacturing: AI and Semantic Web of Things for Industry 4.0," in IEEE Intelligent Systems, vol. 33, no. 4, pp. 79-86, Jul./Aug. 2018.

[5] X. Xu, X. Liang and Z. Guo, "Application of Industrial Robot and Internet of Things in Intelligent Manufacturing System Supported by Software and Hardware," 2022 IEEE International Conference on Electrical Engineering, Big Data and Algorithms (EEBDA), Changchun, China, 2022, pp. 415-418.

[6] R. Kassim, A. Rahmat, H. Mustapa and A. Bakri, "Internet of Things Implementation in Manufacturing Value Chain Process," 2022 International Visualization, Informatics and Technology Conference (IVIT), Kuala Lumpur, Malaysia, 2022, pp. 317-320.

[7] H. Shi, L. Niu and J. Sun, "Construction of Industrial Internet of Things Based on MQTT and OPC UA Protocols," 2020 IEEE International Conference on Artificial Intelligence and Computer Applications (ICAICA), Dalian, China, 2020, pp. 1263-1267.v, Texas, May 2012.

Electrochemical EGFET pH Sensing Performance using ZnO-based Composite Thin Films Sensing Electrode

Nurbaya Zainal
College of Electrical Engineering, School of Engineering, MARA University of Technology,
40450 Shah Alam, Selangor, Malaysia
nurbayazainal@uitm.edu.my

Abdur Rahman
College of Electrical Engineering, School of Engineering, MARA University of Technology,
40450 Shah Alam, Selangor, Malaysia
abdur1012@gmail.com

Nur Syahirah Kamarozaman
College of Electrical Engineering, School of Engineering, MARA University of Technology,
40450 Shah Alam, Selangor, Malaysia
nursyahirahk@uitm.edu.my

Zurita Zulkifli
College of Electrical Engineering, School of Engineering, MARA University of Technology,
40450 Shah Alam, Selangor, Malaysia
zurita101@uitm.edu.my

Sukreen Hana Herman
College of Electrical Engineering, School of Engineering, MARA University of Technology,
40450 Shah Alam, Selangor, Malaysia
hana1617@uitm.edu.my

Abstract—Zinc oxide-based composite films extended gate field effect transistor (EGFET) sensing electrode were fabricated on indium tin oxide, ITO substrate using spin coating method. The sensing performance and structural properties of two different structure sensing electrode; bilayer films and composite films were highlighted in this study. The sensitiveness of the sensing electrode was investigated towards different pH buffer solutions within range pH2 to pH12 and semiconductor measuring unit. Here, commercial reference electrode (RE) was used to complete the electrochemical measurement. As a result, the pH sensitivity and linearity value of the bilayer composite was found highest in this study, that is 45.4 mV/pH and 0.9864 respectively. The results can be observed as correlated with the physical properties of the sensing electrodes characterized by field emission scanning electron microscopy and contact angle.

Keywords—*Zinc Oxide, EGFET, pH sensor, Spin Coating, sensitivity.*

I. INTRODUCTION

Soil biogeochemical processes, which connect biological, chemical, and geological processes, have a strong association with the ecosystem functions of soil [1]. There are factors that can be considered in agricultural soil which include nitrate, soil water content, pH, and temperature. A pH test is undoubtedly the most useful test to determine the characteristics of the soil, specifically the health condition of the soil [2], [3]. Over the last decade, there has been an effort to produce portable equipment for continuous pH determination [4]–[6] . Electrochemical sensors are well-known for their fast reaction times, great sensitivity, and ease of use [7]. Given that an Extended Gate Field Effect Transistor (EGFET) is one of the electrochemical sensor devices that is used for pH detection. The initial discovery of EGFET comes from the construction of Ion-Sensitive Field-Effect Transistor (ISFET). When compared to ISFET, EGFET has various inherent advantages, including a simpler fabrication process, less influence from optical light and operating temperature, and a most importantly disposable gate which makes it more convenient in making a study [8]. EGFET consists of a reference electrode, a sensing electrode, and a MOSFET to function. The sensing electrode is an ion-sensitive layer that will be connected to the gate of the MOSFET to act as the extended gate. Searching the material for sensing electrodes have been captured researchers' interest since decades and mostly metal oxide nanomaterials are famously known for their potentiality in numerous applications of ion detection [9]–[13]. However, composited materials have not yet been fully discovered as sensing electrode for electrochemical pH sensor. This study focuses on the fabrication of sensing electrode comprised by composition of materials; zinc oxide (ZnO), polyvinyl butyral (PVB), and graphene oxide (GOx). For comparison purpose, the sensing electrodes are being prepared with two different structure which is ZnO:GOx:PVB composite films and ZnO:GOx/PVB bilayer composite films.

Various deposition methods mainly chemical techniques were used to prepare the sensing electrodes such as chemical bath deposition [14], spray pyrolysis [15], electro deposition [16], spin coating [17], and others. In comparison to other deposition techniques, spin coating has a number of advantages including flexibility, homogeneity, safety, and economic viability.

II. METHODOLOGY

A. Preparing Solutions

For preparing 50ml of ZnO solution, 4.39 g of zinc acetate powder, $Zn(CH_3CO_2)_2$ is added into a beaker. After that, 1.2 ml monoethanolamine, C_2H_7NO is added to the zinc

979-8-3503-2369-6/23 $31.00 © 2023 IEEE

acetate powder by using a pipette. The mixture is then added with 2-methoxyethanol, $C_3H_8O_2$ until 50 ml of zinc oxide solution. As all the recipe is added, the solution is then stirred using hot plate magnetic stirrer with the setting of 80°C and 300 rpm for 3 hours. After the solution is stirred for 3 hours, the heat is turned off and the solution is left stirred until the next day for ageing process. Once the solution is done, it is ready to be deposited to the ITO substrate by using spin coating method. Polyvinyl butyral, PVB solution is prepared by mixing 0.01g of PVB, ~8ml of ethanol and ~2ml of deionized water inside a beaker. The solution is then stirred at 350 rpm for one hour. As for the preparation of different zinc oxide-based solutions, 5 ml of ZnO, 2.5 ml of PVB, and one drop of Graphene Oxide, GOx is used for each composite solution

B. Fabrication of Sensing Electrode

In spin coting technique, thin film can be present on top of the ITO substrates to act as a sensing membrane. Half of each substrate is covered by kapton tape such that only half of its surface will be deposited with zinc oxide-based solutions as in Fig. 1. As the other half, it will be extended with copper wire to be attached to the gate of MOSFET in the next procedure. Argon gas is supplied in the instrument to vacuum the chamber before spin coating the substrates. Then, 10 droplets are dropped using a pipette on top of the substrate for around 10 seconds at 1500 rpm. It is then left spinning for 50 seconds at 3000 rpm. After all the steps are repeated on each substrate for different solutions, they are then placed inside the oven for drying process for 10 minutes at 100°C. Once the drying process is done, the ZnO-based composite sample will go through annealing process for 10 minutes at 400°C. Fig. 1 shows the cross-sectional schematic diagram of thin films deposited using spin coating technique.

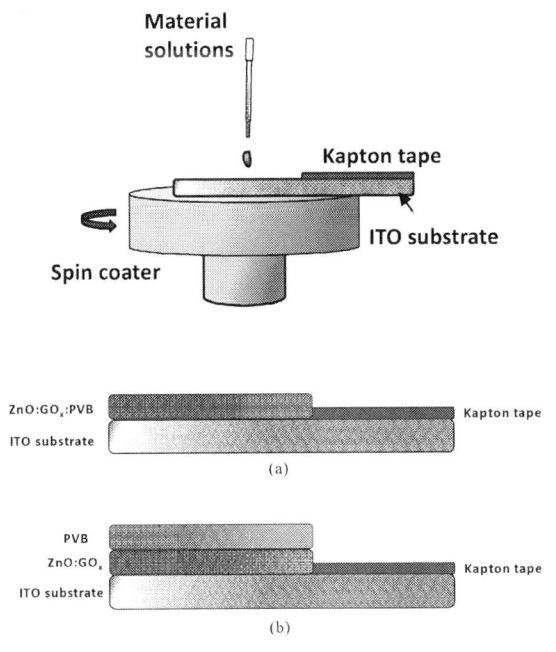

Fig. 1. Thin films deposition process by spin coating technique and cross-sectional view of sample structure (a) ZnO:GOx:PVB composite films, (b) ZnO:GOx/PVB bilayer composite films.

C. EGFET Measurement Setup

As the sensing electrode is fabricated by using spin coating method, the electrical performance of it will be tested by using Semiconductor Parametric Device Analyzer (SDA) Keysight B1500A as the Semiconductor Measuring Unit (SMU) and the physical properties are observed by using the Field Emission Scanning Electron Microscopy (FESEM) and Contact Angle. Firstly, the device needs to be calibrated by testing it with a direct connection to CD4007UB MOSFET where the gate, drain, and source of the MOSFET is connected to SMU1, SMU2, and SMU3 respectively as shown in Fig. 2. Different buffer solutions are used for pH sensing electrode with pH value of 2, 4, 7, 10, 12. An I-V transfer function should appear to indicate that the calibration is correct when the current increases gradually once the gate voltage passed the threshold voltage and will eventually be saturated at a certain point. The surface morphology for the sensing electrode is analyzed by the field emission scanning electron microscope (FESEM) at x10k magnification.

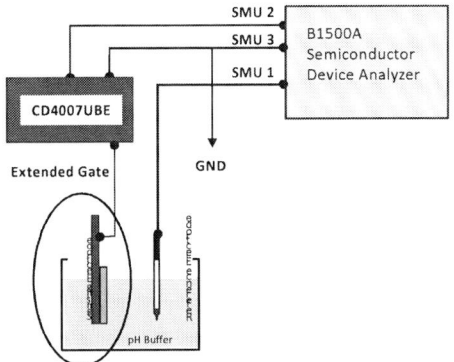

Fig. 2. EGFET measurement setup

As for the contact angle, each substrate is measured by using the Video Contact Angle System (VCA3000S) which can snap the image of the droplet and automatically calculate the tangent lines. The volume of distilled water droplets used is 5μl which will be dropped on each sample prior to capturing the image. The hydrophobicity of the contact angle is referred to as in Fig. 3 where hydrophilic is at an angle less than 90°, hydrophobic is between 90° to 150° and superhydrophobic is more than 150°.

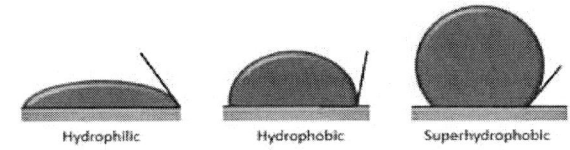

Fig. 3. Classification hydrophobicity

III. RESULTS AND DISCUSSIONS

A. Structural Property

FESEM was used to examine the microstructure and surface characteristics of ZnO-based sensing electrodes in order to determine their structural properties as in Fig. 4. The samples were fabricated using the spin coating technique, and a FESEM with a 10k magnification was used for analysis. The FESEM's high-resolution imaging capabilities made it possible to observe distinguished surface morphology of both ZnO:GOx:PVB composite and ZnO:GOx/PVB bilayer composite films in Fig. 4. (a) and (b) respectively. As can be seen from the FESEM images with the magnification of 10k in Fig. 6, this may be brought on by the material's surface morphology which plays a role in the sensitivity and linearity of the sensing material. The surface of ZnO:PVB:GOx has a porous structure that correlated to highly rough surface. Meanwhile, the bilayer sample's surface morphology represents fibrous surface structure of the PVB layer. Next, the wettability of -based sensing electrodes were analyzed using contact angle measurements as in Fig. 5. The contact angle is a measure of the degree of wetting of a liquid on the surface. The results obtained from the contact angle measurements shows that ZnO:PVB:GOx is hydrophobic (102.80°), whereas ZnO:GOx/PVB is hydrophilic (84.20°). These structural properties provide valuable insights into the sensitivity and linearity for sensing electrode.

B. pH Sensisng Characteristics

The I-V transfer characteristic can be obtained by simulating on Keysight B1500 with each sample used as the sensing electrode. From the transfer curve, the value of the gate to source voltage, V_{GS} at the linear region is recorded specifically when drain to source current, I_{DS} is at $100\mu A$. On the other hand, the output characteristic can be obtained on the same SMU. The output current's value, I_{DS} is recorded when the drain to source voltage, V_{DS} at 2V which is in the saturation region. The average of three readings is calculated to obtain to plot the reference voltage versus pH graph. Sensitivity and linearity were obtained from the linear fit of the graph as in Fig. 6. The sensitivity value of ZnO:GOx/PVB bilayer composite films was found higher than the composite thin films. The average sensitivity value is 45.4mV/pH and linearity of 0.9864 for the bilayer thin films whereas the sensitivity and linearity are 28.8mV/pH and 0.6889 respectively was obtained from the ZnO:GOx:PVB composite films. The reason to poor pH sensitivity performance might be due to the composite material that was not properly dispersed which can lead to clogging or blocking the ion sensing diffusion on the buried site.

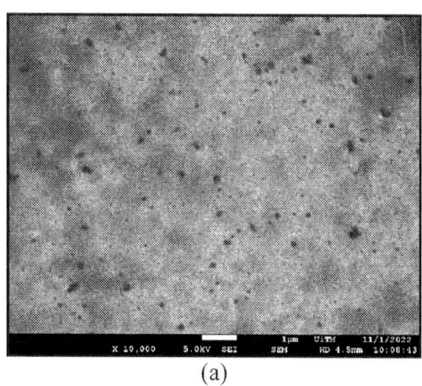

(a)

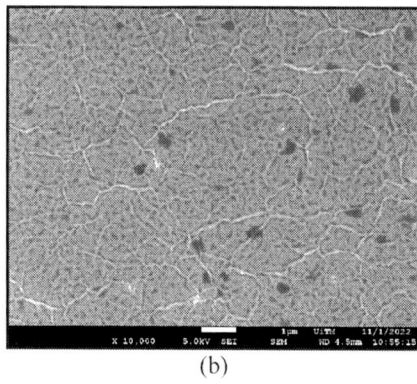

(b)

Fig. 4. FESEM images of (a) ZnO:GOx:PVB composite and (b) ZnO:GOx/PVB bilayer composite films

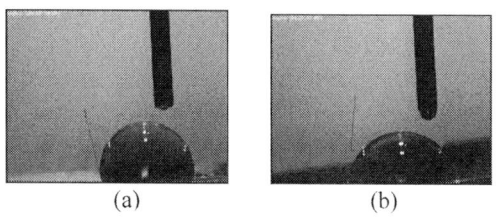

(a)   (b)

Fig. 5. Contact angle of (a) ZnO:GOx:PVB and (b) ZnO:GOx/PVB

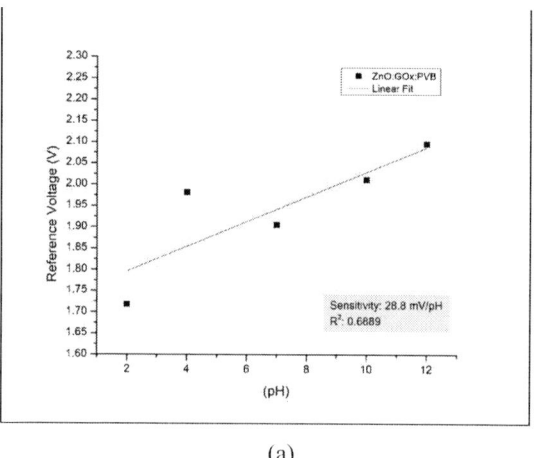

(a)

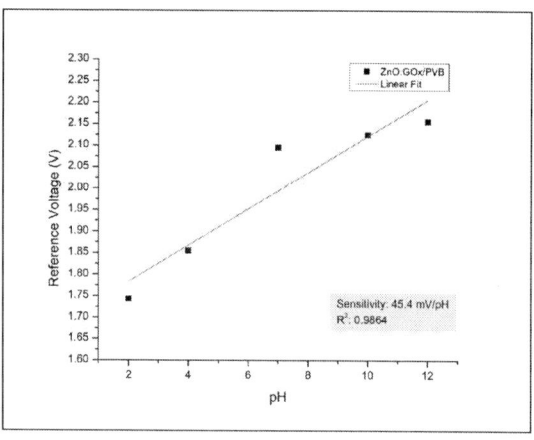

Fig. 6. Reference volgate versus pH value (a) ZnO:GOx:PVB composite and (b) ZnO:GOx/PVB bilayer composite films

TABLE I. SENSITIVITY AND LINEARITY FOR PH SENSING CHARACTERISTICS

No.	Transfer Characteristic		
	Sensing electrode	Sensitivity, mV/pH	Linearity
1.	ZnO:GOx:PVB	28.8	0.6889
2.	ZnO:GOx/PVB	45.4	0.9864

IV. CONCLUSIONS

In conclusion, different results are obtained when the electrical properties of zinc oxide-based sensing electrode are fabricated using the spin coating process on top of an ITO substrate. The EGFET pH sensing performance was led by the ZnO:GOx/PVB bilayer composite films with the sensitivity of 45.4 mV/pH and the linearity value of 0.9864. This thin film was perfectly suited to be used as sensing electrode due to its wettability property proved by the hydrophilicity surface result. Further pH sensing measurement that needs to be taken into account such as hysteresis and drift measurement in next study.

AKNOWLEDGEMENT

This study was partially supported by Research Management Centre, Universiti Teknologi MARA (Project Code: 600-RMC/KEPU 5/3 (007/2021)). Authors acknowledge the technical support from Instrumentation 1 Laboratoray and NANO-ElecTronic Centre (NET), UiTM.

REFERENCES

[1] R. A. Dahlgren, "Biogeochemical processes in soils and ecosystems: From landscape to molecular scale," J. Geochemical Explor., vol. 88, pp. 186–189, 2006.

[2] T. Merl, M. R. Rasmussen, L. R. Koch, J. V. Sondergaard, F. F. Bust, and K. Koren, "Measuring soil pH at in situ like conditions using optical pH sensors (pH-optodes)," Soil Biol. Biochem., vol. 175, pp. 4–6, 2022.

[3] G. W. Thomas, "Soil pH and Soil Acidity," in Methods of Soil Analysis. Part 3. Chemical Method, 1996, pp. 475–490.

[4] L. Manjakkal, K. Cvejin, J. Kulawik, K. Zaraska, and D. Szwagierczak, "A Low-Cost pH Sensor Based on RuO_2 Resistor Material," Nano Hybrids, vol. 5, pp. 1–15, 2013.

[5] R. Zhao, M. Xu, J. Wang, and G. Chen, "A pH sensor based on the TiO_2 nanotube array modified Ti electrode," Electrochim. Acta, vol. 55, pp. 5647–5651, 2010.

[6] S. Nakata, M. Shiomi, Y. Fujita, T. Arie, S. Akita, and K. Takei, "A wearable pH sensor with high sensitivity based on a flexible charge-coupled device," Nat. Electron., vol. 1, pp. 596–603, 2018,

[7] S. Singh et al., "Nitrates in the environment: A critical review of their distribution, sensing techniques, ecological effects and remediation," Chemosphere, vol. 287, p. 131996, 2022,

[8] Y. S. Chiu, C. Y. Tseng, and C. T. Lee, "Nanostructured EGFET pH sensors with surface-passivated ZnO thin-film and nanorod array," IEEE Sens. J., vol. 12, pp. 930–934, 2012.

[9] C. Chen, Y. Zhang, H. Gao, K. Xu, and X. Zhang, "Fabrication of Functional Super-Hydrophilic TiO_2 Thin Film for pH Detection," Chemosensors, vol. 10, pp 1-15, 2022.

[10] H. A. Khizir and T. A. H. Abbas, "Hydrothermal synthesis of TiO2 nanorods as sensing membrane for extended-gate field-effect transistor (EGFET) pH sensing applications," Sensors Actuators A Phys., vol. 333, p. 113231, 2022.

[11] A. K. Mishra, D. J. Kumar, B. Mukherjee, A. Kumar, S. Ratan, and S. Jit, "CuO Nanowire-Based Extended-Gate Field-Effect-Transistor (FET) for pH Sensing and Enzyme-Free/Receptor-Free Glucose Sensing Application," IEEE Sens. J., vol. 20, pp. 5039–5047, 2020.

[12] R. A. Rahman, M. A. Zulkefle, K. A. Yusoff, W. F. H. Abdullah, M. Rusop, and S. H. Herman, "Drying Temperature Dependence of Sol-gel Spin Coated Bilayer Composite ZnO/TiO2 Thin Films for Extended Gate Field Effect Transistor pH Sensor," IOP Conf. Ser. Mater. Sci. Eng., vol. 340, 2018.

[13] P. D. Batista, M. Mulato, C. F. D. O. Graeff, F. J. R. Fernandez, and F. D. C. Marques, "SnO_2 Extended Gate Field-Effect Transistor as pH Sensor," Brazilian J. Phys., vol. 36, pp. 478–481, 2006

[14] A. B. Rosli, S. B. Hashim, R. A. Rahman, S. H. Herman, W. F. H. Abdullah, and Z. Zulkifli, "Correlation of ZnO Surface Morphology and Sensing Performance of EGFET Nitrate Sensor," J. Mech. Eng., vol. 11, pp. 65–80, 2022.

[15] N. Lehraki, M. S. Aida, S. Abed, N. Attaf, A. Attaf, and M. Poulain, "ZnO thin films deposition by spray pyrolysis: Influence of precursor solution properties," Curr. Appl. Phys., vol. 12, pp. 1283–1287, 2012,

[16] S. Rajpal and S. R. Kumar, "Effect of Annealing on Nano crystalline ZnO Thin Film Developed by Electro deposition Method," Mater. Today Proc., vol. 4, pp. 3754–3759, 2017.

[17] J. Charles Babu and V. Jagadeesan, "Impact of pH on the structural characteristics of spin coated zinc oxide thin films," Mater. Today Proc., vol. 66, pp. 2226–2229, 2022.

Fabrication of TiO$_2$-PANI Nanostructure using Electrospray for the pH Sensing Electrode

Aina Syakirah Mohd Masri
School of Electrical Engineering
College of Engineering
Universiti Teknologi MARA
40450 Shah Alam, Malaysia
syakirahh21@gmail.com

Nur Syahirah Kamarozaman
School of Electrical Engineering
College of Engineering
Universiti Teknologi MARA
40450 Shah Alam, Malaysia
nursyahirahk@uitm.edu.my

Nurbaya Zainal
School of Electrical Engineering
College of Engineering
Universiti Teknologi MARA
40450 Shah Alam, Malaysia
nurbayazainal@gmail.com

Zurita Zulkifli
School of Electrical Engineering
College of Engineering
Universiti Teknologi MARA
40450 Shah Alam, Malaysia
zurita101@uitm.edu.my

Sukreen Hana Herman
School of Electrical Engineering
College of Engineering
Universiti Teknologi MARA
40450 Shah Alam, Malaysia
hana1617@uitm.edu.my

Abstract—In this paper, the effect of surface morphology of TiO$_2$-PANI fabricated electrode for pH sensing application using the electrospray method was discussed. ITO glass was used as the substrate for the electrodes. The surface morphology was influenced by the effect of the deposition distance between needle-tip-to-collector at 8, 10, 12 and 14cm. Besides the surface morphology, the pH sensing performance was also investigated. The result from the FESEM images shows that the surface morphological nanocomposite of TiO$_2$-PANI appeared to be nanoflakes structure. The wettability test results show that TiO$_2$-PANI nanocomposite electrodes present high peeling resistance with a contact angle of 102 degrees almost twice from TiO$_2$ nanostructure, which exhibits satisfying hydrophobicity. The polyaniline hydrophobic features contributed to improved electrode insolubility and durability. The experiment constructed an EGFET circuit by connecting the nanocomposite sensing electrode to a commercial MOSFET (CD4007UB) as an evaluation of sensitivity and linearity. TiO$_2$-PANI exhibited the highest sensitivity with 53.75mV/pH and linearity 0.9826 at 8cm deposition distance. Consequently, TiO$_2$-PANI satisfies the criteria for insoluble pH detection, owing to its higher sensitivity.

Keywords—*TiO$_2$-PANI nanocomposite, TiO$_2$ nanostructure, ITO, DMF, pH-sensing, sheet resistance, wettability, sensitivity, linearity, EGFET*

I. INTRODUCTION

The pH sensors are tough and dependable measuring instruments that are suited for submersible sensing devices like stream and lake monitoring, aquaculture research, and biochemical studies [1]. It has a standardized scale for identifying whether the solution is alkaline or acidic and measures the number of hydrogen and hydroxide ions. It is crucial to notice that the hydrogen bonds produced in the simulated system are extremely strong and do not change over time [2], [3]. Today, many chemical and biological processes depend on pH value escalation. The measurement of pH is the most well-known and often performed test in biochemical labs [4]. As a result, it is critical to fabricate precise pH sensing materials with certified pH, insoluble, conductivity, oxidation-reduction potential (Redox) and ion concentration.

TiO$_2$ nanoparticles were used to increase the conductivity of the nanocomposite structure. Due to its great photoactivity, high stability, and popularity as a semiconductor photoactive material, TiO$_2$ has attracted a lot of attention [3]. Titanium dioxide (TiO$_2$) nanocomposites were fabricated by using Polyaniline (PANI) as a synthetic polymer at constant concentrations [5]. Polyaniline (PANI) has been widely explored as a conventional conductive polymer material due to its exceptional optical property, electrical conductivity, corrosion resistance, outstanding environmental and chemical stability, and ease of compounding by chemical and electrochemical processes [6]. A more advanced study was conducted by adding TiO$_2$ nanocomposite with a constant mass of polyaniline polymer to examine conductivity behaviour as a pH-sensing electrode. The electrospray technique was used to fabricate the nanocomposites, the most straightforward and versatile methods for producing TiO$_2$ fine droplets with high porosity, controllable dimensions, designed structures, and adjustable mechanics [7]. A high voltage is applied to the capillary or nozzle, this results in the formation of a cone-shaped protrusion called the cone jet mode [8]. The electrospray process varies by varying deposition distances [9]. As the distance applied needle-tip-to-collector varies, the mass (g) of the polyaniline (PANI) deposition flow rate (ml/h) and other deposition parameters were constant respectively.

EGFET circuit was employed to examine the sensitivity performance of the pH sensing electrode for nanocomposites of TiO$_2$-PANI [10]. The EGFET circuit consists of an EGFET transistor with the gate electrode exposed to the electrolyte solution [11]. Besides that, by implementing optimized process parameters, the surface morphology nanocomposite electrode, wettability, pH sensitivity and linearity of the prepared TiO$_2$-PANI sensing electrode were investigated. In sum, the fabrication of insoluble and high sensitivity pH sensing electrodes via electrospray and the characterizations of TiO$_2$-PANI decorated nanocomposite were aimed to be as promising pH sensor electrodes. The performance of the pH sensing device and material analysis of nanocomposite were also reviewed concerning the effects of deposition distance at 8, 10, 12, and 14 cm.

II. METHODOLOGY

The work can be divided into several phases to achieve the research's main objective.

A. Sample Preparation

The size ITO glass of 1cm x 2cm was used as the substrate. The TiO_2 solution preparation procedure was used. The solution recipe is as Table I.

TABLE I.   CHEMICAL MATERIALS COMPOSITION OF TiO_2 SOLUTION

Beaker A	Beaker B
1.5ml of Titanium (IV) Isopropoxide (TTIP) 23ml of Absolute Ethanol 2.5ml of Glacial Acetic Acid (GAA)	23ml of Absolute Ethanol 1 drop of Triton-X 100 0.2ml of deionized water

Each solution was stirred for an hour at room temperature at speed of 400 rpm. After that, the solutions were combined and swirled at room temperature for another hour at the same speed. Polyaniline powder was weighed using an electronic balance. As for the TiO_2-PANI solution, 10mg of PANI powder was diluted with 1 ml of N-Dimethylformamide (DMF) with a stirring time of six hours, this was done a day before the deposition process [12]. 50 ml of TiO_2 solution was mixed with diluted PANI and stirred for 1 hour. The 50 ml of TiO_2 solution was mixed with diluted PANI then stirred for an hour. All the stirring processes were done at room temperature. A dark green solution of TiO_2-PANI nanocomposite was observed, due to the existence of polyaniline.

B. Deposition Process

The TiO_2-PANI solution is deposited by the electrospray process on the conductive side of the ITO substrates. All the deposition parameters are constant with the voltage used was 15kV, current at 400 µA using 18G gauge at the deposition rate of 1.2ml/h. The TiO_2-PANI solution was deposited using the electrospray method at varying needle-tip-to-collector distances of 8, 10, 12, and 14 cm. Prior research has suggested that the range around 9cm with high applied voltage was critical for achieving a desirable coating morphology [13]. By selecting distances at regular intervals (8cm, 10cm, 12cm, and 14cm), we aimed to cover a diverse set of deposition conditions. This approach allows us to study the effects of varying distances on the surface morphology and provides insights into how the deposition process changes with increasing or decreasing distance. Following electrospray deposition, the nanocomposite substrates were annealed in the furnace for 15 minutes at 300°C.

C. Surface morphology

The samples were observed using a digital microscope by OLYMPUS (U-MSSP4) with different lenses of 20x, and 50x for the surface visualized in the microscale was depicted in the result. In the nanoscale, Field Emission Scanning Electron Microscopy (FESEM JEOL JSM-7600F) was used at magnifications of 20k.

D. pH-Measurement

pH measurement characterization by immersing the deposited substrate in 20 ml pH buffer concentration 2, 4, 7, 10 and 12. The Keysight B1500A Semiconductor Device Analyzer was utilized to contribute accuracy in results. The sensing membrane is connected to the gate of a CD4007UB commercial N-MOSFET to construct the EGFET as shown in Fig.1. The sensing membrane was connected to the gate of a CD4007UB [14].

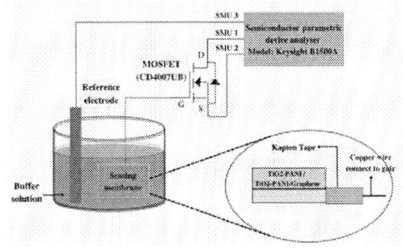

Fig. 1.   EGFET circuit setup.

III. RESULTS

A. Morphology on Digital Microscope

A few microfibers were found in the nanostructure of TiO_2 and TiO_2-PANI has been successfully fabricated using the electrospray method and has been confirmed by Digital Microscope images as depicted in Fig. 2 The digital microscope images represent two distinct nanostructure samples that are prepared under different deposition distances of needle-to-tip-collector of 8, 10, 12 and 14cm. In this study comparison of the transparency of microfibers between two distinct nanostructures was observed. At 20x and 50x magnification indicates that TiO_2 and TiO_2-PANI microfibers possess a transparent structure. It is found that the distribution is not uniform in size, there are several agglomerations of TiO_2 nanoparticles resulting in large lumps. The aggregated TiO_2 revealing the transparent and moderately rough TiO_2 phase [15]. The visualization of surface morphology on a digital microscope was influenced by several factors including the solution composition, polymer solubility, and viscosity. Viscosity and surface tension were influenced by molecular weight. The morphology of the fibers changes as the polymer's molecular weight increases, including a decrease in the number of beads and an inconsistent shape [16]. Thus, research revealed that altering the viscosity and concentration of sol-gel solutions could regulate the formation of the microfiber's diameter and accumulation could regulate the formation of the microfiber's diameter and accumulation.

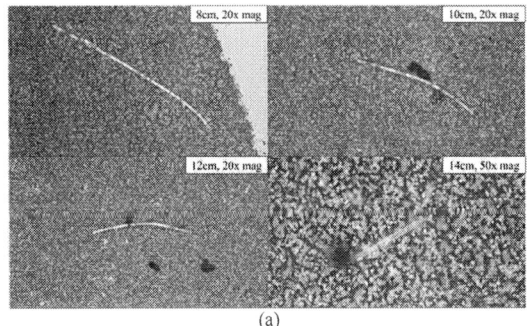

(a)

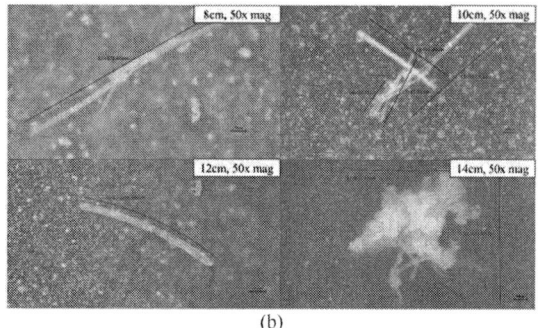

(b)

Fig. 2. Digital microscope image of (a) TiO₂ and (b) TiO₂-PANI.

B. FESEM

The FESEM images revealed the nanostructure synthesized by the electrospray technique with a distinct deposition distance, as shown in Fig. 2. According to past research, for smooth nanocomposite preparation of sample TiO₂-PANI, nanoflakes and nanorods are formed on a large scale in FESEM image [17]. As depicted, 20k magnification is chosen for FESEM analysis. The FESEM image of both nanocomposites clearly shows the formation of the nanoflakes and irregular nanostructure with minimal agglomeration. The nanostructure rough surface has the potential to provide hydrophobicity. The size of the nanoflakes in both samples seems identical as they range between 500 nm and below in size. The images depicted for the TiO₂-PANI sample, decreasing in deposition distance produced a higher density of nanoflakes. This happens because the evaporation rate is influenced by the strength of the electric field. The distance should be sufficient to allow the solvent to evaporate and extend before being deposited on the collector [18]. In this study, shorter distance allows for more efficient deposition nanoflakes on the substrate's surface.

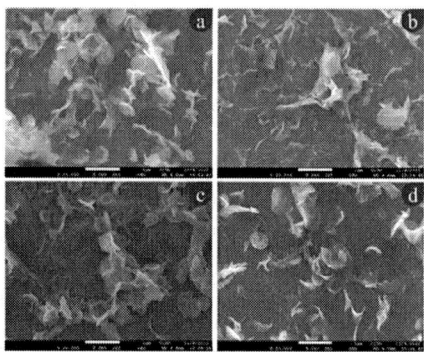

Fig. 3. FESEM images of TiO₂-PANI with different deposition distances a) 8cm b) 10cm c) 12cm d) 14cm.

C. Contact angle

The optical contact angle measurement of the three samples, TiO₂ and TiO₂-PANI wetting behaviour were observed and the mean of the angle was calculated and shown in Table II. The TiO₂ substrate presents the lowest contact angle of 69.27° and exhibits gratifying hydrophilicity surface. The contact angle for TiO₂-PANI was the highest value with 102°, which confirmed the chemical stability of the contact angle output as its exhibit's properties of hydrophobicity nanocomposites. Since then, it has been confirmed that TiO₂-

PANI nanocomposite promotes strong durability and insoluble properties for pH sensing applications.

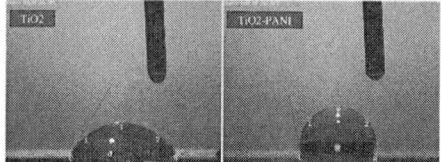

Fig. 4. Sessile drop lying on the surface of the sensing membrane.

TABLE II.   THE CONTACT ANGLE FOR 8CM DEPOSITION OF TiO₂ AND TiO₂-PANI

Nanocomposite	Contact angle (°)			Mean (°)
TiO₂	62.50	59.20	86.10	60.27
TiO₂-PANI	101.30	102.00	102.70	102.00

D. pH-Measurement

The characterization of the structural, morphological, optical, and electrical properties of the synthesized TiO₂-PANI nanostructures confirms their enhanced stability and improved conductivity by annealing. Fig. 5. depicted information on reference voltage (V) versus pH value. The sensitivity of the fabricated pH sensor was measured to determine its sensing properties for varying pH buffer concentrations of 2, 4, 7, 10 and 12. Sensor response for pH sensitivity data was obtained on the Vds linear region at 100uA. The sensitivity of the pH sensing membrane refers to the change in the output reference voltage per unit change in pH, and linearity indicates how well the sensor response fits a straight line. The TiO₂-PANI nanocomposite in Table III, shows the highest sensitivity with a value of 53.75mV/pH at 8cm with linearity of 0.9826. The study shows at lower deposition distance improved the performance of the pH-sensing membrane. Moreover, at 8 cm deposition distance, the TiO₂ nanostructure in Table III, shows a sensitivity value of 53mV/pH and linearity of 0.9642. The sensitivity has enhanced by 1.415 percent in the TiO₂-PANI nanocomposite compared to TiO₂ nanostructure. A minor increment was observed, attributed to the composition ratio of polyaniline (PANI) in the sensing membrane. Therefore, it was believed that adding 10mg of PANI to a 50ml TiO₂ solution would lead to synergistic effects, enhancing the pH sensitivity of the electrode.

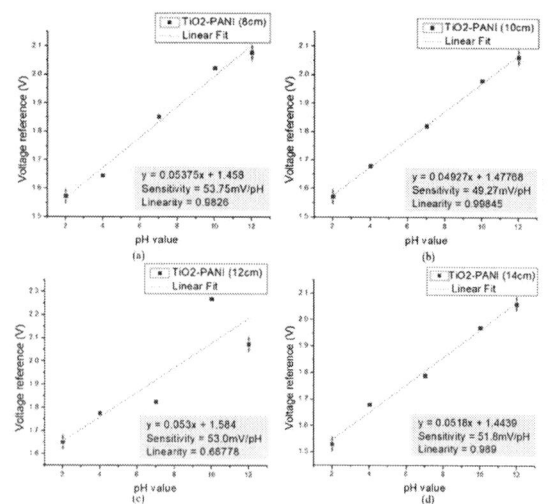

979-8-3503-2369-6/23 $31.00 © 2023 IEEE

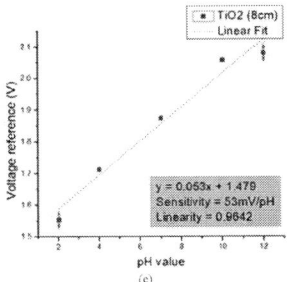

Fig. 5. Sensor response pH sensitivity and linearity of TiO$_2$-PANI nanocomposites layers deposition distance of (a) 8cm, (b) 10cm, (c) 12cm, (d)14cm and TiO$_2$ nanostructure (e) at 8cm deposition distance.

TABLE III.   PH MEASUREMENT SENSITIVITY AND LINEARITY OF TiO$_2$-PANI DEPOSITION DISTANCE OF 8, 10, 12 AND 14CM NANOCOMPOSITE AND TiO$_2$ NANOSTRUCTURE AT 8CM DEPOSITION DISTANCE

Nanocomposite	TiO$_2$-PANI				TiO$_2$
Deposition distance (cm)	8	10	12	14	8
Sensitivity (mV/pH)	53.75	49.27	53.00	51.80	53.00
Linearity	0.9826	0.9985	0.6878	0.989	0.9642

IV. CONCLUSION

In conclusion, TiO$_2$-PANI nanocomposites were made using the electrospray process and placed onto an ITO glass substrate for the pH sensor's sensing electrodes. The morphology of the TiO$_2$-PANI nanocomposites demonstrates the ability of nanoflakes to increase surface energy. The contact angle data shows that hydrophobic surface features combined with polyaniline improve electrode insoluble and durability performance. The pH sensitivity results for TiO$_2$-PANI nanocomposite showed a high pH sensitivity membrane behaviour compared to TiO$_2$ nanostructure. Hence, TiO$_2$-PANI sensing membrane shows almost twice performance in contact angle value compared to TiO$_2$ sensing membrane, demonstrating that it is not degrading the sensitivity performance. Current research shows that nanocomposite TiO$_2$-PANI, despite higher sensitivity, may satisfy the requirement for insoluble pH detection.

ACKNOWLEDGMENT

This work was partially supported by the Research Management Centre, Universiti Teknologi MARA (Project Code: 600-RMC/KEPU 5/3 (007/2021)). Authors acknowledge the technical support from NANO-ElecTronic Centre (NET), UiTM.

REFERENCES

[1] Dmitry S. Koktysh, "Ratiometric pH sensor using luminescent CuInS2/ZnS quantum dots and fluorescein," *Material Research Bulletin*, 2020.

[2] M. Dmitrenko et al., "Modification strategies of polyacrylonitrile ultrafiltration membrane using TiO2 for enhanced antifouling performance in water treatment," *Sep Purif Technol*, vol. 286, no. January, p. 120500, 2022.

[3] L. Fan, G. Liang, W. Yan, Y. Guo, Y. Bi, and C. Dong, "A highly sensitive photoelectrochemical aptasensor based on BiVO4 nanoparticles-TiO2 nanotubes for detection of PCB72," *Talanta*, vol. 233, Oct. 2021.

[4] H. A. Khizir and T. A. H. Abbas, "Hydrothermal synthesis of TiO2 nanorods as sensing membrane for extended-gate field-effect transistor (EGFET) pH sensing applications," *Sens Actuators A Phys*, vol. 333, Jan. 2022.

[5] S. W. H. M. D. Titi Istirohah, "Fabrication of Aligned PAN/TiO2 Fiber using Electric Electrospinning (EES)," *Science Direct*, pp. 1–6, 2018.

[6] C. Hu, K. Kwan, X. Xie, C. Zhou, and K. Ren, "Superhydrophobic polyaniline/TiO2 composite coating with enhanced anticorrosion function," *React Funct Polym*, vol. 179, Oct. 2022.

[7] X. Zhang et al., "UV/TiO2/periodate system for the degradation of organic pollutants – Kinetics, mechanisms and toxicity study," *Chemical Engineering Journal*, vol. 449, Dec. 2022.

[8] J. Y. Kim, S. J. Lee, G. Y. Baik, and J. G. Hong, "Viscosity Effect on the Electrospray Characteristics of Droplet Size and Distribution," *ACS Omega*, vol. 6, no. 44, pp. 29724–29734, Nov. 2021.

[9] B. M. Marsh, K. Iyer, and R. G. Cooks, "Reaction Acceleration in Electrospray Droplets: Size, Distance, and Surfactant Effects," *J Am Soc Mass Spectrom*, vol. 30, no. 10, pp. 2022–2030, Oct. 2019.

[10] T.-M. P. Chih-Wei Wang, "Structural properties and sensing performances of CoNxOy ceramics film for EGFET pH sensors," *Ceram Int*, pp. 1–9, 2021.

[11] R. K. Sahoo, S. Rani, V. Kumar, and U. Gupta, "Zinc oxide nanoparticles for bioimaging and drug delivery," *Nanostructured Zinc Oxide: Synthesis, Properties and Applications*, pp. 483–509, Jan. 2021.

[12] B. Mandal et al., "Supercapacitor performance of nitrogen doped graphene synthesized via DMF assisted single-step solvothermal method," *FlatChem*, vol. 34, Jul. 2022.

[13] P. Y. Lin, Y. Y. Chen, T. F. Guo, Y. S. Fu, L. C. Lai, and C. K. Lee, "Electrospray technique in fabricating perovskite-based hybrid solar cells under ambient conditions," *RSC Adv*, vol. 7, no. 18, pp. 10985–10991, 2017.

[14] T. A.-H. A. Hersh Ahmed Khizir, "Sensors and Actuators: A. Physical," *Hydrothermal synthesis of TiO2 nanorods as sensing membrane for extended-gate field-effect transistor (EGFET) pH sensing applications* , pp. 1–9, 2022.

[15] T.-F. M. C. M. S. A. T.-M. H. T. Wan-Ting Chiu, "Electrocatalytic activity enhancement of Au NPs-TiO2 electrode via a facile redistribution process towards the non-enzymatic glucose sensors," *Sens Actuators B Chem*, pp. 1–12, 2020.

[16] A. Al-Abduljabbar and I. Farooq, "Electrospun Polymer Nanofibers: Processing, Properties, and Applications," *Polymers*, vol. 15, no. 1. MDPI, Jan. 01, 2023.

[17] K. A. K. Rahman Kazi Hasibur, "Titanium-dioxide (TiO2) concentration-dependent optical and morphological properties of PAni-TiO2 nanocomposite," *Mater Sci Semicond Process*, pp. 1–11, 2020.

[18] N. Z Al-Hazeem, "Effect of the Distance between the Needle Tip and the Collector on Nanofibers Morphology," *Nanomedicine & Nanotechnology Open Access*, vol. 5, no. 3, 2020.

Determination of the Aptamer Probe Density by Double Layer and Redox Capacitance of CNT-Based Electrochemical DNA-Aptasensors

Mohammad Al Mamun
Nanotechnology and Catalysis Research Centre,
Universiti Malaya,
50603 Kuala Lumpur, Malaysia
Department of Chemistry,
Jagannath University,
Dhaka-1100, Bangladesh
zithrox@gmail.com

Yasmin Abdul Wahab
Nanotechnology and Catalysis Research Centre,
Universiti Malaya,
50603 Kuala Lumpur, Malaysia
yasminaw@um.edu.my

M. A. Motalib Hossain
Nanotechnology and Catalysis Research Centre,
Universiti Malaya,
50603 Kuala Lumpur, Malaysia
motalib123@yahoo.com

Abu Hashem
Nanotechnology and Catalysis Research Centre,
Universiti Malaya,
50603 Kuala Lumpur, Malaysia
Microbial Biotechnology Division,
National Institute of Biotechnology,
Ganakbari, Ashulia, Savar,
Dhaka-1349, Bangladesh
hashemnib04@yahoo.com

Mohd Rafie Johan
Nanotechnology and Catalysis Research Centre,
Universiti Malaya,
50603 Kuala Lumpur, Malaysia
mrafiej@um.edu.my

Nurul Ezaila Alias
School of Electrical Engineering,
Universiti Teknologi Malaysia,
Johor Bahru, Malaysia
ezaila@fke.utm.my

Hanim Hussin
Faculty of Electrical Engineering,
Universiti Teknologi MARA,
Shah Alam, Malaysia
hanimh@uitm.edu.my

Maizan Muhamad
Faculty of Electrical Engineering,
Universiti Teknologi MARA,
Shah Alam, Malaysia
maizan@uitm.edu.my

Abstract—Electrochemical DNA-Aptasensors are gaining increasing popularity for the recognition of numerous analytes ranging from tiny molecules to entire microorganisms due to their multiple benefits over antibody-based biosensors. However, the reliable determination of aptamer probe density in DNA-Aptasensors is a potential barrier to ensuring their quality and performance. In this regard, an innovative, simple, and reliable electrochemical method has been introduced by exploring the redox capacitance (CR) using $[Ru(NH_3)_6]^{3+/2+}$ as a model redox species to determine the DNA-Aptamer probe density on a CNT (carbon nanotube) modified *gc* (glassy carbon) electrode surface. Moreover, the results were compared with the probe density calculated by the existing method using double layer capacitance (CD). Despite a lack of reproducibility caused by the non-specifically adsorbed aptamers on CNT, an excellent coincidence was found between the results obtained following those two methods. Therefore, the new method will accelerate the commercialization of electrochemical DNA-Aptasensors by confirming their quality through assessing the reliable DNA-Aptamer probe density.

Keywords—DNA-Aptasensors, electrochemical, double layer, redox capacitance, probe density

I. INTRODUCTION

Aptamer or synthetic single stranded DNA (ssDNA) or RNA (ssRNA) based electrochemical biosensors have shown tremendous scope for the detection of not only small molecules but also entire microorganisms [1, 2]. Integration of nanomaterials (NMs) (such as carbon nanotubes, graphene and gold nanomaterials) as the signal transducing elements has added further benefits to building a more stable, robust, selective, and highly sensitive bio-sensing surface [1, 3]. The incorporation of NMs in DNA biosensors increases the surface area dramatically [2, 3]. As a result, more recognition elements, such as DNA aptamer probes, can be adopted on the biosensor surface, which enhances the scope of interaction between the targeted analytes and the recognition layers during the electrochemical detection in solution. In other words, the higher the probe density, the higher the sensitivity towards the analytes to be examined [4]. Therefore, accurate probe density determination is a crucial issue to optimize the quality of Aptasensors' surface. However, the precise calculation of aptamer surface charge density mainly depends on accurate measurements of the surface area of the signal transduction layer. There are a number of methods for the calculation of electrode surface area in Electrochemistry [5-7]. Among them, using electrical double layer capacitance (DLC), CD has become popular in the case of electrochemical biosensors due to its simplicity and rapidity [5, 6]. But the problem is that no ideal polarizable electrode (IPE) or ideal double-layer capacitor over the entire potential window is available in solution. Even in the case of a porous electrode surface (like a CNT surface), charge can cross the IPE surface if the potential through it is changed. As a result, the nature of the electrode-electrolyte interface is not exactly equivalent to that of an ideal capacitor. Furthermore, the charging of functionalized CNT electrode surfaces (such as carboxylated, esterified, aptamer functionalized, etc.) is more complex than that of bare *gc* surfaces or non-functionalized CNT surfaces since it is also governed by pH through surface proton exchange [8]. However, it is noted that there is no available method or reference to confirm or validate the reliability of the determined probe density of electrochemical DNA-Aptasensors using CD, which is necessary to ensure the quality of the DNA biosensors' surface.

In these circumstances, an alternative with simple method is urgently needed to validate the surface area determined by the DLC method and the calculated surface charge density using DLC-based surface area. In this regard, we have proposed an efficient alternative electrochemical method by calculating redox capacitance (CR) using $[Ru(NH_3)_6]^{3+}$ as a model redox species for results comparison. This technique might introduce a reliable platform to ensure the precise determination and optimization of probe density on electrochemical DNA biosensors surface to confirm its quality.

II. MATERIALS AND METHODS

A. Chemicals and Reagentes

Analytical grade chemicals and reagents were used. SWCNT (single walled carbon nanotube) were obtained as bulk form (>90% purity) with 1.4–1.5 nm diameter and 150 mm of mean length. 5′-end aminated 15-mer (5′-GGT TGG TGT GGT TGG-3′) with a 6-carbon spacer thrombin specific aptamers (TBA) were purchased from Eurogentec (Madrid, Spain). The reagents prepared from NHS (N-hydroxysuccinimide), EDC (1-ethyl-3-(3-dimethylaminopropyl) carbodiimide hydrochloride), and (MES) (2-(N-morpholino) ethanesulfonic acid), CTAB (Cetyl trimethylammonium bromide), $[Ru(NH_3)_6]Cl_3$ complex (98%), DMF (Dimethyl Formamide), Tris-(hydroxymethyl)aminomethane, KCl (potassium chloride), NaCl (sodium chloride), Na_2HPO_4 (di-sodium hydrogen phosphate) and NaH_2PO_4 (sodium di-hydrogen phosphate) were purchased from Sigma-Aldrich (Spain). All the soultions were prepared with deionized water (>18.2 MΩ.cm) obtained from a Millipore deionization water plant.

B. Construction of Electrochemical DNA-Aptasensors

There are 3 steps involved in building the electrochemical DNA-Aptasensors surface on a *gc* (glassy carbon) electrode, as illustrated in Scheme 1.

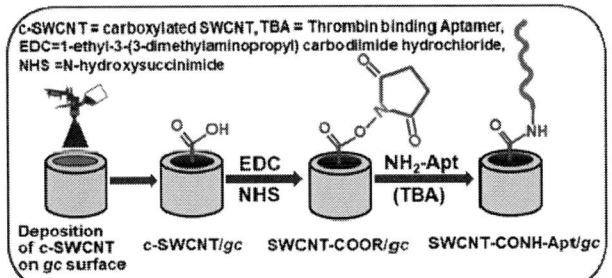

Scheme 1: Construction steps of the CNT-based electrochemical DNA-Aptasensors by the covalent functionalization of a 15-base aptamer (TBA) (5′-end modified by the NH_2-$(CH_2)_6$ group) associating the activation of carboxylic groups of c-SWCNT using NHS and EDC following the covalent immobilization of TBA on the activated surface.

In the first step, after cleaning, polishing, and sonicating the *gc* electrode surface, the carboxylated and pretreated 1 mg/mL SWCNTs suspension in DMF was deposited by spraying (35 times) while being dried [6], and the surface is denoted as c-SWCNT/*gc*. Then, the cSWCNT/*gc* electrode was dipped into a mixture of 50 mM NHS and 200 mM EDC prepared in 50 mM of MES buffer with pH 5.0) for the esterification of the surface to yield a SWCNT-COOR/*gc* surface. In the final step, the TBA aptamers (1 μM) were covalently immobilized onto the layer of esterified SWCNTs surface via the activation of the surface following a recognized carbodiimide-facilitated wet-chemistry method [9]. In this approach, the amide bonds are formed between the carboxylic species of the SWCNT and the amine spacer [10, 11] of TBA. The resulting DNA-aptasensors are denoted as SWCNT-CO-NH-Aptamer/*gc*.

C. Instrumental setup and Measurements

Both the C_D and C_R were determined by the CV (Cyclic Voltammetry) technique using 3 electrodes inserted in a single cell, where the DNA-Aptasensor, Ag/AgCl, and glassy carbon (*gc*) rod were used as working electrode (WE), reference electrode (RE) and counter electrode (CE), respectively. 5 mL of 10 mM tris-buffer (pH = 7.4) was used as an electrolyte medium. Before each measurement, N_2 gas was purged into the solution for about 15 min with constant stirring to avoid the O_2 interference. All the CVs were taken at ambient temperature and pressure.

Experimental conditions (C_D) :Tris-buffer (pH 7.4, 10 mM) as an electrolytic medium, scanning potential window selected as -550 mV to -650 mV, CVs taken at the scan rates of 5 mV s^{-1}, 10 mV s^{-1}, 15 mV s^{-1}, 20 mV s^{-1}, 25 mV s^{-1}, and 30 mV s^{-1} in the absence of redox species in the electrolyte medium to avoid the Faradaic current.

Experimental conditions (C_R): Tris-buffer (pH 7.4, 10 mM) as an electrolytic medium, scanning potential window selected as 500 mV to -1000 mV, CVs taken at the scan rates of 80 mVs^{-1}, 100 mVs^{-1}, 200 mVs^{-1} and 300 mVs^{-1}. Faradic current within this potential window has been considered, in the presence of redox species $[Ru(NH_3)_6]^{3+}$ added in the electrolyte medium until saturation of the peak current has been attained.

III. RESULTS AND DISCUSSION

Double Layer Capacitance, DLC (C_D)

According to the theoretical concept illustrated in Fig. 1, if the over-all charge density obtained from a precisely adsorbed ion in the IHP (Inner Helmholtz Plane) is σ^i and the excess charge density of non-specifically adsorbed ions in the OHP or diffuse layer is σ^d, the total extra charge density on the solution side of the electrified interface, σ^s can be expressed as:

$$\sigma^s = \sigma^i + \sigma^d = -\sigma^M \qquad (1)$$

Where σ^M = the charge density on the electrode surface. Considering only the double layer charging process, the slope of the straight line of current (*i*) *vs* scan rate (dE/dt) plot gives the total value of the interfacial differential capacitance can be expressed by the following equation (2)

$$C_D = dq/dE = i \, dt/dE = i /(dE/dt) \qquad (2)$$

Where dq/dE expresses the differential capacitance, *i* is the current at the middle point of the CV taken between the non-Faradaic potential window and dE/dt = scan rate.

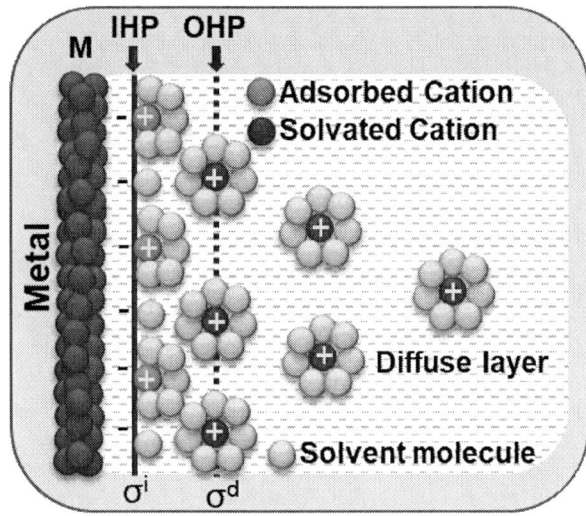

Fig. 1: Electrochemical double layer through the electrode-solution interface. OHP = Outer Helmholtz Plane, IHP = Inner Helmholtz Plane, σ = Charge Density = C_D/A (Double Layer Capacitance /Surface Area).

In order to determine the double layer capacitance, the CV curves were taken in a narrow potential window (within the non-Faradic region) following the prescribed experimental conditions as described in Section II, and illustrated in Fig. 2 The current in the middle point of the potential range (at -0.601 V) is then plotted against the scan rates to calculate the C_D from the slope of the regression line (Fig. 2b).

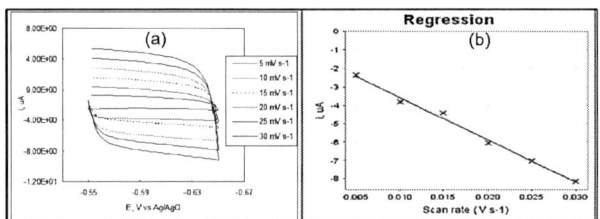

Fig. 2 : (a) CVs of 10 mM tris-buffer of pH 7.4 on the electrochemical DNA-Aptasensors surface (SWCNT-CO-NH-Aptamer/*gc*) between the potential window of -550 mV and -650 mV *vs* Ag/AgCl reference electrode at different scan rates and the corresponding C_D calculation plots, *i* (μA) *vs* scan rates (Vs⁻¹).

From the CVs of catholytes (Fig. 2) (in absence of redox markers), it was observed that there was a significant quantity of capacitive current on the surface of Aptasensors (Aptamer-$(CH_2)_6$-NH-CO-SWCNT/*gc*), indicating the high charge density and surface area of the electrode surface as the integration of nanomaterials (CNT) as a signal transduction layer [3]. Five similar measurements were carried out on different DNA-Aptasensors and then the calculated C_D was found to be 268±30 μF. Then, the electrode surface area was calculated as 26.8±2.9 cm² using the formula, A= C_D/C^*, where C^* is the reference value of specific capacitance for c-SWCNT ($C^* = 10$ μF/cm²) [5].

Redox capacitance (C_R)

Redox capacitance was determined on the basis of the redox reaction that occurred at the electrode solution interface, as illustrated in Fig. 3. When Ru^{3+} ions are added in PBS buffer (pH =7.4) of Aptasensors containing 3-electrode electrochemical cell, Ru^{3+} ions are adsorbed on the negatively charged phosphate backbone of TBA (thrombin binding aptamer) due to the electrostatic interaction, and at saturation, all the negatively charged sites on the backbone are occupied by Ru^{3+} ions (Fig. 3 A).

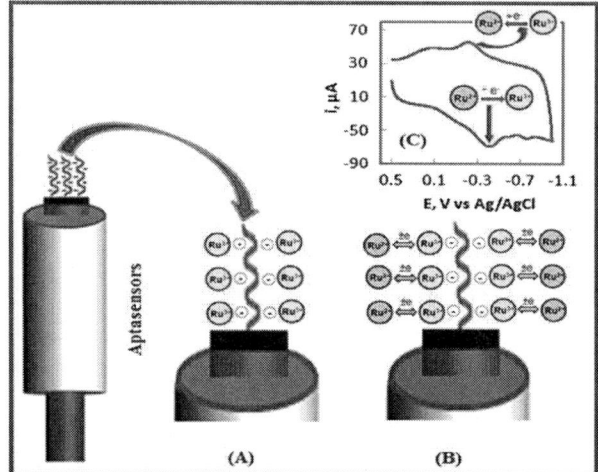

Fig. 3: Electrochemical mechanism for the measurement of redox capacitance (C_R).

The charge transfer redox reaction undergoes while scanning between the potential window of 500 mV and -1000 mV at 100 mVs⁻¹ scan rate, and at ambient temperature (Fig. 3B). The corresponding cyclic voltammogram (CV) (Fig. 3C) indicates the respective oxidation and reduction peaks obtained from the redox charge transfer processes. The cathodic peak current, i_{pc}, for the reduction of Ru^{3+} to Ru^{2+}, was used for the calculation of the redox capacitance, C_R from the CVs of the Ru^{3+} saturated DNA-Aptasensors surface using the equation (3) at different scan rates and tabulated in Table 1.

$$C_R = i_s/(dE/dt) = i/v \qquad (3)$$

Where i_s is the saturation current relating to the cathodic peak of the CV and v is the voltage scan rate in Vs⁻¹.

Table 1 : Calculated redox capacitance, electrode surface area, area charge, Ru^{3+} surface density and aptamer probe density at different scan rates.

Scan rate (mV s⁻¹)	I_s (A) x e⁵	C_R (F) x e⁶	A (cm²)	Q (C) x e⁵	$\Gamma_{Ru} = Q/nFA$ (Mole cm⁻²) x e¹¹	$\Gamma_A = \Gamma_{Ru}$ (z/m)NA (molecules cm⁻²) x e⁻¹²
80	2.63	328.75	32.9	7.88	2.48	2.99
100	3.11	311.00	31.1	7.40	2.47	2.96
200	5.14	257.00	25.7	6.63	2.67	3.21
300	6.45	215.00	21.5	5.78	2.79	3.36

I_s = Saturation Current, C_R= Redox Capacitance, A = Electrode Surface Area, Q = Area Charge, Γ_{Ru}= Ru^{3+} Surface Density, Γ_A = Aptamer Probe Density

From Table 1, it is clear that the saturation current has increased with the increase in scan rate, which is a common electrochemical phenomenon that happens as the rate of diffusion increases due to the increase of scan rate [12]. On the other hand, the C_R decreases with increasing scan rate, which reflects the electron transfer process is controlled by the adsorbtion of Ru^{3+} on the Aptasensors surface [13]. For the same reason, the corresponding surface area has been decreased.

Determination of the DNA-aptamer probe density

We have calculated the DNA-Aptamer probe density, Γ_A (molecules cm^{-2}) using the following equation 4 [6]:

$$\Gamma_A = \Gamma_{Ru} (z/m) N_A \qquad (4)$$

Where, Γ_{Ru} = Q/nFA (mole cm^{-2}), Q = area charge, n = number of electron transfer in the redox reaction, F= Faraday constant, A = surface area, z = the number of charges in the $[Ru(NH_3)_6]^{3+}$ complex, m = the number of nucleotides in the TBA, and N_A = Avogadro's number. In terms of double layer capacitance, the DNA-Aptamer probe density was calculated as $3.30 \pm 0.33 \times 10^{12}$ molecules cm^{-2}. On the contrary, the calculated DNA-Aptamer probe density using the redox capacitance was found between 2.96×10^{12} molecules cm^{-2} and 3.36×10^{12} molecules cm^{-2} at different scan rates (Table 1). It is interesting to note that those results are much comparable to the DNA-Aptamer probe density calculated by using C_D. However, if we focus on the reproducibility issue, both methods do not satisfy the targeted reproducibility. This may be due to the inhomogeneity of the Aptasensors surface caused by the presence of non-specifically adsorbed aptamers on the surface. Because the non-specifically adsorbed DNA-Aptamers have strong supramolecular interactions on the SWCNT surface, it is not possible to completely remove them from the surface by just buffer washing. That is why it is essential to explore an alternative method for the complete removal of non-covalently adsorbed DNA-aptamers from the surface to achieve a reproducible and homogeneous surface. That reproducible and homogeneous surface could overcome the limitations of those methods regarding the irreproducibility issue.

IV. CONCLUSION

To conclude, the new method for the determination of DNA-Aptamer probe density on the electrochemical DNA-Aptasensor surface using the redox capacitance (C_R) might be a reliable method due to its significant match to the probe density associated with the double layer capacitance (C_D). The computed probe density using the double layer capacitance was found to be $3.30 \pm 0.33 \times 10^{12}$ molecules cm^{-2}, whereas that using the redox capacitance was found to be between 2.96×10^{12} molecules cm^{-2} and 3.36×10^{12} molecules cm^{-2} at different scan rates. However, more exploration is needed regarding how to improve the reproducibility of the Aptasensors while maintaining the stability of the electrode surfaces to get more precise results from the DNA-aptamer probe density.

ACKNOWLEDGMENT

The work is financially supported by the Ministry of Higher Education Malaysia (MOHE) via Fundamental Research Grant Scheme (FRGS/1/2022/TK09/UM/02/27). The authors also extend their appreciation to the Universiti Malaya (Grant No. ST055-2022) and Bangabandhu Science and Technology Fellowship, Bangladesh for the financial support during this research.

REFERENCES

[1] M. Al Mamun, Y. A. Wahab, M. M. Hossain, A. Hashem, and M. R. Johan, "Electrochemical biosensors with Aptamer recognition layer for the diagnosis of pathogenic bacteria: Barriers to commercialization and remediation," *TrAC Trends in Analytical Chemistry,* vol. 145, p. 116458, 2021.

[2] M. Al Mamun, Y. A. Wahab, M. M. Hossain, A. Hashem, and M. R. Johan, "DNA-Aptamer–Based Electrochemical Biosensors for the Detection of Thrombin: Fundamentals and Applications," in *Functional Nanomaterials for Sensors*: CRC Press, 2023, pp. 201-221.

[3] A. Hashem, M. M. Hossain, M. Al Mamun, K. Simarani, and M. R. Johan, "Nanomaterials based electrochemical nucleic acid biosensors for environmental monitoring: A review," *Applied Surface Science Advances,* vol. 4, p. 100064, 2021.

[4] M. Al Mamun and A. S. Ahammad, "Characterization of Carboxylated-SWCNT Based Potentiometric DNA Sensors by Electrochemical Technique and Comparison with Potentiometric Performance," *Journal of Biosensors & Bioelectronics,* vol. 5, no. 3, p. 1, 2014.

[5] S. Shiraishi, H. Kurihara, K. Okabe, D. Hulicova, and A. Oya, "Electric double layer capacitance of highly pure single-walled carbon nanotubes (HiPco™ Buckytubes™) in propylene carbonate electrolytes," *Electrochemistry Communications,* vol. 4, no. 7, pp. 593-598, 2002.

[6] M. Al Mamun and A. Ahammad, "Characterization of Carboxylated-SWCNT Based Potentiometric DNA Sensors by Electrochemical Technique and Comparison with Potentiometric Performance," *Journal of Biosensors & Bioelectronics,* vol. 5, no. 3, p. 1, 2014.

[7] M. Łukaszewski, M. Soszko, and A. Czerwiński, "Electrochemical methods of real surface area determination of noble metal electrodes–an overview," *Int. J. Electrochem. Sci.* vol. 11, no. 6, pp. 4442-4469, 2016.

[8] A. Daghetti, G. Lodi, and S. Trasatti, "Interfacial properties of oxides used as anodes in the electrochemical technology," *Materials Chemistry and Physics,* vol. 8, no. 1, pp. 1-90, 1983.

[9] C. Furtado, U. Kim, H. Gutierrez, L. Pan, E. Dickey, and P. C. Eklund, "Debundling and dissolution of single-walled carbon nanotubes in amide solvents," *Journal of the American Chemical Society,* vol. 126, no. 19, pp. 6095-6105, 2004.

[10] S. S. Wong, E. Joselevich, A. T. Woolley, C. L. Cheung, and C. M. Lieber, "Covalently functionalized nanotubes as nanometre-sized probes in chemistry and biology," *Nature,* vol. 394, no. 6688, pp. 52-55, 1998.

[11] D.-H. Jung *et al.*, "Covalent attachment and hybridization of DNA oligonucleotides on patterned single-walled carbon nanotube films," *Langmuir,* vol. 20, no. 20, pp. 8886-8891, 2004.

[12] M. Al Mamun, O. Ahmed, P. Bakshi, and M. Ehsan, "Synthesis and spectroscopic, magnetic and cyclic voltammetric characterization of some metal complexes of methionine:[(C5H10NO2S) 2MII]; MII= Mn (II), Co (II), Ni (II), Cu (II), Zn (II), Cd (II) and Hg (II)," *Journal of Saudi Chemical Society,* vol. 14, no. 1, pp. 23-31, 2010.

[13] M. Al Mamun *et al.*, "Electrochemistry of Green Ag Nanoparticles Modified Electrode Surface," in *2022 IEEE International Conference on Semiconductor Electronics (ICSE),* 2022: IEEE, pp. 37-40.

Enhancing Sensitivity of Thermal Biosensors through Vanadium Dioxide (VO$_2$) Thin Films

Abdelkader Hassein-Bey
LPCMIA, Physics Department
University of Blida1
Blida, Algeria
ahassei@hotmail.fr

Leila Asmaa Hassein-Bey
Physics Department
University of Blida1
Blida, Algeria
asmaahasseinbey@gmail.com

Slimane Lafane
Ionized Medium and Laser Division, CDTA
Algiers, Algeria
slafane@cdta.dz

Samira Abdelli-Messaci
Ionized Medium and Laser Division, CDTA
Algiers, Algeria
messaci@cdta.dz

Burhanuddin Yeop Majlis
IMEN, Universiti Kebangsaan Malaysia
Bangi Selangor, Malaysia
burhan@ukm.edu.my

Abstract— There is a growing interest in thin-film thermistors due to their compatibility with standard microfabrication processes in microelectronics. This greatly simplifies the integration of on-chip biosensors during the fabrication stage. Vanadium dioxide VO$_2$ exhibits notable sensitivity and a significant temperature coefficient of resistance (TCR), rendering it an interesting material for thermal sensors with a wide range of potential applications. This paper explores the utilization of vanadium dioxide (VO$_2$) thin films to enhance the sensitivity of thermal biosensors. The thermo-resistive properties of VO$_2$ thin films are investigated, focusing on improving the performance and sensitivity of micro-thermistor for microcalorimetric biosensors. The VO$_2$ layers are fabricated using laser ablation on Au substrates. The VO$_2$/Au thin films exhibit a highly sensitive response with an interesting temperature coefficient of resistance (TCR) value of 19%/°C, significantly surpassing existing VOx-based thermal biosensors. The improved sensitivity enables the detection of low concentrations of biological substances, making VO$_2$/Au thin films promising candidates for micro-thermistor in thermal biosensors applications, facilitating early disease diagnosis with tuned characteristics through electrical biasing.

Keywords— *VO$_2$ thin films; PLD; VO$_2$ Voltage tuned MIT; TCR; Micro-thermistor; Thermal-sensitive transducer element; Microcalorimetric biosensor.*

I. INTRODUCTION

Vanadium dioxide, also known as VO$_2$, is a type of material that goes through a phase change when exposed to heat. Specifically, it transitions from a metallic phase to an insulating phase when it reaches a temperature of around 68°C [1]-[3]. The temperature at which this occurs can be adjusted using external stimuli. During this phase change, VO$_2$ undergoes significant changes in its electrical [3]-[12], optical [13]-[17], and mechanical [18]-[20] properties. Due to these unique characteristics, there are many potential applications for VO$_2$. Its high-temperature coefficient of resistance (TCR) also makes it an ideal material for thermal sensors [21]-[23].

The TCR value for bulk VO$_2$ is generally above 70%/°C around 68°C [24]. This high-temperature range makes it impratical for use in biosensors. However, thin film VO2, which has been extensively studied, exhibits lower TCR values compared to bulk VO$_2$. These values remain relatively promising for enhancing sensitivity in thermal detection across various applications [13]. During the deposition process, thin film VO$_2$ is optimized through methods such as doping or other parameters. Research studies have shown TCR values of 4.5%/°C for chromium (Cr)-doped VO$_2$ [25] and 3.36%/°C using excimer laser irradiation deposited VO$_2$ for infrared sensors [26]. Other studies have achieved TCR values as high as 10%/°C with tungsten (W) doping [27] and up to 11.9%/°C with co-doping of chromium and niobium (Nb). These results are obtained at temperatures close to room temperature, but unfortunately, the TCR is localized around a fixed temperature value. However, for the VO$_2$/Au thermo-sensitive layer, the TCR optimization step can be made independent of the material's fabrication process by the application of an electric voltage during its use. This work proposes an innovative thermo-sensitive element based on VO$_2$/Au, which offers tunable TCR. This means that sensitivity can be adjusted according to our needs through applied electrical polarization. This allows us to achieve our main objective of obtaining thin film VO$_2$ for more practical applications operating at temperatures closer to room temperature, particularly in the field of thermal biosensors.

Thermal biosensors are able to detect and measure biological interactions or the concentration of analytes by detecting temperature changes, providing label-free detection. The transducer in a thermal biosensor is able to convert temperature changes resulting from biological interactions or analyte binding into measurable signals [28]. In order to enhance the sensitivity of the transduction element of the biosensor, it is necessary to develop materials with high thermal sensitivity. This work introduces a VO$_2$/Au thin layer as a highly sensitive transducer element for microcalorimetric biosensors, offering advantages for heat detection. The VO$_2$ layers are produced on Au substrate, using the laser ablation technique (PLD) [29], [30]. In addition, this work proposes a microfluidic biosensor chip design. The study highlights the potential of VO$_2$/Au as an innovative heat-sensitive element, which makes it a great choice fot thermal transducers in biological sensors (microcalorimetric biosensors). There is a growing interest in

979-8-3503-2369-6/23 $31.00 © 2023 IEEE

thin-film thermistors due to their compatibility with standard microfabrication processes in microelectronics [31]. This makes it easier to integrate on-chip biosensors during the fabrication stage.

II. PERFORMANCE OF VO₂/AU AS THERMISTOR: (TCR)

Thermistors and thermocouples are commonly used as transducers in thermal biosensors and microcalorimeters [32]. In contrast to thermistors (semiconductor-based for sensitive temperature measurement) and thermocouples (using dissimilar metals for high-temperature applications), RTD (Resistance Temperature Detector) measures temperature by tracking changes in electrical resistance within specific materials. VO_2, on the other hand, behaves like a thermistor. Its resistance varies with temperature fluctuations, as shown in Fig. 1 [28]. This makes VO_2 a suitable material for thermal sensing applications due to its rapid and significant resistance changes within narrow temperature range.

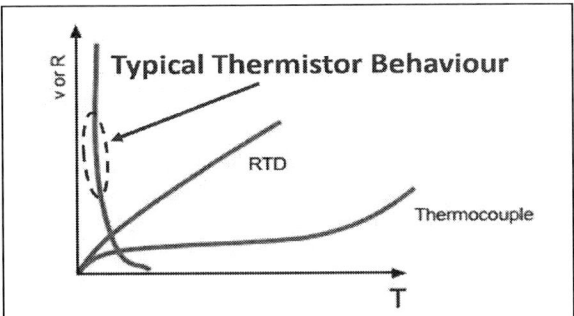

Fig. 1. Temperature-Dependent Resistance Variations in Metal Oxide Thermistors: Irregular and Sudden Changes within a Narrow Range

The fabricated VO_2/Au thin films exhibit notable variations in electrical resistance around the transition zone at temperatures closer to ambient. This makes the VO_2/Au layer a great choice for future fabrication of thermistors in highly sensitive microcalorimetric biosensors based on metal oxides (Fig. 2). By leveraging the key electrical parameters of this phase-transition material, namely electrical resistance and temperature coefficient of resistance (TCR), it becomes possible to develop ultra-sensitive and thermal high-performance biosensors.

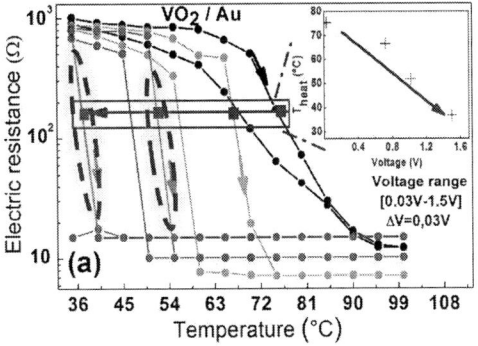

Fig. 2. Thermistor-like Behavior in the VO_2/Au Transition Zone

VO_2 has high sensitivity expressed by the temperature coefficient of resistance (TCR), making it a promising material for thermal sensors [24]. The TCR serves as a performance indicator for thermal detection, quantifying the relative change in resistance per degree of temperature variation (1) [33].

$$TCR = (1/R) \Delta R / \Delta T \qquad (1)$$

When using thermistors, it's important to keep in mind that the resistance value of metal oxides will change as temperature fluctuates. These variations in resistance can be quite irregular or sudden within a narrow temperature range. The most accurate empirical expression describing the R-T relationship in a thermistor is the Steinhart-Hart (2) [31]. For narrow temperature ranges around the transition, the relationship can be approximated by (3).

$$1/T = A + B (\ln R_T) + C (\ln R_T)^3 \qquad (2)$$

$$R_T = R_{T0} \exp [\beta (1/T - 1/T_0) \qquad (3)$$

However, the TCR value for bulk VO_2 is impractical for use in biosensors due to the high temperature range. VO_2 in thin layers has lower TCR values but can be optimized through doping or other parameters. Research has shown TCR values ranging from 3.36%/°C to 11.9%/°C for temperatures close to ambient, but with a localized TCR around a fixed temperature value [25]-[27].

It is interesting to note that TCR optimization for the thermo-sensitive layer can be separated from the material development stage by applying an electrical voltage during use or post-deposition characterization. This means that the TCR is not fixed and can actually be modulated by the electrical bias, as shown in Fig. 3. This approach provides greater flexibility in optimizing the performance of this heat-sensitive layer specifically for thermal detection applications.

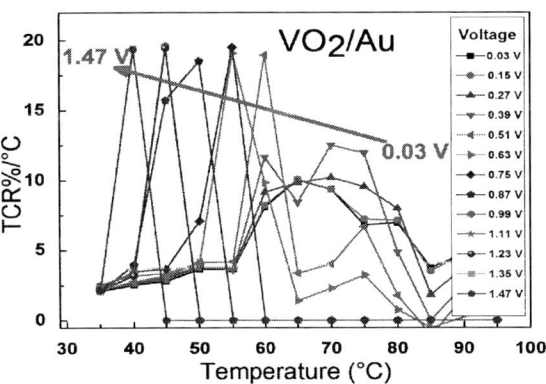

Fig. 3. Experimental results of temperature coefficients of resistance (TCR) modulated by electrical bias.

The experimental results show that the TCR of the VO_2/Au thin film can be significantly modulated by an electrical bias (Fig. 3). Specifically, the TCR of the material was measured to be 0.87%/°C at zero electrical bias, but it increased to 2.24%/°C when an electrical bias of 3 volts was applied.

The TCR continued to increase with higher applied electrical biases. In fact, it reached a maximum value of 3.6%/°C at 5 volts (Fig. 3). These results demonstrate the potential for the VO_2/Au thin film to be used as a highly sensitive and tunable temperature element in microcalorimetric biosensing applications. Moreover, the

TCR values we obtained were comparable to existing ones, but significantly increased from 0.51 V to 1.47 V, reaching 19%/C°. This is very promising and compares favorably to other studies that have been conducted in this area [34]-[36].

III. INTEGRATION OF VO₂/AU THERMAL-SENSITIVE ELEMENTS IN MINIATURIZED BIOSENSOR DESIGN

We propose an innovative thermo-sensitive element based on VO₂/Au with a modulated TCR. This design allows for voltage-tunable sensitivity through electrical polarization, which makes it ideal for use in biosensors. One of the most interesting possibilities this opens up is the potential for real-time TCR control in future thermal biosensors that integrate VO₂/Au-based micro-thermistors.

In biosensors, the heat-sensitive layer must be set to an operating point close to the temperatures used in medical analysis laboratories. To achieve this, the heat-sensitive layer can be controlled in a similar way to a transistor (Fig. 4). To target a specific TCR value, there are two key parameters that can be adjusted: electrical polarization and temperature. The integration of VO₂/Au as a heat-sensitive element in a biosensor involves defining an operating point and integrating it into a microfluidic chip with electronic acquisition and control circuits.

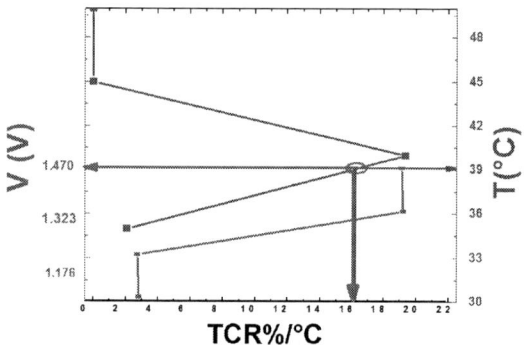

Fig. 4. Operating point selection for a VO₂/Au heat-sensitive element.

The TCR is significant coefficient that indicates the performance of temperature sensor materials. It is great to know that this can be accurately controlled and adjusted through electrical biasing in the intelligent and innovative VO₂/Au material. With this innovation, it is possible to finely tune the biosensor's operating point by manipulating the electrical bias value.

We propose the integration of this intelligent material with its innovative properties as a micro-thermistor in future microfluidic chips for microcalorimetric detection (Fig. 5). It is impressive to see that the thermo-sensitive element developed has a much higher TCR compared to the ones reported in the literature [36] [37].

Additionally, the fact that the TCR of VO₂/Au element is modulable through the bias voltage, is a unique feature that can be advantageous for designing future biosensors. Morever, the simplicity of the fabrication process of VO₂ layers, compared to other more complicated processes described in the literature, is also a positive point in favor of our approach. With these features, VO₂/Au thermo-sensitive layer can be a promising candidate for various thermal biosensing applications.

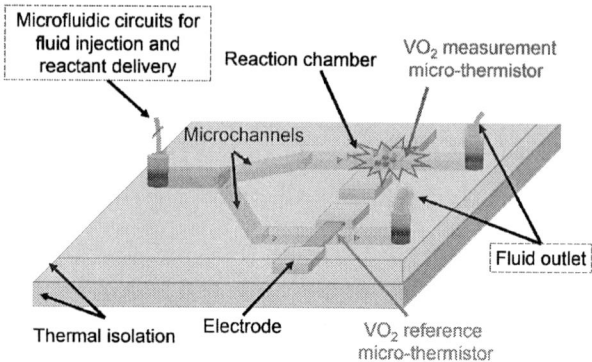

Fig. 5. Proposed 3D Schematic Diagram of a Microfluidic Biosensor with VO2/Au Micro-Thermistor for Calorimetric Detection

IV. CONCLUSION

Indeed, the potential applications of VO₂-based biosensors are numerous and promising, especially in the healthcare sector. With the ability to detect a biological substances at low concentrations and with higher sensitivity, the diagnosis and treatment of deseases could be greatly improved. The incorporation of VO₂/Au thin film as an ultra-sensitive transduction element in biosensors is suggested. This micro-thermistor based on VO₂/Au could detect even the smallest temperature changes. To further enhance the performance of calorimetric biosensors, it would be worthwhile to explore the integration of VO₂/Au with other materials and technologies, as well as the development of specific biosensor the detection of different biomolecules and diseases. Ultimately, the optimization of the thermo-sensitive properties of VO₂ is a key step towards the development of innovative thermal biosensors with high sensitivity.

ACKNOWLEDGMENT

The authors would like to express our sincere appreciations to the Ionized Media and Laser Division and Microelectronics and Nanotechnology Division of the Advanced Technology Development Research Center (CDTA, Algeria) for their invaluable assistance and support throughout the project. We are also deeply grateful to the Nano Physics Group (Physics department) at the (University of Blida 1, Algeria) for their helpful advice and assistance. The authors wholeheartedly appreciate the collective efforts of all individuals involved in this project for their guidance and support.

REFERENCES

[1] Aurélien Didelot,"Films d'oxydes de vanadium thermochromes dopés aluminium obtenus après un recuit d'oxydation-cristallisation pour applications dans le solaire thermique," Université de Lorraine, Nancy cedex, 2017.

[2] Grandi, F., Amaricci, A., and Fabrizio, M.,"Unraveling the Mott-Peierls intrigue in vanadium dioxide," , Physical Review Research, V. 2, n° 1, (March 2020).

[3] Morin, F.J.,"Oxides which show a metal-to-insulator transition at the neel temperature," , Physical Review Letters, V. 3, n° 1, (July 1959), 34–36.

[4] Ruzmetov, D., Zawilski, K.T., Senanayake, S.D., Narayanamurti, V., and Ramanathan, S.,"Infrared reflectance and photoemission spectroscopy studies across the phase transition boundary in thin film

vanadium dioxide," , Journal of Physics Condensed Matter, V. 20, n° 46, (November 2008).

[5] Villeneuve, G., Drillon, M., and Hagenmuller, P.,"Contribution a l'etude structurale des phases V1-xCrxO2," , Materials Research Bulletin, V. 8, n° 9, (September 1973), 1111–1121,

[6] Phillips, T.E., Murphy, R.A., and Poehler, T.O.,"Electrical studies of reactively sputtered Fe-doped VO2 thin films," , Materials Research Bulletin, V. 22, n° 8, (August 1987), 1113–1123.

[7] Drillon, M., and Villeneuve, G.,"Diagramme de phases du systeme V1-xAlxO2," , Materials Research Bulletin, V. 9, n° 9, (September 1974), 1199–1207.

[8] Chae, B.G., and Kim, H.T.,"Effects of W doping on the metal-insulator transition in vanadium dioxide film," , Physica B: Condensed Matter, V. 405, n° 2, (January 2010), 663–667.

[9] F. Béteille, R. Morineau, J. Livage, M.N.,"Switching properties of V1−xTixO2 thinfilms deposited from alkoxides," , Mater. Res. Bull., V. 32, n° no 8, (1997), 1109‑1117.

[10] Mai, L.Q., Hu, B., Hu, T., Chen, W., and Gu, E.D.,"Electrical property of mo-doped VO2 nanowire array film by melting-quenching sol-gel method," , Journal of Physical Chemistry B, V. 110, n° 39, (October 2006), 19083–19086.

[11] Holman, K.L., McQueen, T.M., Williams, A.J., Klimczuk, T., Stephens, P.W., Zandbergen, H.W., Xu, Q., Ronning, F., and Cava, R.J.,"Insulator to correlated metal transition in V1-x Mox O2," , Physical Review B - Condensed Matter and Materials Physics, V. 79, n° 24, (June 2009).

[12] Orlianges, J.C., Leroy, J., Crunteanu, A., Mayet, R., Carles, P., and Champeaux, C.,"Electrical and optical properties of vanadium dioxide containing gold nanoparticles deposited by pulsed laser deposition," , Applied Physics Letters, V. 101, n° 13, (September 2012).

[13] Ke, Y., Wang, S., Liu, G., Li, M., White, T.J., and Long, Y.,"Vanadium Dioxide: The Multistimuli Responsive Material and Its Applications," , Small, V. 14, n° 39, (September 2018), 1802025.

[14] Londos, C.A., Sgourou, E.N., and Chroneos, A.,"Defect engineering of the oxygen-vacancy clusters formation in electron irradiated silicon by isovalent doping: An infrared perspective," , Journal of Applied Physics, V. 112, n° 12, (December 2012).

[15] Stefanovich, G., Pergament, A., and Stefanovich, D.,"Electrical switching and Mott transition in VO2," , Journal of Physics Condensed Matter, V. 12, n° 41, (October 2000), 8837–8845.

[16] Chen, C., Zhao, Y., Pan, X., Kuryatkov, V., Bernussi, A., Holtz, M., and Fan, Z.,"Influence of defects on structural and electrical properties of VO 2 thin films," , Journal of Applied Physics, V. 110, n° 2, (July 2011), 023707.

[17] Peralta, X.G., Brener, I., Padilla, W.J., Young, E.W., Hoffman, A.J., Cich, M.J., Averitt, R.D., Wanke, M.C., Wright, J.B., Chen, H.T., O'Hara, J.F., Taylor, A.J., Waldman, J., Goodhue, W.D., Li, J., and Reno, J.,"External modulators for TeraHertz Quantum Cascade Lasers based on electrically-driven active metamaterials," , Metamaterials, V. 4, n° 2–3, (August 2010), 83–88.

[18] Leroy, J., Crunteanu, A., Bessaudou, A., Cosset, F., Champeaux, C., and Orlianges, J.C.,"High-speed metal-insulator transition in vanadium dioxide films induced by an electrical pulsed voltage over nano-gap electrodes," , Applied Physics Letters, V. 100, n° 21, (May 2012).

[19] akai, J.,"High-efficiency voltage oscillation in VO2 planer-type junctions with infinite negative differential resistance," , Journal of Applied Physics, V. 103, n° 10, (May 2008).

[20] Marezio, M., McWhan, D.B., Remeika, J.P., and Dernier, P.D.,"Structural aspects of the metal-insulator transitions in Cr-doped VO2," , Physical Review B, V. 5, n° 7, (April 1972), 2541–255.

[21] Wang, L., Sipe, D.M., Xu, Y., and Lin, Q.,"A MEMS thermal biosensor for metabolic monitoring applications," , Journal of Microelectromechanical Systems, V. 17, n° 2, (April 2008), 318–327.

[22] Gurung, N., Ray, S., Bose, S., and Rai, V.,"A broader view: Microbial enzymes and their relevance in industries, medicine, and beyond," , BioMed Research International, V. 2013, (2013).

[23] Mehrotra, P.,"Biosensors and their applications - A review," , Journal of Oral Biology and Craniofacial Research, V. 6, n° 2, (May 2016), 153–159.

[24] Takami, H., Kawatani, K., Kanki, T., and Tanaka, H.,"High Temperature-Coefficient of Resistance at Room Temperature in W-Doped VO 2 Thin Films on Al 2 O 3 Substrate and Their Thickness Dependence," , Japanese Journal of Applied Physics, V. 50, n° 5R, (May 2011), 055804.

[25] Lu, J., Kittiwatanakul, S., Sauber, N., Cyberey, M., Lichtenberger, A., and Weikle, R.,"Tuning of TCR in poly-crystalline VO2 for enhanced IR detection," in Image Sensing Technologies: Materials, Devices, Systems, and Applications V, May 2018, (May 2018), 9.

[26] Ishizaki, H., Nakajima, T., Shinoda, K., Tohyama, S., Kurashina, S., Miyoshi, M., Sasaki, T., and Tsuchiya, T.,"Improvement of temperature coefficient of resistance of a VO2 film on an SiN/polyimide/Si substrate by excimer laser irradiation for IR sensors," , Japanese Journal of Applied Physics, V. 53, n° 5 SPEC. ISSUE 1, (May 2014).

[27] Takami, H., Kawatani, K., Kanki, T., and Tanaka, H.,"High temperature-coefficient of resistance at room temperature in W-doped VO2 thin films on Al2O3 substrate and their thickness dependence," , Japanese Journal of Applied Physics, V. 50, n° 5 PART 1, (May 2011).

[28] Naresh, V., and Lee, N.,"A review on biosensors and recent development of nanostructured materials-enabled biosensors," Sensors (Switzerland), V. 21, n° 4, (February 2021). 1–35, Feb. 05, 2021.

[29] Hassein-Bey, A.L.S., Tahi, H., Lafane, S., Djafer, A.Z.A., Hassein-Bey, A., and Belgroune, N.,"Substrate effect on electrical properties of vanadium oxide thin film for Memristive device applications," in IEEE International Conference on Semiconductor Electronics, Proceedings, ICSE, Aug. 2016, V. 2016-Septe, (August 2016), 240–243.

[30] Hassein-Bey, A.L.S., Tahi, H., Lafane, S., Hassein-Bey, A., Abdelli-Messaci, S., and El-Amine Benamar, M.,"Voltage tuning effect on metal-insulator phase transitions in vanadium dioxide thin film," , ASM Science Journal, V. 12, n° SpecialIssue4, (2019), 19–28.

[31] Xie, B., and Danielsson, B.,"Thermal Biosensor and Microbiosensor Techniques," in Handbook of Biosensors and Biochips, (March 2008), Chichester, UK: John Wiley & Sons, Ltd, 2008.

[32] Lee, W., Lee, and Koh,"Development and applications of chip calorimeters as novel biosensors," , Nanobiosensors in Disease Diagnosis, V. 1, (April 2012), 17,

[33] Wood, R.A.,"Chapter 3 Monolithic Silicon Microbolometer Arrays," in Semiconductors and Semimetals, V. 47, n° C, (1997), 1997.

[34] Inomata, N., Pan, L., Wang, Z., Kimura, M., and Ono, T.,"Vanadium oxide thermal microsensor integrated in a microfluidic chip for detecting cholesterol and glucose concentrations," , Microsystem Technologies, V. 23, n° 7, (July 2017), 2873–2879.

[35] Miyazaki, K., Shibuya, K., Suzuki, M., Sakai, K., Fujita, J.I., and Sawa, A.,"Chromium-niobium co-doped vanadium dioxide films: Large temperature coefficient of resistance and practically no thermal hysteresis of the metal-insulator transition," , AIP Advances, V. 6, n° 5, (May 2016).

[36] Y. Yorozu, M. Hirano, K. Oka, and Y. Tagawa, "Electron spectroscopy studies on magneto-optical media and plastic substrate interface," IEEE Transl. J. Magn. Japan, vol. 2, pp. 740–741, August 1987 [Digests 9th Annual Conf. Magnetics Japan, p. 301, 1982].

[37] M. Young, The Technical Writer's Handbook. Mill Valley, CA: University Science, 1989.

Fabrication of Flexible and Printable Organic Thin-Film Transistor-based Sensor

Fazliyatul Azwa Md Rezali
Department of Electrical Engineering, Faculty of Engineering
Universiti Malaya
Kuala Lumpur, Malaysia
Centre of Printable Electronics
Universiti Malaya
Kuala Lumpur, Malaysia
s2027584@siswa.um.edu.my

Norhayati Soin
Department of Electrical Engineering, Faculty of Engineering
Universiti Malaya
Kuala Lumpur, Malaysia
Centre of Printable Electronics
Universiti Malaya
Kuala Lumpur, Malaysia
norhayatisoin@um.edu.my

Sharifah Fatmadiana Wan Muhamad Hatta
Department of Electrical Engineering, Faculty of Engineering
Universiti Malaya
Kuala Lumpur, Malaysia
Centre of Printable Electronics
Universiti Malaya
Kuala Lumpur, Malaysia
sh_fatmadiana@um.edu.my

Siti Nabila Aidit
Centre of Printable Electronics
Universiti Malaya
Kuala Lumpur, Malaysia
nabilaidit@um.edu.my

Abstract—As the next generation technology, organic thin-film transistor (OTFT) is fully desired in manufacturing flexible electronics due to its' mechanically strong property and low-temperature processing at large-scale production. In this work, OTFT device using poly(3,4-ethylene dioxythiophene) polystyrene sulfonate (PEDOT:PSS) is successfully fabricated on a flexible film by screen printing and drop casting technique at maximum process temperature of 80 °C. A For such straightforward deposition techniques, the device demonstrated a potentially good reproducibility and repeatability. Moreover, a significant property is observed on the device's current-voltage (I-V) performance that shows the average output source current is larger by ~4.4 times at maximum when dielectric reduce from ~24.04 μm to ~13.87 μm. The implementation of extended gate and reference electrode further enable detection in phosphate buffered saline (PBS) with small hysteresis. Eventually, the flexibility and printability of the fabricated OTFT raised the opportunity of low-cost application such as disposable biosensor to detect specific analyte at a small volume of solution.

Keywords— PEDOT:PSS, screen print, drop cast, reproducibility, repeatability

I. Introduction

Smart sensors based on organic thin-film transistor devices fabricated on thin plastic film substrates have garnered significant consideration in research and development [1], [2]. The OTFTs made by various printing method are fully desired to reduce fabrication steps and cost for large-scale manufacturing at low temperature [3], [4] as compared to conventional wafer fabrication method which uses lithographic patterning and vacuum deposition. Some of ongoing works of OTFT are using poly(3,4-ethylene dioxythiophene)-poly(styrene sulfonate) (PEDOT:PSS), poly(3-hexylthiophene-2,5- diyl) (P3HT), pentacene, and many more as the organic semiconductors (OSC) [5]–[7]. These OTFTs have been adapted for targeted application to integrated circuits and sensors such as to detect pH, gas, and biomolecules. Unlike the standard electrode sensor, the OTFT configuration optimization brings inherent signal amplification capability and scalability advantages in manufacturing reliable and high-performance sensor [8], [9].

Herein, a flexible and printable OTFT was proposed as biosensor by using straightforward screen printing and drop casting technique. To accomplish this, exclusive use of screen masking patterns specifically for gate and source/drain region are required. While most of the studies focused on improving the sensitivity of sensor, their performance stability under the same processing condition, has rarely been considered. Moreover, sensors of specific applications are more concerning to attain a high yield percentage of device during manufacturing rather than high carrier mobility. Despite many advances on OSC materials, the PEDOT:PSS is still popularly used in sensor application as it can yields good uniformities and stable performance [4], [10]. In order to validate the reproducibility and repeatability of the device in measurement, fabrication was carried out by batches with each batch consist of 4 arrays of OTFT device. Since screen printing is limited by the accuracy control on thickness and micrometre length [11], study was also conducted on different thickness of dielectric layer to observe their effect on I-V characteristics of the device. This paper also aims to explore the sensing performance of the device in detecting low sample volume. PBS is selected in this work since the solution is commonly applied as an electrolyte for biological study as ion concentration and osmolality of PBS are similar to the human body fluid [5].

II. Methodology

The OTFT structure in this work composed of multi-layer components with the configuration of extended-gate type sensor. At first, a 150-μm-thick polyethylene terephthalate (PET) film was wiped with isopropyl alcohol (IPA), followed by acetone and deionized water to thoroughly cleaned the substrate. A polyester screen-printing stencil in 200-mesh size was utilized with each pattern consist of 4 arrays. A flexible conductive silver (Ag) paste and polymer dielectric grey paste based on titanium dioxide (TiO₂), purchased from Sigma-Aldrich, were used to screen print gate electrode and dielectric layer respectively. After that, the substrate was baked in a convection oven for 30 minutes at

---

This work was supported in part by the Collaborative Research in Engineering, Science and Technology Center (CREST).

80 °C for each layer. To vary the thickness, the screen printing of dielectric was performed by single printing and double printing. The PEDOT:PSS (CLEVIOS™ P VP AI 4083) solution from Heraeus was filtered through a 0.45 μm filter before used as organic semiconductor. For semiconductor deposition, 3 μL of PEDOT:PSS ink was drop casted on top of dielectric layer and dried for 10 minutes at 80 °C. For source/drain pattern, the channel length of the device was screen printed with the initial gap of 500 μm using Ag paste at then heated at 80 °C for 30 minutes. In order to protect the semiconductor layer, the device was encapsulated with 60-μm-thick non-sticky PET tape. For sensing application, Ag-based electrode coated with Ag/AgCl ink, acquired from JE Solutions Consultancy Ltd, was screen printed as reference electrode and heated at 80 °C for 30 minutes [12]. Total devices fabricated were 20 devices, in which 12 devices with single layer dielectric, while another 8 devices with double layer dielectric

The I-V characterizations of fabricated OTFT were carried out by Keithley 4200 semiconductor characterization system (SCS) and required PBS electrolyte for sensor testing. The thickness of dielectric layer was measured using Bruker Dektak XT stylus profilometer, while digital microscope was used to capture image of fabricated device and its' dimensions. To perform I-V measurement, voltage was applied on gate (V_g) and source (V_s) while drain (V_d) was grounded. The output source current (I_s) produced was evaluated and threshold voltage (V_{th}) was presented for analysis. During sensing measurement, the gate voltage is applied through reference electrode (V_{ref}) instead and 10 μL of PBS was dropped on sensing area before measurement was taken.

III. RESULTS AND DISCUSSION

Fig. 1(a) shows the microscopic image of fabricated OTFT with reference electrode for sensing application. During fabrication, the OTFT dimensions, particularly dielectric thickness (T_{die}), channel length (L_{ch}) and width (W_{ch}), are utmost concern in order to generate uniform batch characterization for sensor [4]. Fig. 1(b) represents a surface profile obtained for the dielectric thickness layers, in which the average thickness was found to be ~13.87 μm and ~24.04 μm for single and double printed layer respectively. Meanwhile, the channel width-to-length ratio obtained was 3.47 with the average L_{ch} was 427 μm ± 37 μm based on all 20 devices. In screen printing, it was expected that the dimensions of the device to be different compare to the initial design due to the surface tension and viscosity of the paste when deposited on PET film. Moreover, the spreading of paste on the substrate hugely dependent on the mesh and particle size to squeeze very high viscosity paste through the mask [13].

Fig. 2(a) shows the transfer curve (I_s-V_g) for two different T_{die} when sweeping V_g from 2 V to -5 V at a fixed V_s of -10 mV. Six devices prepared for single and double dielectric layer were assessed in terms of I_s and V_{th}. The I_s-V_g curve obtained shows consistent transistor behavior for every devices indicating considerably good reproducibility [14]. The I_s performance was highest with thinner T_{die} and the average I_s was calculated to be greater by ~4.4 times for single layer dielectric compared to double layer dielectric at

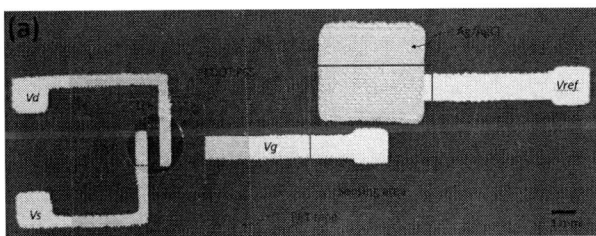

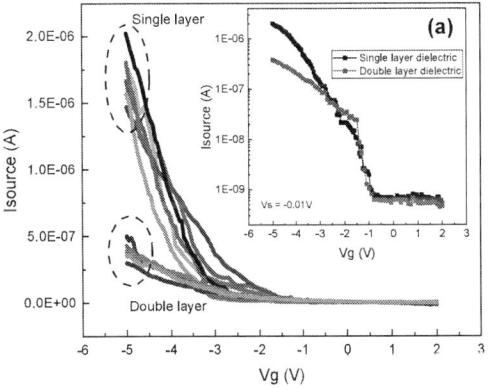

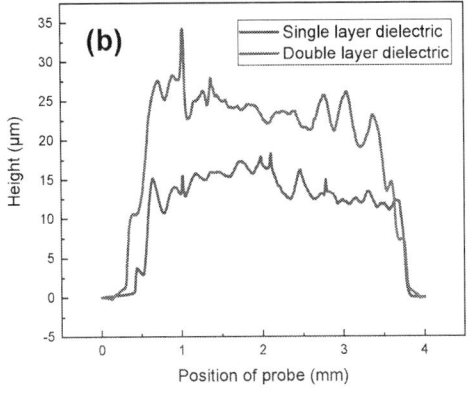

Fig. 1. Structure characterization of OTFT. (a) Microscopic image of OTFT fabricated using screen print and drop casting technique. The red rectangular area represents the sensing area where the electrolyte was dropped. (b) Surface profile obtained for thickness of a single and double dielectric layer by screen printing.

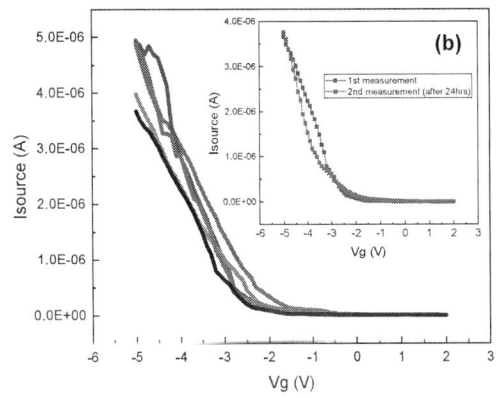

Fig. 2. The reproducibility and repeatability of fabricated OTFTs based on I-V graph at V_s of (a) -0.01 V (b) 1 V.

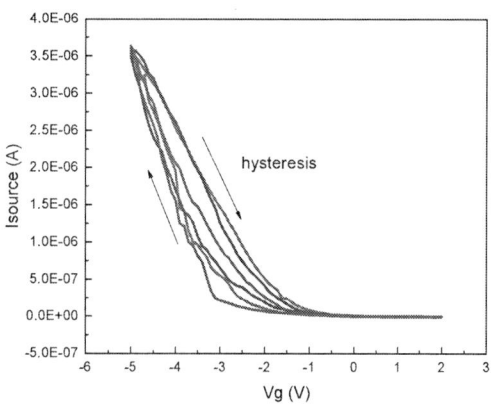

Fig. 3. The sensing characteristic of three fabricated OTFT devices in electrolyte PBS.

V_g of -5V. The average V_{th} measured was (-2.82 ± 0.57) V for single layer dielectric, in which such variations may resulted from the high roughness of dielectric layer [15]. The performance degradation for double layer dielectric can be associated with the dielectric capacitance ($C = k\varepsilon/T_{die}$), where k is dielectric constant relative to vacuum, ε [16]. Thicker T_{die} reduces the dielectric capacitance, followed with lowering transconductance and increasing operating voltage, hence result in lower I_s value [17]. Next, I_s-V_g plot from six devices of single layer dielectric at V_s of 1 V was presented in Fig. 2(b). which shows uniform performance at much higher V_s. The measurement was repeated after 24 hours to confirm its' stability. The results suggested good repeatability as the I_s performance during subthreshold does not significantly changed before and after 24 hours of measurement. A slight discrepancy observed between the two measurements within 24 hours may due to environmental exposure during measurement and when storing the device [18].

To enable detection in electrolyte, the extended gate configuration was adapted onto the OTFT, where sensing performance can be evaluated based on changes in its V_{th} or I_s level [19]. Fig. 3 shows the I_s-V_g curve when PBS was applied on the sensing area of OTFT with V_{ref} swept between 2 V to -5 V at a fixed V_s of 1V. Based on the graph, minimal hysteresis observed between forward and reverse V_{ref}, resulted from high ionic mobility [20]. Such sensing behavior is favorable to ensure reliable measurement. Moreover, similar hysteresis characteristic was obtained when measured on another two samples. In practice, the extended-gate type OTFT-based sensor can employ biorecognition layer on its extended gate to amplify output signal based on sample concentration [11]. Hence, further implementation of sensing layer such as enzyme could allow detection of specific analyte for targeted application.

IV. CONCLUSION

This paper has demonstrated a facile deposition process to fabricate OTFT devices on flexible substrate by screen printing and drop casting technique at maximum processing temperature of 80 °C. The I-V measurements for the OTFT device show standard transistor behaviour with the thin dielectric thickness is preferable for greater output current. The single printed dielectric layer resulted in maximum of ~4.4 times higher in I_s average value compared to double printed dielectric layer. In terms of reproducibility and repeatability, the device had acceptable I-V performance with consistent response, thus allowing it to be used for sensing application. The extended gate configuration and Ag/AgCl-based reference electrode was utilized to enable detection in electrolyte. Although the demonstration may be limited to laboratory demonstrations, the OTFT device show potential for various sensing applications and is appealing towards printable and flexible technologies.

REFERENCES

[1] C. Sun, X. Wang, M. A. Auwalu, S. Cheng, and W. Hu, "Organic thin film transistors-based biosensors," *EcoMat*, vol. 3, no. 2, p. e12094, 2021.

[2] N. S. Yusof, M. F. P. Mohamed, N. A. Ghazali, M. F. A. J. Khan, S. Shaari, and M. N. Mohtar, "Evolution of solution-based organic thin-film transistor for healthcare monitoring– from device to circuit integration: A review," *Alexandria Eng. J.*, vol. 61, no. 12, pp. 11405–11431, 2022.

[3] J. Kim *et al.*, "Enhanced performance and reliability of organic thin film transistors through structural scaling in gravure printing process," *Org. Electron.*, vol. 59, no. April, pp. 84–91, 2018.

[4] M. Zabihipour *et al.*, "High yield manufacturing of fully screen-printed organic electrochemical transistors," *npj Flex. Electron.*, vol. 4, no. 1, p. 15, 2020.

[5] J. Fan, A. A. Forero Pico, and M. Gupta, "A functionalization study of aerosol jet printed organic electrochemical transistors (OECTs) for glucose detection," *Mater. Adv.*, vol. 2, pp. 7445–7455, 2021.

[6] S. Sagar and B. C. Das, "Highly-sensitive full-scale organic pH sensor using thin-film transistor topology," *Org. Electron.*, vol. 111, no. September, p. 106654, 2022.

[7] B. Shao *et al.*, "Crystallinity and grain boundary control of TIPS-pentacene in organic thin-film transistors for the ultra-high sensitive detection of NO2," *J. Mater. Chem. C*, vol. 7, no. 33, pp. 10196–10202, 2019.

[8] I. Gualandi, E. Scavetta, F. Mariani, D. Tonelli, M. Tessarolo, and B. Fraboni, "All poly(3,4-ethylenedioxythiophene) organic electrochemical transistor to amplify amperometric signals," *Electrochim. Acta*, vol. 268, pp. 476–483, 2018.

[9] K. Fukuda *et al.*, "Printed Organic Transistors with Uniform Electrical Performance and Their Application to Amplifiers in Biosensors," *Adv. Electron. Mater.*, vol. 1, no. 7, 2015.

[10] S. Gupta, R. Datt, A. Mishra, W. C. Tsoi, A. Patra, and P. Bober, "Poly(3,4-ethylenedioxythiophene):Poly(styrene sulfonate) in antibacterial, tissue engineering and biosensors applications: Progress, challenges and perspectives," *J. Appl. Polym. Sci.*, vol. 139, no. 30, pp. 1–21, 2022.

[11] K. N. Agamine *et al.*, "Printed Organic Transistor-based Biosensors for Non-invasive Sweat Analysis," *Anal. Sci.*, vol. 36, pp. 291-302, 2020.

[12] T. Minamiki, S. Tokito, and T. Minami, "Fabrication of a flexible biosensor based on an organic field-effect transistor for lactate detection," *Anal. Sci.*, vol. 35, no. 1, pp. 103–106, 2019.

[13] Y. Zhang *et al.*, "Ink formulation, scalable applications and challenging perspectives of screen printing for emerging printed microelectronics," *J. Energy Chem.*, vol. 63, pp. 498–513, 2021.

[14] G. Kim, C. Fuentes-Hernandez, X. Jia, and B. Kippelen, "Organic Thin-Film Transistors with a Bottom Bilayer Gate Dielectric Having a Low Operating Voltage and High Operational Stability," *ACS Appl. Electron. Mater.*, vol. 2, no. 9, pp. 2813–2818, 2020.

[15] M. Geiger *et al.*, "Effect of the Degree of the Gate-Dielectric Surface Roughness on the Performance of Bottom-Gate Organic Thin-Film Transistors," *Adv. Mater. Interfaces*, vol. 7, no. 10, 2020.

[16] X. Guo *et al.*, "Current Status and Opportunities of Organic Thin-Film Transistor Technologies," *IEEE Trans. Electron Devices*, vol. 64, no. 5, pp. 1906–1921, 2017.

[17] P. Mittal, S. Yadav, and S. Negi, "Advancements for organic thin film transistors: Structures, materials, performance parameters, influencing factors, models, fabrication, reliability and applications," *Mater. Sci. Semicond. Process.*, vol. 133, p.

105975, 2021.

[18] M. Nikolka et al., "High operational and environmental stability of high-mobility conjugated polymer field-effect transistors achieved through the use of molecular additives," *Nat. Mater.*, vol. 16, pp. 356–362, 2017.

[19] G. M. Ali, "Interdigitated Extended Gate Field Effect Transistor Without Reference Electrode," *J. Electron. Mater.*, vol. 46, no. 2, pp. 713–717, 2017.

[20] M. Singh et al., "The double layer capacitance of ionic liquids for electrolyte gating of ZnO thin film transistors and effect of gate electrodes," *J. Mater. Chem. C*, vol. 5, no. 14, pp. 3509–3518, 2017.

Smoothing Sensor Data in a Controlled IoT Framework with Moving Averages

Akmal Mustaffa Zulhakim
School of Electrical Engineering
College of Engineering
Universiti Teknologi MARA
Shah Alam, Selangor
akmalmustaffa99@gmail.com

Wan Fazlida Hanim Abdullah
School of Electrical Engineering
College of Engineering
Universiti Teknologi MARA
Shah Alam, Selangor
wanfaz@uitm.edu.my

Ili Shairah Abdul Halim
School of Electrical Engineering
College of Engineering
Universiti Teknologi MARA
Shah Alam, Selangor
shairah@uitm.edu.my

Robaiah Binti Haji Mamat
School of Electrical Engineering
College of Engineering
Universiti Teknologi MARA
Shah Alam, Selangor
robaiah_hjmamat@yahoo.com

Muhammad Izzat Alif Muslan
School of Electrical Engineering
College of Engineering
Universiti Teknologi MARA
Shah Alam, Selangor
izzat.alif@hotmail.com

Ahmad Zaki Abu Bakar
Mimos Berhad
Technology Park Malaysia
Kuala Lumpur
zaki.bakar@mimos.my

Abstract—This research presents a study on the effect of incorporating two moving averaging algorithms to an ESP32 microcontroller which was connected to an Extended Gate Field Effect Transistor (EGFET) sensor and an IoT framework via Wi-Fi. The collected data parameters are shown on an analytical platform which was installed in the cloud server and saved inside the server's database. The Simple Moving Average (SMA) algorithm was used to remove noise in the data by taking an average of values over a period, but this method is prone to having large delays especially with larger sampling rates. Therefore, the study proposes an enhanced version of the SMA algorithm called Enhanced Simple Moving Average (ESMA) which incorporates a second period to perform the averaging twice. The results show that ESMA significantly reduces the initialization period while keeping the sensor values stable.

Keywords— data averaging, SMA, EGFET, Internet of Things, ESP32, Arduino IDE, Node-Red, Influx DB, Grafana

I. INTRODUCTION (*HEADING 1*)

Wireless sensor networks and the Internet of Things (IoT) have becoming increasingly popular due to a wide variety of applications in agriculture for monitoring different parameters such as soil moisture, temperature, and humidity [1]–[3]. However, sensor data are prone to several sources of errors [4] resulting in noisy data that affects the data integrity and reliability. One way to improve the data integrity is by implementing averaging techniques to the sensor data. Data will be smoothed out before being sent to the IoT framework to avoid inaccuracies and unreliable statistics. The typical approach to remove noise is by using moving average algorithm where an average of values is taken over a period [5] called the Simple Moving Average (SMA). This algorithm comes with many variations, another popular type is the Exponential Moving Average (EMA) and a Weighted Moving Average (WMA) [6] which works the same way but focuses and weighs more at the recent part of the period taken.

Amongst the three algorithms, SMA are the simplest algorithm to be implemented but are also prone to have large delays especially with larger sampling rate because they are bounded to past values [7]. This research took the SMA algorithm and improves the delay by introducing a second period within the original period to perform the averaging twice. EMA and WMA were not included because of their high sensitivity to changes with new values. In addition, EMA algorithm is inflexible as it depends on the value of alpha (α), which ranges from 0-1 thus the number of samples taken each second is uncertain and cannot be controlled.

This research presents the effect of incorporating two moving averaging algorithms to an ESP32 microcontroller which was connected to an Extended Gate Field Effect Transistor (EGFET) sensor and an IoT framework via Wi-Fi. The EGFET sensor comprising of a sensing and a working electrode, and an interfacing circuitry that are used to capture the ion reactions in the target solution. The collected data parameters will be shown on an analytical platform which was installed in the cloud server and saved inside the server's database. A Message Queuing Telemetry Transport (MQTT) protocol are used as messaging protocol between the microcontroller and the server.

II. METHODOLOGY

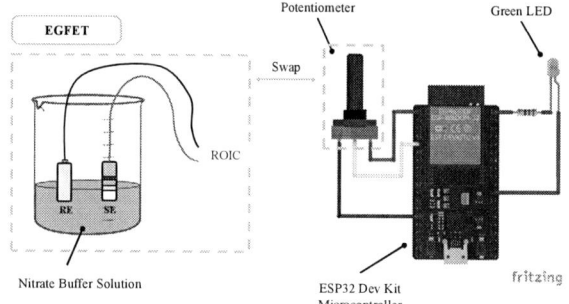

Fig. 1. Schematic diagram of the system.

This research was conducted by implementing data averaging techniques on sensor data which was paired with an IoT architecture for the research's scalability using SMA and an enhanced version of the SMA, ESMA. Two

979-8-3503-2369-6/23 $31.00 © 2023 IEEE

experimental works will be conducted; the first by simulation using a potentiometer, which represented the sensor in the Arduino IDE, and the second is by replacing the potentiometer with EGFET as shown in the schematic diagram in Fig. 1. EGFET will be dipped inside a nitrate buffer solution of 20 ppm concentration during the experimentation.

A. Simple Moving Average (SMA) Algorithm

SMA algorithm works by taking the arithmetic mean of given set of values over a series of time periods. When new values are introduced into the set, the old value drops out and the averaging cycle continues along the timescale [8].

$$SMA = \frac{P_M + P_{M-1} + \cdots + P_{M-(n-1)}}{n} \qquad (1)$$

In the formula above, n represents the sampling rate or the number of data points, and P_M represents the data point value at time M. When calculating successive values, the formula takes the following form:

$$SMA_{now} = \frac{P_M}{n} + SMA_{prev} + \frac{P_{M-n}}{n} \qquad (2)$$

B. Enhanced approach of Simple Moving Average (ESMA) Algorithm

The enhanced approach of the SMA is done by introducing a second period prior to the intended period. The approach was called Enhanced of Simple Moving Average (ESMA). To put it simply, to calculate a moving average of 10 data points (n=10) for every 10 seconds, the second interval can be calculated using the following formula.

$$second\_interval, t_2 = \frac{first\_interval, t_1}{sampling\_rate, n} \qquad (3)$$

In the span of 10 seconds, there are 10 moving averages that occur as the value of second interval is 1 second ($t_2=1s$), which will greatly diminish the delay from the main interval ($t_1=10s$). To sum up, the process of ESMA starts with performing the first moving average using interval t_2 followed by the second moving average using interval t_1.

C. IoT System Architecture

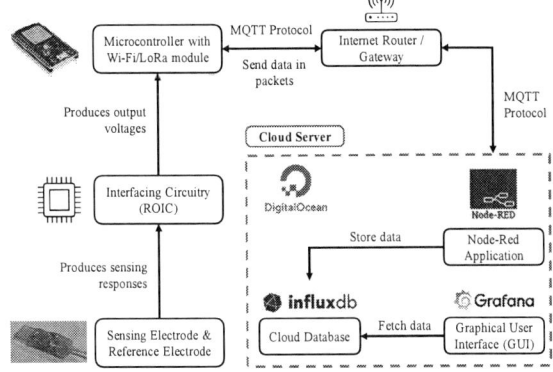

Fig. 2. IoT Architecture of the system.

The IoT framework comprises of three different subsystems: sensor node devices, a gateway, and a cloud server that includes several more applications, such as a flow-based development tool, a MQTT broker, a database, and a dashboard as shown in Fig. 2. The study incorporates Wi-Fi as the connectivity protocol because it is easier to set up with the microcontroller and the cloud server.

The sensor nodes used in this research consist of an EGFET for data acquisition and a microcontroller for data averaging. EGFET was fabricated and assemble by the research team while the microcontroller coding was completed and assessed in the Arduino IDE, where all the necessary libraries and boards were installed. The gateway in this context refers to the router that manages the network traffic between the microcontroller and the cloud server. In continuation to that, a flow-based development tool call Node-Red, is used to manage, filter, and converting data into various formats. The overall Node-Red process is shown in Fig. 3. Node-Red uses MQTT for messaging protocol, which is a simple and lightweight messaging protocol that relies on a Pub-Sub client for messaging.

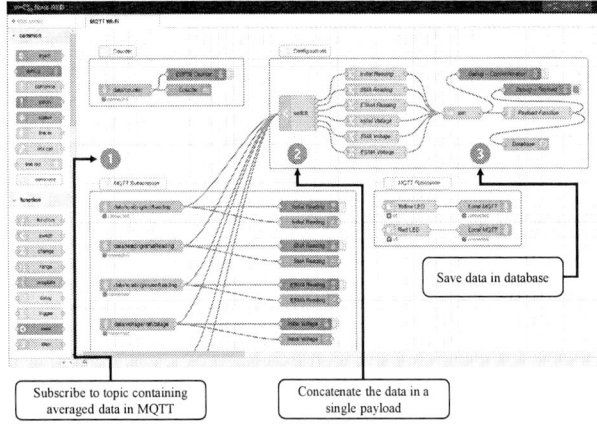

Fig. 3. Node-Red overall workflow.

The microcontroller publishes data in a topic, with Node-Red subscribing to the topic to receive the data. The sensor data from MQTT is then concatenated into a single payload string file, stored inside the Influx DB database, and displayed onto the Grafana dashboard. As Grafana fetched the data from the database, some configurations are made to enhance visualization to users. On a side note, the output of each building block in Node-Red is monitored to ensure that the data formatting is correct.

D. Flow of the System

The system starts when the microcontroller initializes and the built-in LED blinks. The microcontroller will then try to connect to the Internet using the Wi-Fi credentials programmed. If the LED stops blinking and stays ON, the microcontroller is now connected to the internet. The microcontroller will then attempt to connect to MQTT and publish a topic every second after a successful connection. The published topic will contain a message sequence consisting of a message counter, the initial nitrate reading in bit format, the current nitrate reading in bit format, and the nitrate voltage value. Node-Red will consistently monitor that specified topic to receive data. Once subscribed, the data will be displayed on the configured dashboard and stored in the database. The displayed data will be differentiated between the initial and the current reading of nitrate voltage.

III. RESULTS AND DISCUSSION

The research findings and a detailed discussion of the results will be presented in two sections: simulation results in

Arduino IDE and results in Grafana dashboard. For the simulation results, two sampling rates were used: 10 samples/second and 30 samples/second, with a 1-second delay between each sample. In the latter part of the experiment, the microcontroller was tested to compute 3000 samples/second. In the Grafana dashboard, a sampling rate of 300 samples/second was used, with the same 1-second delay between each transferred data. The EGFET sensor was used to provide data to determine the voltage value in a 20-ppm nitrate buffer solution.

A. Simulation result in Arduino IDE

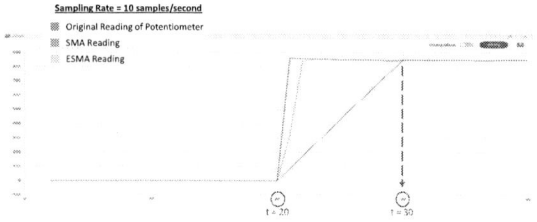

Fig. 4. Initialization period of sampling rate of 10 samples/second.

Fig. 4 above shows the simulation results in the Arduino Serial plotter when the sampling rate is 10 samples/second. The blue line (value 1) indicates the original reading of the potentiometer, with the red line (value 2) being the reading after processed through SMA, and the orange line (value 4) being the reading of ESMA. At the beginning, the value remains at 0 until the 20th second (t=20) when the original reading spiked to around 850. As shown in the figure, ESMA caught up with the original reading in only 2 seconds, while SMA took 10 seconds as indicated by the t=30 on the figure. This clearly shows that with ESMA, the initialization period was reduced by a significant amount.

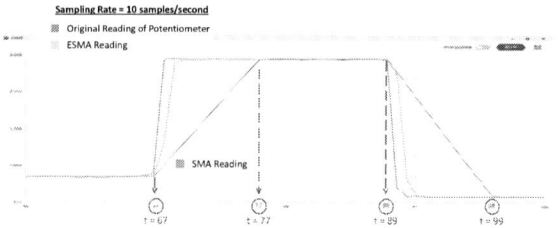

Fig. 5. Testing the algorithm for large changes in sensor value at SR=10.

To further strengthen the statement, the initial reading's value was raised from 800 bits to approximately 2400 bits before dropping back down to around 600 bits, as illustrated in Fig. 5. Despite the drastic changes in value, SMA still required 10 seconds (t=67 to t=77 & t=89 to t=99) to adjust, while ESMA still follows the original value with only a slight delay of a few seconds.

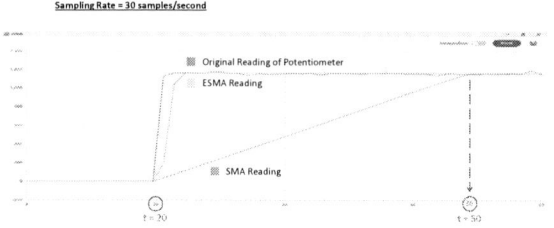

Fig. 6. Initialization period of sampling rate of 30 samples/second.

The investigation that explores a higher sampling rate. The main disadvantage of SMA is illustrated in Fig. 6 above. When the sampling rate was changed to 30 samples/second, SMA took 30 seconds to catch up with the original value, from t=20 to t=50. In contrast, ESMA caught up with the original value quickly and still performed the averaging. From t=20 onwards, the original value became noisy, but ESMA continued to average as intended.

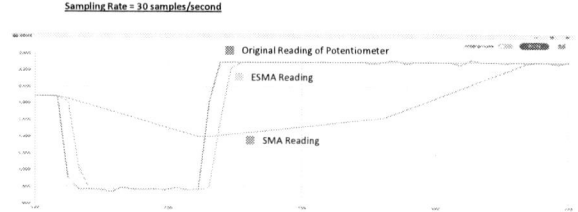

Fig. 7. Testing the algorithm for large changes in sensor value at SR=30.

This Fig. 7, the values of the potentiometer was changed from 1900 bits to 800 bits, then changed again to 2300 bits. The figure clearly shows that ESMA outperformed SMA in adapting to drastic changes in value. This figure also demonstrates that with a higher sampling rate, SMA will experience longer delays before reaching the stage where averaging is done properly. ESMA, on the other hand, performs exceptionally well in smoothing the data even when the sampling rate is increased.

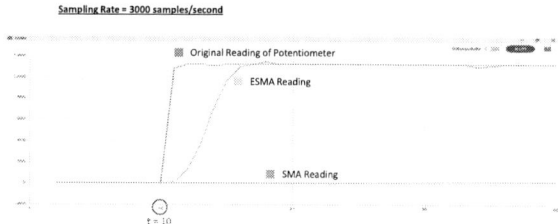

Fig. 8. Initialization period of sampling rate of 3000 samples/second.

The Fig. 8 above shows the results when the sampling rate was changed to 3000 samples/second, which is the current limit of the ESP32 microcontroller with the code written in the Arduino IDE. If the sampling rate is set higher than 3000, the microcontroller will run out of memory and cannot compile or run the code. It may be possible to push the limit higher with code optimization, but this will not be covered in this research. When the sampling rate was set to the maximum, ESMA still experienced a slight delay of around 7 seconds while SMA would take 3000 seconds (50 minutes) to reach the original reading. If the original reading fluctuates significantly due to human error during this time, SMA would reset, and another 50 minutes would be needed to provide a new stable value. These results clearly prove the limitations of SMA at high sampling rates.

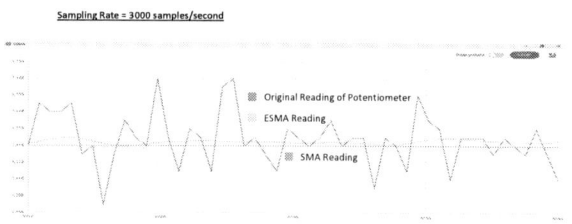

Fig. 9. Close-up comparison between the algorithms after stable.

979-8-3503-2369-6/23 $31.00 © 2023 IEEE

Fig. 9 shows a close-up view of the original signal with a sampling rate of 3000. The value of the original signal fluctuates between 1107-1122 bits. Once the SMA stabilizes, the value becomes more concrete, which in this case is 1114. ESMA values have minimal fluctuation between 1114-1115 bits but considering the advantages of minimized delay with high sampling rate, it becomes acceptable.

B. Data in Grafana Dashboard

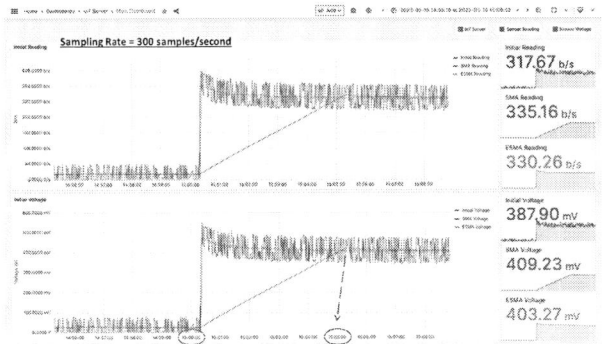

Fig. 10. Graph of sensor data before and after dipping in buffer solution.

The graph in Fig. 10 displays the sensor value before and after being dipped in a 20-ppm nitrate buffer solution, with a sampling rate of 300 samples/second. The upper graph shows the sensor reading in bits, while the lower graph displays the sensor reading converted to voltage values. The blue line represents the initial sensor reading, the red line represents the reading after being averaged with SMA, and the orange line represents the reading after being averaged with ESMA. When EGFET is dipped in the nitrate solution, the sensor data remained noisy, with an increase in reading but still accompanied by noise. If the user were to take 5-second data samples and manually average the data, false information might be received. In contrast, SMA and ESMA perfectly averages the sensor reading according to the initial reading, producing more dependable, stable, and accurate results. Based on the figure, SMA took 300 seconds (5 minutes) to have a stable reading from time 15:00:00 (3.00 p.m.) to 15:05:00 (3.05 p.m.).

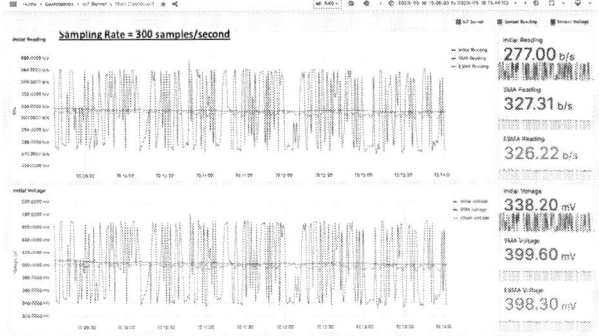

Fig. 11. Close-up comparison between the algorithms after stable.

Assuming time is not a concern, the Fig. 11 shows a comparison of the initial sensor reading, the reading of the SMA, and the ESMA after stable. Looking at the lower graph for voltage value, the original signal is very noisy, ranging from 340mV to 470mV. The SMA provides greater stability than the ESMA, with a value of 399.60mV. The

ESMA has a value of 398.30mV, which only differs by 0.33%. Keep in mind that this is only with a sampling rate of 300 and ESMA can be programmed for a higher sampling rate. Then, ESMA can achieve greater stability with a minimal delay.

IV. SUMMARY

Results from both the simulation and the actual experiment show that the implemented SMA and ESMA algorithms in the microcontroller can significantly reduce noise from the EGFET sensor. The final averaging values of SMA and ESMA are very similar, with a marginal error of only 0.33% when using 300 samples/second as the sampling rate. However, ESMA's sampling rate can be further increased up to 3000, limited only by the microcontroller's capability. This makes ESMA more reliable and produces more accurate values compared to SMA, which depends on its previous value and causes delays at higher sampling rates. Therefore, ESMA is a better option for a real-time IoT system that continuously sends data to the server.

ACKNOWLEDGMENT

The research project is funded by the Ministry of Higher Education (MOHE) under the Fundamental Research Grant Scheme (FRGS)(FRGS/1/2022/TK07/UITM/02/38) and supported by the College of Engineering, Universiti Teknologi MARA.

REFERENCES

[1] M. E. E. Alahi, N. Pereira-Ishak, S. C. Mukhopadhyay, and L. Burkitt, "An Internet-of-Things Enabled Smart Sensing System for Nitrate Monitoring," IEEE Internet Things J, vol. 5, no. 6, pp. 4409–4417, Dec. 2018, doi: 10.1109/JIOT.2018.2809669.

[2] M. R. Nithya, P. Lakshmi, J. Roshmi, R. Sabana, and R. U. Swetha, "Machine Learning and IoT based Seed Suggestion: To Increase Agriculture Harvesting and Development," 2023 International Conference on Sustainable Computing and Data Communication Systems (ICSCDS), pp. 1202–1207, Mar. 2023, doi: 10.1109/ICSCDS56580.2023.10104981.

[3] N. Shanmugasundaram, G. Santhip Kumar, S. Sankaralingam, S. Vishal, and N. Kamaleswaran, "Smart Agriculture Using Modern Technologies," 2023 9th International Conference on Advanced Computing and Communication Systems (ICACCS), pp. 2025–2030, Mar. 2023, doi: 10.1109/ICACCS57279.2023.10113059.

[4] E. Elnahrawy and B. Nath, "Cleaning and Querying Noisy Sensors," 2nd ACM International Workshop on Wireless Sensor networks and Applications, WSNA 2003, pp. 78–87, 2003, doi: 10.1145/941350.941362.

[5] M. H. Prami Swari, I. P. Susila Handika, and I. K. Susila Satwika, "Comparison of Simple Moving Average, Single and Modified Single Exponential Smoothing," Proceedings - 2021 IEEE 7th Information Technology International Seminar, ITIS 2021, 2021, doi: 10.1109/ITIS53497.2021.9791516.

[6] Y. Zhuang, L. Chen, X. S. Wang, and J. Lian, "A weighted moving average-based approach for cleaning sensor data," Proc Int Conf Distrib Comput Syst, 2007, doi: 10.1109/ICDCS.2007.83.

[7] M. E. Haque, M. N. S. Khan, and M. R. I. Sheikh, "Smoothing control of wind farm output fluctuations by proposed Low Pass Filter, and Moving Averages," ICEEE 2015 - 1st International Conference on Electrical and Electronic Engineering, pp. 121–124, Mar. 2016, doi: 10.1109/CEEE.2015.7428234.

[8] S. Hansun, "A new approach of moving average method in time series analysis," 2013 International Conference on New Media Studies, CoNMedia 2013, 2013, doi: 10.1109/CONMEDIA.2013.6708545.

Morphology and Electrical Properties of Pristine and Composite Rice Husk Ash Nano/Micro Particles Thick Films for Gas Sensing Applications

Jamila lamido Sumaila
Department of Physics
Yusuf Maitama Sule University
kano
Kano,Nigeria
l j.lamido@yahoo.com

Dahiru Sani Shu'aibu
Department of Electrical
Engineering
Bayero University Kano
Kano, Nigeria
dsshuaibu.ele@buk.edu.ng

Mohd Nizar Hamidon
Institute of Nanoscience and
Nanotaechnology
Universiti Putra Malysia
Serdang, Malaysia
mnh@upm.edu.my

Zainab Yunusa
Department of Electrical
Engineering
Hafr Al Batin Al Jamiah
Hafr Al Batin, Saudi Arabia
zee2yunusa@gmail.com

Nuraddeen Magaji
Department of Electrical
Engineering
Bayero University Kano
Kano, Nigeria
nmagaji.ele@buk.edu.ng

Azlinda Abubakar
Institute of Nanoscience and
Nanotaechnology
Universiti Putra Malysia
Serdang, Malaysia
azlinda@unikl.edu.my

Farah Nabilah Shafiee
Institute of Nanoscience and
Nanotaechnology
Universiti Putra Malysia
Serdang, Malaysia
farahnabilahshafiee@gmail.com

Sulaiman Babani
Institute of Nanoscience and
Nanotaechnology
Universiti Putra Malysia
Serdang, Malaysia
sbabani2000@gmail.com

Abstract— **In this work, we investigate the morphology and electrical properties of thick film made from Rice husk ash (RHA) and RHA mixed with biochar of an invasive plant from Northern Nigeria called Prosopis Africana char (PAC). The RHA and the RHA+PAC composite were processed into microns and nanoparticles, linseed oil was incorporated as a vehicle binder for the thick film formulation. The developed thick films were characterized to study their morphology and evaluate their electrical properties. FESEM, EDX characterizations analysis were carried out on the films and a two-point probe meter and Keithley source meter were used to measure the resistance, sheet resistance, and I-V characteristics respectively. Finally, we were able to calculate the resistivity and conductivity of each thick film developed. Thick films made from nanoparticles shoes relatively low conductivity compared to its microns' counterparts.**

Keywords— *Resistivity, conductivity, sheet resistance, thick film*

I. INTRODUCTION

The need for gas sensors is on the rise due to increased gas emissions into our environment. This has led to global warming due to large amounts of greenhouse gasses released into the atmosphere, such as CO_2, NO_2, etc. Health hazards include lung diseases and renal failures due to inhalation of toxic gasses such as ammonia gas. Other major disasters are fire outbreaks and explosions which can be caused by H2 gas and other highly inflammable gases. It is, therefore, imperative to monitor these gases so as to ensure they are within the allowable range to reduce the danger posed to living beings[1].

As a result of its simplicity, speed, and low cost, the thick film fabrication technique is one of the most utilized methods in electronic applications. Also, the layers of the thick film can be deposited layer by layer to accommodate the needed applications, making it possible for the miniaturization of electronics[2]. Thick film technology date back to the 1950s, making it one of the oldest microelectronics-enabling technologies. It offered an alternative to printed circuit board technology and the potential to build small, integrated, durable circuits at the time. Since the 1960s, it has primarily existed in the shadow of silicon technology. The films are deposited using screen printing (stenciling), a method of graphic replication that dates back to the great Chinese dynasties approximately one thousand years ago. In fact, there is evidence that even the earliest Paleolithic cave paintings from approximately 15,000 B.C. were likely created using primitive stenciling techniques. With the introduction of surface-mounted electronic devices in the 1980s, thick film technology regained popularity as it permitted the manufacture of circuits without connecting components [3]. Generally speaking, there are two types of thick films: ceramic-metallic and polymer. Ceramic-metallic thick films, also known as cermet thick films, are composite materials composed of ceramic and metal. This sort of thick film requires glass frits, such as boron oxide (B2O3), titanium oxide (TiO_2), and lead oxide (PbO), which are ceramic-based inorganic binders. Incorporating glass frits requires a somewhat high temperature (over 450°C) for it to operate as a binder. Polymer substances such as ethyl cellulose (boiling point: 250°C) and acrylic-based adhesives (boiling point: 160°C) are often used as the binder in carbon-based polymer thick films, resulting in a comparatively low firing temperature[4]. Moreover, polymer thick film can be utilized to fabricate flexible thick film, which has gained considerable interest, particularly for flexible and portable device applications [5]. Recent research has demonstrated that linseed oil can be used as a binder in the production of thick films. Linseed oil is a drying oil due to its high alpha-linoleic acid concentration (about 60 percent). It is extensively used as a binder in the painting industry due to its ability to dry and create a film, making it an excellent coating substance [6]. The specifications for manufacturing microelectronic circuits call for a printer with a distinct type of screen material. A typical thick-film screen will consist of a mesh woven from stainless steel, polyester, or nylon. This is put under tension on a metal frame and covered with an ultraviolet-sensitive emulsion upon which a photographic circuit pattern can be produced. The desired pattern can be printed through the open mesh sections of the completed screen. In a screen-printing machine, the screen is held at a distance of approximately 0.5 mm from the substrate's surface. The thick-film paste, which is commonly resistive, conductive, or dielectric, is poured onto the top surface of the stencil, and a squeegee is used to apply pressure to the screen. This motion drives the paste through the open places by bringing the screen into touch with the substrate. Therefore, the appropriate pattern is deposited on the substrate[7]. Drying the printed film is the next step in the process. Various organic solvents are present in all pastes in order to generate the correct viscosity for screen printing. These can be eliminated by drying the film in a furnace. After drying, the films are relatively resistant to smearing and retain a solid pattern on the substrate. In certain instances, an additional layer can be printed directly onto a dry film, but in most cases, the films must be annealed at high temperatures. The majority of thick-film furnaces give the operator control over a variety of parameters, such as throughput speed, peak temperature, and dwell time. After the firing process, the film securely adheres to the substrate, and more screen-printed layers can be applied as necessary[8].

Investigations on morphology and electrical properties of biochar thick films were carried out, which inferred that biochar films, which appear chaotic and granular based on morphological characteristics are fairly conductive. Owing to its inexpensive cost make appropriate for uses in variety of applications[9]. Rice husk thick film used for Humidity gas sensing application in[10] with a particle size of $45\mu m$ dispalayed a high a impedance value of about $19M\Omega$. In [11] films of Rice husk cellulose nano fiber were developed for gas sensing application, the electrical property measurement indicated a resistace value ranging from $5k\Omega$ to $10M\Omega$.

In this paper, thick films were developed using RHA, and RHA mixed with char obtained from an invasive plant called Prosopis Africana using a 50:50 rati0, the mixed powder was divide into two different parts first part was sieved using 20 microns sieve and the second part milled for 6hrs. Furthermore, we study each film's morphology and electrical properties based on particle size. The developed films would subsequently be used for gas-sensing applications.

II. MATERIAL AND METHODS

A. Formulating the thick film paste

Biochars were manually ground, one part sieved to 200 microns, and another milled for six hours using SPEX 800m with Vial. were used for milling. 40wt% of biochar was combined with 60wt% organic binder based on linseed oil. The mixtures were stirred for 24hrs after which was allowed to rest for 1hr. the paste was printed onto an alumina substrate consisting of four different films sample. The samples were named RH1, RH2 for pristine rice husk ash micro and Nano particles respectively and RP1, RP2 for composite micro and Nano particles respectively. All samples were annealed at 350°C for 1hr:30mins. The thick film were then characterized using an FEI NOVA Field-Emission Scanning Electron Microscope. Two-point probe approach measured thick film electrical resistance.

III. RESULTS AND DISCUSSIONS

FESEM images in the figures below show large particles of micro sizes in figures 1 and 3. Some defects in the particle shapes can also be observed. Moreover, the longer milling process in RH2 and RP2 yielded smaller particles agglomerated together with more defects on the particle. It could be observed that in all the micrographs, the particles are binded together due to the of linseed oil binder used during thick film formulation. It could be observed in micrographs of Fig 1 and Fig 3 the existence of cracks in the films which is largely due to the binder drying up during the annealing process. While in fig 2 and fig4 micrographs, a significant amount of the binder could be observed in the film. This would likely results in low conductivity value of the films as the binder behaves like as an insulator.

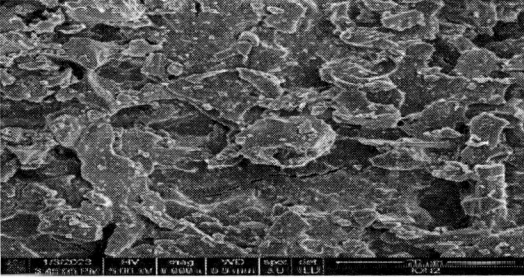

Fig.1. Fesem Micrograph of sample RH1

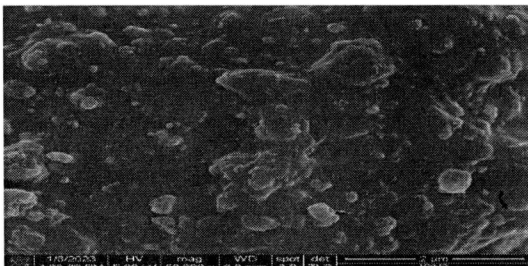

Fig. 2. Fesem Micrograph of sample RH2

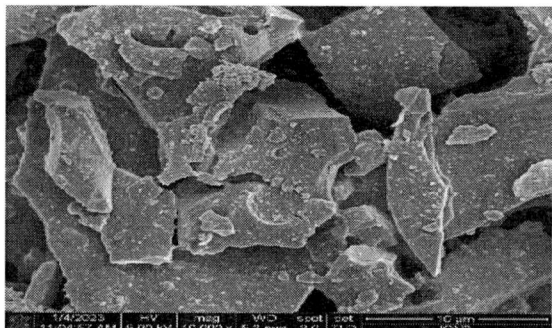

Fig. 3. Fesem Micrograph of sample RP1

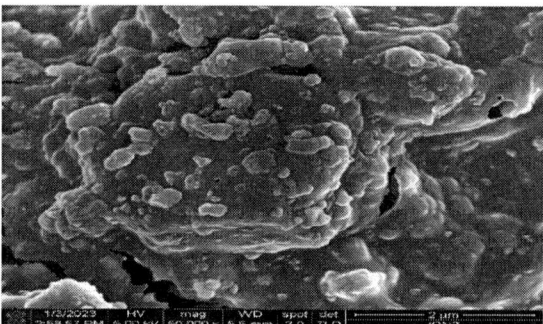

Fig. 4. Fesem Micrograph of sample RP2

The figures below depict the EDX spectrums of the films, which show the elemental constituents of each of the samples. It could be observed in RH1 and RH2 there is a high amount of silicon followed by Oxygen and then carbon with traces of Potassium which may be due to contamination. The linseed oil binder may have contributed to the high oxygen content of the samples while it it obvious that rice husk is a rich source of silica [12]]. Furthermore, spectrums of RP1 and RP2 indicates relatively low oxygen content with high amount of carbon. The silicon content also depicted in the spectrums is moderately low. This was due to the nature of 50:50 weight ratio used with RHA and PAC respectively. Traces of iron in the content was from the ball used during milling process of the chars.

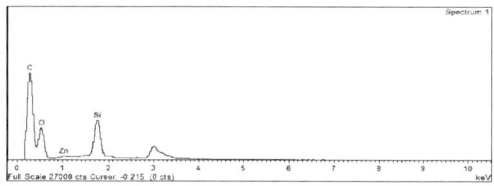

Fig. 5. EDS spectrum of sample RH1

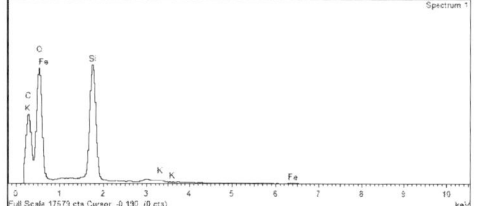

Fig. 6. EDS spectrum of sample RH2

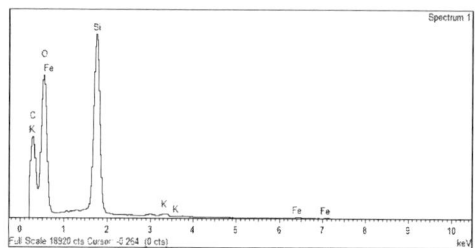

Fig. 7. EDS spectrum of sample RP1

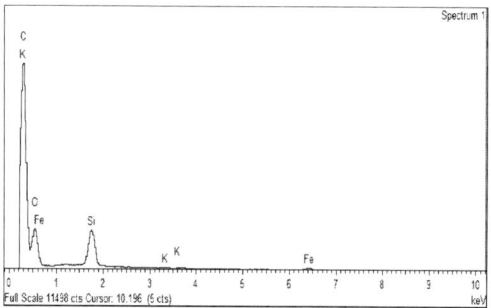

Fig. 8. EDS spectrum of sample RP2

A. Electrical Measurements

Using a two-point probe, the electrical resistance of each film was measured. Applying the formula below we calculated the sheet resistance, resistivity, and conductivity.

$$R_s = R * R_{cf} \quad (1)$$
$$\rho = R_s * t \quad (2)$$
$$\delta = \frac{1}{\rho} \quad (3)$$

Where R_s is the sheet resistance in Ω/sq, R_{cf} resistance correction factor, ρ resistivity in Ω/m, t is the thickness of the film in μm and δ conductivity in S/m. It could be observed that thick films made from micro particles size displayed high conductivity when compared to thick films of nanoparticles. This could be related to the amount of binder removed during the annealing process and the particles size, shape, and particles interconnection.

TABLE I. Electrical properties

S/N	THICK FILM	R(Ω)	R_s(Ω/sq.,)	ρ(Ω/m)	δ (S/m)
1.	RH1	0.45E6	3.80E7	556.8	1.8E-3
2.	RH2	Overload	-	-	-
3.	RP1	303	2.766E5	4.42	2.26E-1
4.	RP2	4.6k	3.0E7	418	2.39E-3

979-8-3503-2369-6/23 $31.00 © 2023 IEEE

IV. CONCLUSIONS

Morphology and electrical properties of thick film developed using rice husk has been successfully investigated. It could be observed that the thick film based on micro particle size displayed high conductivity due to low resistance value compared to thick film based on nanoparticle sizes. The milling process install defect in the particle size and shape which vehemently affect the conductivity of the film.

Acknowledgment

The authors would like to acknowledge the management of Yusuf Maitama Sule University, Kano.

REFERENCES

[1] A. M. Al-Diabat, N. A. Algadri, N. M. Ahmed, A. Abuelsamen, and S. A. Bidier, "A high-sensitivity hydrogen gas sensor based on carbon nanotubes fabricated on SiO2 substrate," *Nanocomposites*, vol. 7, no. 1, pp. 172–183, 2021, doi: 10.1080/20550324.2021.1977063.

[2] G. Korotcenkov, *Handbook of Gas Sensor Materials*, vol. 2. 2014. [Online]. Available: http://link.springer.com/10.1007/978-1-4614-7388-6

[3] N. White, "Thick Films," pp. 707–721, 2017, doi: 10.1007/978-3-319-48933-9.

[4] M. Masat, H. K. Sağlam, H. Korul, and M. Ertuğrul, "TiO2 Thick Film Gas Sensor Fabricated by Screen Printing Method for Airplanes," *AIP Conf. Proc.*, vol. 2506, no. 1, pp. 26–30, 2022, doi: 10.1063/5.0084102.

[5] R. Alrammouz, J. Podlecki, P. Abboud, B. Sorli, and R. Habchi, "A review on flexible gas sensors: From materials to devices," *Sensors Actuators, A Phys.*, vol. 284, pp. 209–231, 2018, doi: 10.1016/j.sna.2018.10.036.

[6] F. N. Shafiee, "Effect of nanometric and micronic particles size on physical and electrical properties of graphite thick film Mohd Nizar Hamidon * Mohd Haniff Wahid Abdul Halim Shaari Mehmet Ertugrul Nor Hapishah Abdullah and Mohd Asnawi Mohd Kusaimi Muhammad Syazwan Mus," vol. 17, pp. 825–839, 2020.

[7] S. A. M. Chachuli *et al.*, "Effects of MWCNTs/graphene nanoflakes/MXene addition to TiO2 thick film on hydrogen gas sensing," *J. Alloys Compd.*, vol. 882, p. 160671, 2021, doi: 10.1016/j.jallcom.2021.160671.

[8] M. A. Kusaimi, M. N. Hamidon, and S. Azhari, "Importance of Annealing Temperature on the Electrical Conductivity of Screen Printed Graphite Organic Paste OPTICAL DETECTION SYSTEM FOR FREE FATTY ACIDS IN CRUDE PALM OIL BASED ON ENZYMATIC METHOD View project CNTs-PDMS Nanocomposite View project Importa," no. September 2017, pp. 2016–2019, 2016, [Online]. Available: https://www.researchgate.net/publication/319416636

[9] M. Yasir, P. Zaccagnini, G. Palmara, F. Frascella, N. Paccotti, and P. Savi, "Graphene and Biochar Thick Films," 2021.

[10] D. Ziegler, F. Boschetto, E. Marin, P. Palmero, G. Pezzotti, and J. Tulliani, "Sensors and Actuators : B . Chemical Rice husk ash as a new humidity sensing material and its aging behavior," *Sensors Actuators B. Chem.*, vol. 328, no. June 2020, p. 129049, 2021, doi: 10.1016/j.snb.2020.129049.

[11] N. Shahi, E. Lee, and B. Min, "Rice Husk-Derived Cellulose Nanofibers : A Potential Sensor for Water-Soluble Gases," 2021.

[12] L. W. O. Soares, R. M. Braga, J. C. O. Freitas, R. A. Ventura and D. M. A. Melo, "The effect of rice husk ash as pozzolan," *J. Pet. Sci. Eng.*, vol. 131, pp. 80–85, 2015, doi: 10.1016/j.petrol.2015.04.009.

Effect of the Electrodeposition Cycle of RGO Towards Glucose Detection

Muhammad Haziq Ilias
School of Electrical Engineering,
College of Engineering,
Universiti Teknologi MARA,
Shah Alam, Selangor, Malaysia
muhammadhaziq0403@gmail.com

Norhazlin Khairudin
School of Electrical Engineering,
College of Engineering,
Universiti Teknologi MARA,
Shah Alam, Selangor, Malaysia
norhazlin380@uitm.edu.my

Ahmad Sabirin Zoolfakar
School of Electrical Engineering,
College of Engineering,
Universiti Teknologi MARA,
Shah Alam, Selangor, Malaysia
ahmad074@uitm.edu.my

Maizatul Zolkapli
School of Electrical Engineering,
College of Engineering,
Universiti Teknologi MARA,
Shah Alam, Selangor, Malaysia
maizatul544@uitm.edu.my

Zainiharyati Mohd Zain
Faculty of Applied Sciences,
Universiti Teknologi MARA,
Shah Alam, Selangor, Malaysia
zainihar@uitm.edu.my

Noor Fitrah Abu Bakar
School of Chemical Engineering,
College of Engineering,
Universiti Teknologi MARA,
Shah Alam, Selangor, Malaysia
fitrah@uitm.edu.my

Rozina Abdul Rani
School of Mechanical Engineering,
College of Engineering,
Universiti Teknologi MARA,
Shah Alam, Selangor, Malaysia
rozina7370@uitm.edu.my

Azrif Manut
School of Electrical Engineering,
College of Engineering,
Universiti Teknologi MARA,
Shah Alam, Selangor, Malaysia
ayeh77@gmail.com

Abstract—**The author report on the synthesis of reduced graphene oxide (RGO) using electrodeposition method on screen printed gold electrode (SPGE). The RGO were deposited on the screen-printed gold electrode (SPGE) using cyclic voltammetry technique with different cycles of 3, 5 and 7 of voltage range from -1.35V to -2.0V voltage applied. The surface morphology of RGO was being characterized using field-emission scanning electron microscopy (FESEM) and Raman spectroscopy. The intensity ratio of Raman spectra was being calculated in order to investigate the amount of disordered phase for each cycle. The cyclic voltammetry (CV) analysis was used in the detection of glucose and the linear curve plot was obtained. The sensitivity for each electrode was calculated from the slope obtained from the linear obtained from each electrode. The increase in the number of cycles increase the sensitivity of the sensor. The highest sensitivity in the detection of glucose concentration (1mM – 10mM) was discovered from the RGO 7 cycles of electrodeposition.**

Keywords—*Reduced Graphene Oxide (RGO), Screen-printed Gold Electrode (SPGE), Glucose detection, Electrochemical sensor.*

I. INTRODUCTION

One of the most important global public health problems, diabetes mellitus has a substantial impact on both socioeconomic development and public health globally. Diabetes prevalence has increased in most other emerging and developed countries during the past several decades, even though the incidence of the disease has started to drop in certain countries. In the Western Pacific area and among the highest in the world, diabetes is most prevalent in Malaysia and costs the country over US$600 million annually. In Malaysia, 3.6 million persons (18 and older) had diabetes in 2019, yet 49% of those cases (3.7 million) lacked a diagnosis [1]. According to the World Health Organization's (WHO) recommendations, 10 to 25 grammes of sugar should be consumed per day. People who consume more sugar than is advised over an extended period are far more likely to develop diabetes mellitus. Patients who had diabetes may experience symptoms including nausea, continuous hunger, and excessive perspiration. Patients run the danger of more serious results, such as death and seizure, if the issue is not handled [2].

Electrochemical sensors are a type of chemical sensor in which an electrode serves as a transducer element in the presence of an analyte. Physical, chemical, or biological characteristics may all be detected by modern electrochemical sensors using a variety of attributes. Electrochemical sensors are widely used due to their numerous advantages, including their ability to achieve low detection limits (as low as picomoles), their fast response time, and the use of cost-effective equipment for sensing purposes. These sensors are available in various forms, ranging from benchtop setups to fully integrated wearable devices. In electrochemical sensing, the electrical signal generated by the interaction between the target analyte and the recognition layer provides the analytical information. Depending on the specific analyte, sample matrix characteristics, and desired sensitivity or selectivity, different types of electrochemical devices can be utilized for environmental monitoring applications [3].

In recent times, there has been a growing interest in nanomaterials due to the increasing need to control specific molecules present in the human body and the environment. Nanomaterials consist of nanoparticles (NPs) that have a size of less than 100 nm in at least one dimension. The field of "nanotechnology" focuses on materials at the nanoscale. The synthesis and manipulation of nanomaterials require knowledge from various disciplines such as physics, chemistry, electronics, computer science, biology, engineering, agriculture, and more. This interdisciplinary approach can pave the way for the development of innovative and multifunctional nanotechnologies. Semiconductor materials like graphene, zinc oxide, and titanium oxide are widely employed in biosensors. These materials have demonstrated the ability to provide high sensitivity for detecting very small amounts of analytes.

979-8-3503-2369-6/23 $31.00 © 2023 IEEE

However, current research suggests that graphene-based biosensors outperform others in terms of offering low-noise and highly sensitive detection capabilities. Graphene and its derivatives possess exceptional electrical characteristics and possess a unique structural geometry. Graphene can be obtained through various methods, including mechanical and chemical exfoliation, as well as epitaxial growth. Another approach involves the reduction of solution-based graphene oxide (GO) to form reduced graphene oxide (rGO) through chemical or thermal means. rGO exhibits high electrical and thermal conductivity, substantial carrier mobility, mechanical strength, and optical properties. These distinctive features make graphene and its derivatives an exciting class of nanomaterials for applications in biosensing [4]. Reduced Graphene Oxide had been used for other sensing device such as humidity and UV sensor [5] [6] [7]

In this research, electrodeposition method was used to fabricate the sample on screen-printed gold electrode (SPGE). Electrodeposition is a process that is governed by kinetics and involves the selective formation of nuclei followed by the growth of metal nanoparticles (NPs) on a suitable electrode surface [8]. Electrodeposition is a kinetic-controlled process that involves the preferential nucleation and subsequent development of metal nanoparticles on an appropriate electrode surface. Deposition of pure metal and alloys at the cathode or of oxides and hydroxides at the anode/cathode can be used for electrodeposition. By adjusting process factors including current density, applied potential, electrolyte chemistry, temperature, pH, stirring, and particle loading in the bath, the electrodeposition technique makes it simple to produce graded coatings [9].

In this work, reduced graphene oxide (RGO) was being deposited on the screen-printed gold electrode (SPGE) using electrodeposition method. The aim of this study was to analyse the effect of cycle of RGO in glucose detection.

II. METHODOLOGY

A. Apparatus and Materials

The graphene oxide (GO) powder was acquired from GO Advanced Solution. The phosphate buffer saline (PBS) (pH 7) was acquired from R&M Chemicals. Sodium hydroxide (NaOH) granules and D-glucose was acquired from Sigma Aldrich.

The fabrication process was being done using Autolab PGSTAT204 from Metrohm Autolab (Utrecht, Netherlands). The cyclic voltammetry (CV) analysis was done with PalmSens4 (Netherlands) for the investigation of the electrode in glucose detection.

B. Fabrication of RGO on SPGE

Graphene oxide (GO) solution was produced by adding 0.15g of GO powder to a phosphate buffer saline (PBS) and sodium hydroxide solution, all of which were at PH 9. The electrodeposition process was being done using cyclic voltammetry (CV) using AUTOLAB PGSTAT101. 50 ml of GO solution is poured into a beaker and immersed into the water bath at 40°C throughout the process. The window parameters for the CV staircase were set up -1.35V for start and stop potential. The upper vertex potential was set at -2V and low vertex potential was set at -1.35V. The number of

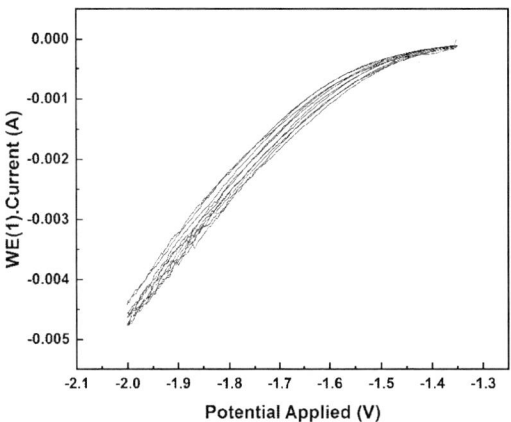

Fig. 1. Cyclic Voltammetry (CV) of electrodeposition of RGO

scans was being varied at 3, 5 and 7 cycles with scan rate at 0.005 V/s.

C. Characterization of RGO/SPGE

The morphology surface of the RGO/SPGE was being studied using field emission scanning electron microscopy (FESEM) and Raman spectroscopy.

D. Electrocatalytic performance of RGO/SPGE sensor

The electrocatalytic performance of the sensor was being studied using cyclic voltammetry (CV) and differential pulse voltammetry (DPV). The CV analysis was being done with different concentration of glucose ranging from 1 mM to 10 mM (scan rate 01 V/s).

III. RESULT AND DISCUSSION

A. Characterization of RGO/SPGE

The electrodeposition of RGO was being done with cyclic voltammetry (CV). Fig. 1 shows the cyclic voltammetry (CV) plot which shows the reduction of graphene oxide (GO) into reduced graphene oxide (RGO). The windows parameters of RGO were chosen from -1.35 V to -2.0 V due to the location of functional group in GO plane and carboxyl functional groups are often at the edge and the other groups such as epoxy and hydroxyl are on the basal

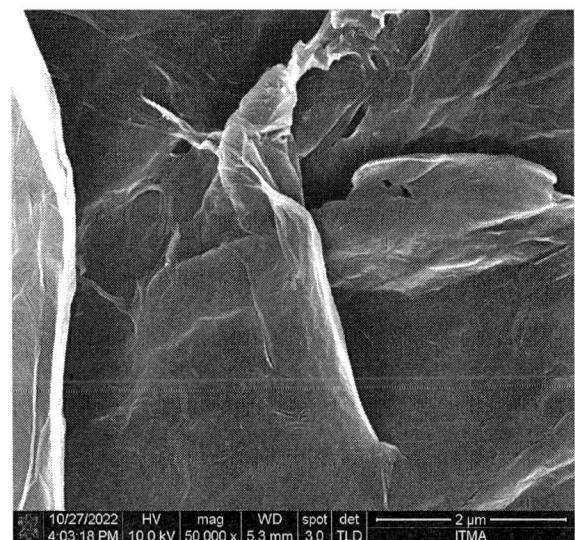

Fig. 2. FESEM image of RGO formation on SPGE.

plane of GO. Most of non-carboxylic groups are removed in potentials smaller than -1.0 V, whereas the carboxylic groups need potentials more negative than -1.0 V. By raising the negative potential value from -1.35 to -2.0 V, large amounts of carboxylic groups are removed from the GO planes [10]. The CV plot shows that the GO had been successfully reduced and form RGO.

The surface morphology for the formation of RGO on the SPGE was being studied using field emission scanning electron microscopy (FESEM). At magnifications ranging from 10x to 300,000x and with a nearly infinite depth of field, field emission scanning electron microscopy (FESEM) delivers topographical and elemental information. Field emission scanning electron microscopy (FESEM) delivers clearer, less electrostatically distorted images with spatial resolution down to 1 1/2 nanometers - three to six times better than conventional scanning electron microscopy (SEM) [11]. The morphological surface of RGO following the electrodeposition process was shown in Fig. 2. In the dry stage, RGO form dense agglomerates with layered structure, and the nanosheets have curved/wrinkled morphology, as seen by the FESEM image. The curved form prevents graphene sheets from stacking on top of one another and adds to the mesoporous characteristic of the material. Mesopores and wrinkles on the surface of graphene sheets may decrease ion diffusion lengths and enable full utilization of graphene nanosheets through better electrolyte ion transport [12].

To further study the morphology of RGO, Raman spectroscopy analysis was employed as displayed in Fig. 3. The major properties of carbon-based materials (the D and G bands), both coming from sp²-hybridized carbon atom vibrations, emerge approximately 1350 cm⁻¹ and 1600 cm⁻¹, respectively. Raman spectroscopy is an effective method for determining the amount of graphene layers and the change in RGO crystal structure. The D band and the G band are the two main bands. The D band is associated with graphene

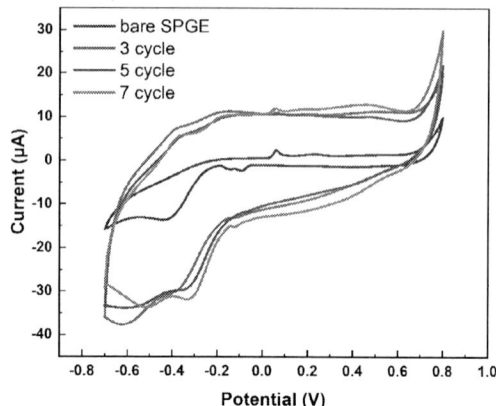

Fig. 4. CV of comparison between bare SPGE and different cycles of RGO.

disorders and imperfections. The G band is the Raman active for sp² hybridized carbon-based material [13]. The ratio of D to G band intensities defines the quality factor of RGO structures [14]. Based on Table 1, the intensity ratio (I_D/I_G) of RGO increase as the number of cycle increase. The increased intensity of the D band in comparison to the G band suggests an increase in the amount of disordered phase in the RGO. The intensity of the D band is greater than that of the G band, which is connected to the production of sp³ hybridized bonds as a result of graphite oxidation [15].

B. Cyclic voltammetry (CV) analysis of RGO on SPGE.

The cyclic voltammetry (CV) plot of bare SPGE and different cycle of RGO/SPGE was being compared in the presence of 5mM of glucose concentration as shown in Fig. 4. It shows that the redox reaction increases in glucose oxidation along with the number of cycles. The increase in redox reaction in the detection of glucose can be related to the change in surface area which may affect the interaction between RGO and glucose [16].

The sensor performance was best to be studied in glucose concentration. Cyclic voltammetry (CV) was performed in different concentration of glucose at a scan rate of 0.1 V/s shown in Fig. 5 for 5 cycles of RGO/SPGE. Based on the cyclic voltammetry (CV) plot, the oxidation peak decrease significantly as the concentration increase (1 mM – 10 mM).

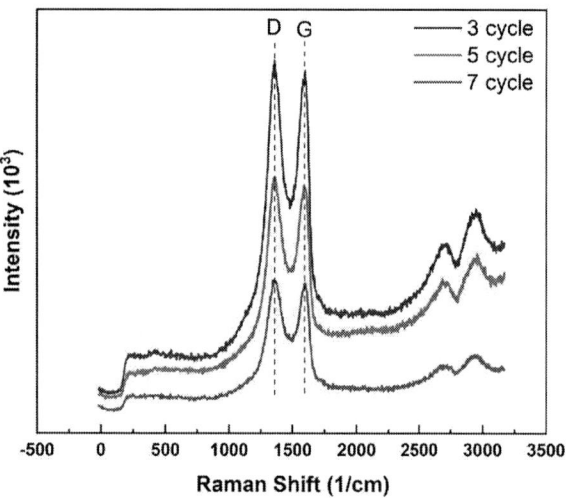

Fig. 3. Raman Spectra of RGO deposited with different cycle on SPGE.

Table 1. Intensity ratio (ID/IG) obtained from Raman Spectra

Number of Cycle	Intensity Ratio (I_D/I_G)
3	1.45
5	1.57
7	1.74

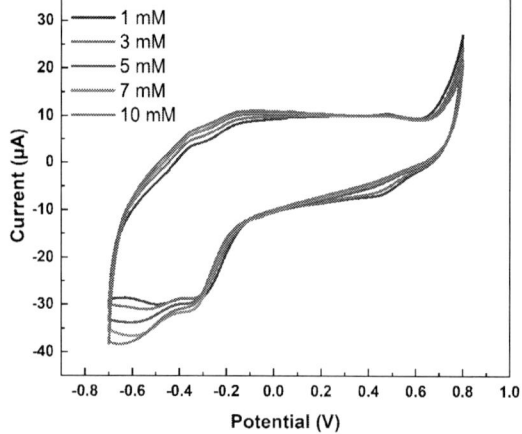

Fig. 5. CV for 5 cycles in different concentration of glucose at a scan rate of 0.1 V/s.

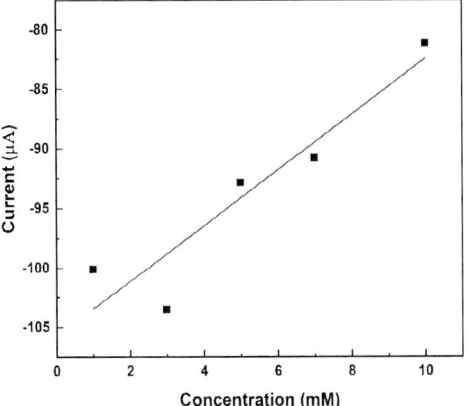

Fig. 6. Linear curve obtained from the CV curve for 5 cycles.

Table 2. Sensitivity calculated from the slope of the linear curve.

Number of Cycle	Sensitivity ($\mu A.mM^{-1} cm^{-2}$)	R^2
3	0.8302	0.1942
5	2.3332	0.8732
7	3.4053	0.7505

The glucose sensor's analytical response was proven by creating a linear curve between oxidation current and glucose concentration (1 mM – 10 mM), as illustrated in Fig. 6. The sensitivity of the sensor calculated from the slope was 2.3332 $\mu A.mM^{-1} cm^{-2}$ ($R^2 = 0.8732$). The sensitivity for other cycle was summarized in Table 2. It was notified that the sensitivity of the sensor increases as the number of cycle increase. This is due to the RGO surface's ability to ionize more electrons from more oxygen-containing functional groups as the number of sensor layers rises, which boosts sensor sensitivity [17].

IV. CONCLUSIONS

For conclusions, the reduced graphene oxide was successfully fabricated on screen-printed gold electrode using electrodeposition method. The electrode was modified with different cycle of RGO which are 3, 5 and 7 cycle. The characterization of the sample was being done using FESEM and Raman spectroscopy. The morphology of RGO shows that the RGO had wrinkle structure. In addition, the characterization of RGO was being done using Raman spectroscopy which shows that the intensity ratio increases as the number of cycles increase. The sample was continued being studied for the detection of glucose. The CV analysis was recorded in 5mM of glucose concentrations. It shows that the oxidation of glucose increases as the number of cycles increase. The sensitivity of the sensor was calculated from the slope of linear curve. The sensitivity increases as the number of cycle increase. It was due to the high oxygen-containing functional groups that increase the sensitivity.

ACKNOWLEDGMENT

The author thanks the MIMOS Berhad and Faculty of Applied Science for their hospitality and help in completing this work.

REFERENCES

[1] S. Akhtar, J. A. Nasir, A. Ali, M. Asghar, R. Majeed, and A. Sarwar, "Prevalence of type-2 diabetes and prediabetes in Malaysia: A systematic review and meta-analysis," *PLoS One*, vol. 17, no. 1 January, Jan. 2022, doi: 10.1371/journal.pone.0263139.

[2] K. L. Leong, M. Y. Ho, X. Y. Lee, and M. S. L. Yee, "A Review on the Development of Non-Enzymatic Glucose Sensor Based on Graphene-Based Nanocomposites," *Nano*, vol. 15, no. 11. World Scientific, Nov. 01, 2020. doi: 10.1142/S1793292020300042.

[3] J. Baranwal, B. Barse, G. Gatto, G. Broncova, and A. Kumar, "Electrochemical Sensors and Their Applications: A Review," *Chemosensors*, vol. 10, no. 9. MDPI, Sep. 01, 2022. doi: 10.3390/chemosensors10090363.

[4] D. Kadadou *et al.*, "Optimization of an rGO-based biosensor for the sensitive detection of bovine serum albumin: Effect of electric field on detection capability," *Chemosphere*, vol. 301, Aug. 2022, doi: 10.1016/j.chemosphere.2022.134700.

[5] Zainor F.A., Zoolfakar A.S., Rani R.A., Syono M.I., Manut A., Mamat M.H., Zain A.M., Omar N. (2019) Proceedings of the 2019 IEEE Regional Symposium on Micro and Nanoelectronics, RSM 2019, art. no. 8943558, pp. 83 – 86. DOI: 10.1109/RSM46715.2019.8943558

[6] Khairudin N., Ilias M.H., Rani R.A., Ahmad M.Z., Manut A., Burham N., Nour M., Zoolfakar A.S., (2022) International Journal of Integrated Engineering, 14 (3), pp. 215 – 228, DOI: 10.30880/ijie.2022.14.03.024

[7] Mohd M.A.H., Rani R.A., Omar N., Zain A.M., Zolkapli M., Manut A., Khairudin N., Burham N., Mamat M.H., Zoolfakar A.S., (2020) IEEE International Conference on Semiconductor Electronics, Proceedings, ICSE, 2020-July, art. no. 9166882, pp. 108 – 111, DOI: 10.1109/ICSE49846.2020.9166882

[8] M. Li, X. Bo, Z. Mu, Y. Zhang, and L. Guo, "Electrodeposition of nickel oxide and platinum nanoparticles on electrochemically reduced graphene oxide film as a nonenzymatic glucose sensor," *Sens Actuators B Chem*, vol. 192, pp. 261–268, 2014, doi: 10.1016/j.snb.2013.10.140.

[9] S. Paul, "Nanomaterials synthesis by electrodeposition techniques for high-energetic electrodes in fuel cell," *Nanomaterials and Energy*, vol. 4, no. 1. ICE Publishing, pp. 80–89, Jan. 01, 2015. doi: 10.1680/nme.14.00031.

[10] A. A. Sehat, A. A. Khodadadi, F. Shemirani, and Y. Mortazavi, "Fast Immobilization of Glucose Oxidase on Graphene Oxide for Highly Sensitive Glucose Biosensor Fabrication," 2014. [Online]. Available: www.electrochemsci.org

[11] M. K. Patil, "Synthesis and Characterization of Graphene Oxide, and Reduced Graphene oxide composites with Conducting Polymer," *International Research Journal of Engineering and Technology*, 2021, [Online]. Available: www.irjet.net

[12] Y. Bai, R. B. Rakhi, W. Chen, and H. N. Alshareef, "Effect of pH-induced chemical modification of hydrothermally reduced graphene oxide on supercapacitor performance," *J Power Sources*, vol. 233, pp. 313–319, 2013, doi: 10.1016/j.jpowsour.2013.01.122.

[13] D. Ickecan, R. Zan, and S. Nezir, "Eco-Friendly Synthesis and Characterization of Reduced Graphene Oxide," in *Journal of Physics: Conference Series*, Institute of Physics Publishing, Oct. 2017. doi: 10.1088/1742-6596/902/1/012027.

[14] M. S. Roslan, K. T. Chaudary, Z. Haider, A. F. M. Zin, and J. Ali, "Effect of magnetic field on carbon nanotubes and graphene structure synthesized at low pressure via arc discharge process," in *AIP Conference Proceedings*, American Institute of Physics Inc., Mar. 2017. doi: 10.1063/1.4978843.

[15] R. Muzyka, S. Drewniak, T. Pustelny, M. Chrubasik, and G. Gryglewicz, "Characterization of graphite oxide and reduced graphene oxide obtained from different graphite precursors and oxidized by different methods using Raman spectroscopy," *Materials*, vol. 11, no. 7, Jun. 2018, doi: 10.3390/ma11071050.

[16] E. Gacka, Ł. Majchrzycki, B. Marciniak, and A. Lewandowska-Andralojc, "Effect of graphene oxide flakes size and number of layers on photocatalytic hydrogen production," *Sci Rep*, vol. 11, no. 1, Dec. 2021, doi: 10.1038/s41598-021-95464-y.

[17] A. Al-Hamry *et al.*, "Layer-by-Layer Deposited Multi-Modal PDAC/rGO Composite-Based Sensors," *Foods*, vol. 12, no. 2, Jan. 2023, doi: 10.3390/foods12

2023 IEEE Regional Symposium on Micro and Nanoelectronics (RSM)

Characterization and Optimization of Ion-Sensitive Field Effect Transistor (ISFET) with Different Gate Dielectric and Thickness

Suhana Mohamed Sultan
Faculty of Electrical Engineering
University Teknologi Malaysia
Johor, Malaysia
suhanasultan@utm.my

Jason Kong Kai Seng
Faculty of Electrical Engineering
University Teknologi Malaysia
Johor, Malaysia
jasonkongkai@graduate.utm.my

Abstract— **The ion-sensitive field effect transistor (ISFET) is one of the emerging chemical sensors with advantages in miniaturization, low-cost manufacture and short response time. Previous studies primarily focused on the types of gate dielectric materials of ISFET and their respective sensitivity in terms of gate voltage. This paper will analyze the effect of the gate dielectric materials and their respective thickness on pH sensing. To achieve this, an ISFET model is developed using COMSOL Multiphysics software. Gate dielectric plays a significant role as sensing film, which has a binding site to detect the ions present in the electrolyte solution. Therefore, gate dielectric materials and their thickness are critical in determining the sensing performance of ISFET. In this study, Y_2O_3, Ta_2O_5, HfO_2 and TiO_2 with 20 nm, 30 nm and 40 nm thicknesses are used as gate dielectric materials. The simulation uses an electrolyte solution with pH values of 3, 7 and 11. The sensitivity of ISFET is determined in terms of the reference voltage and drain current. The ISFET with the highest voltage sensitivity is obtained with TiO_2 as the gate dielectric material at 20nm thickness, which shows a sensitivity of 56.25 mV/pH and 22.62 mA/pH, respectively.**

Keywords—ISFET, pH Sensor, dielectric material, thickness, sensitivity of ISFET

I. INTRODUCTION

The pH measurement in solutions is crucial as the change in pH values is a reliable indicator to track the chemical processes for biomedical and chemical applications. The most common pH measurement method is through the pH glass electrode, which consists of a doped glass membrane that is sensitive to hydrogen. However, the glass electrode has limitations such as low resistance to high temperatures, low durability, complex manufacturing and miniaturization problem [1]. The Ion-Sensitive Field-Effect Transistor (ISFET) was first introduced in 1970. It has obtained much attention for decades due to its outstanding advantages over existing biosensing technologies in sensitivity, production cost, size and lab-on-a-chip integration [2].

ISFET has a structure similar to conventional Metal-Oxide-Semiconductor Field-Effect Transistor (MOSFET) and has three terminals: a source, drain and gate. However, the gate metal of ISFET is replaced by the electrolyte solution with the reference electrode for sensing applications. The gate dielectric acts as the sensing film to interact with

the hydrogen ions in the electrolyte solution to form an electrolyte-insulator interface. The change in the concentration of hydrogen ions in the electrolyte solution will change the pH values. The charge distribution at the surface of the gate insulator will change as more hydrogen ions are available for binding. The surface charge density of the gate insulator will change, and the changes can be reflected in the threshold voltage and the drain current of the ISFET.

The high sensitivity of ISFET has significantly contributed to the use of biosensors. Many attempts have been made to improve the sensitivity of the ISFET close to the Nernst limit. The sensitivity of the ISFET can be determined by relating the potential drop at the gate insulator surface to the changes in the pH of the electrolyte solution [1]. The surface potential at the gate dielectric surface, known as threshold voltage, can compensate for the potential drop at the electrolyte-insulator interface. Therefore, the change in reference voltage at fixed drain current per pH unit change is determined. Fig. 1 illustrates the ISFET structure, which consists of a source, drain, gate insulator and reference electrode.

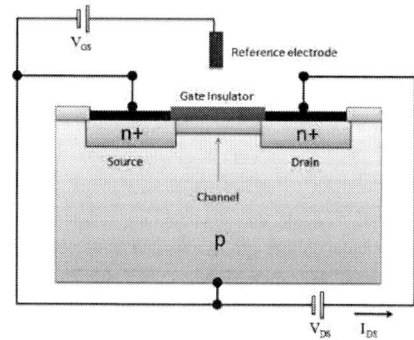

Fig. 1. An ISFET structure which consists of source, drain, gate insulator and reference electrode [3].

In this work, the ISFET structure is modelled using COMSOL Multiphysics using Y_2O_3, Ta_2O_5, HfO_2 and TiO_2 as gate dielectric materials with thickness of 20 nm, 30 nm and 40 nm. The electrical characteristics of ISFET at pH values of 3, 7 and 11 are determined. The sensitivity in terms of reference voltage of the ISFET for each case will be measured. The ISFET with the most optimized gate

979-8-3503-2369-6/23 $31.00 © 2023 IEEE

dielectric materials and its respective thickness will be determined and discussed. The electrical signal captured corresponds to the hydrogen ions concentration in the electrolyte solution will determine the pH values of the electrolyte solution.

II. METHODOLOGY

The ISFET will be modelled using the COMSOL Multiphysics, a simulation platform with fully coupled capabilities for single-physics and multiphysics modelling. After developing the ISFET model, the simulation is conducted by using Y_2O_3, Ta_2O_5, HfO_2 and TiO_2 as gate dielectric materials with thickness of 20 nm, 30 nm and 40 nm. The parameters used to simulate the ISFET are listed in Table I. The relative permittivity for each type of dielectric material is known from the previous study [4]. The wider thickness range of the gate dielectric materials are used by referring to the previous study. The electrical characteristics of ISFET are obtained and the sensitivity of ISFET is extracted.

The 2D geometry of the ISFET is divided into 2 dimensions, which are substrate dimension and electrolyte dimension. For substrate dimension, 3 μm of length is set in the X-axis and 0.7 μm of height is set in the Y-axis. For electrolyte dimension, the length is set to 3 μm and 1.6 μm in the upper X-axis and lower X-axis respectively. The height is set to 1 μm in the Y-axis for electrolyte dimension. The model for the ISFET structure and the mesh applied to the model are illustrated in Fig. 2(a) and Fig. 2(b), respectively. The simulation is executed to generate the transfer characteristics graph of the ISFET at pH = 3, 7 and 11. The sensitivity of the ISFET for each case is determined based on the graphs in the computation of COMSOL Multiphysics.

TABLE I.   PARAMETERS FOR ISFET SIMULATION

Parameters	Values
T	298.15 K
L	1.6 μm
N_A	1×10^{17} cm^{-3}
K_B	1×10^{5} mol m^{-3}
N_D	1×10^{20} cm^{-3}
K_A	0.001 mol m^{-3}

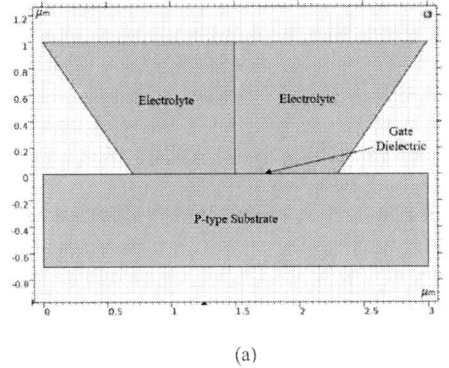

(a)

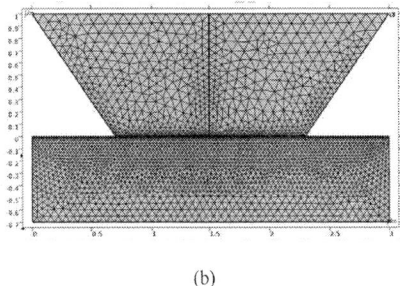

(b)

Fig. 2. (a) ISFET model (b) Mesh applied in the ISFET model.

III. RESULTS AND DISCUSSION

The ISFET is modified from the structure of the conventional MOSFET at which the metal gate is replaced by the electrolyte solution with reference electrode. The reference electrode used is normally silver electrode submerged in the electrolyte solution. The gate insulator that is exposed to the electrolyte solution acts as sensing film that provides the binding site to detect the ions. In pH detection, the concentration of hydrogen ions, H^+ and hydroxide ions, OH^- are the indicator of the pH values of the solution. The pH of the solution can be determine using the expression [5]:

$$pH = -\log[H_3O^+] \qquad (1)$$

where H_3O^+ is the concentration of hydronium ions in the solution. The gate insulators are metal oxides that have hydroxyl groups which can react with the hydrogen ions (protons). This causes the protonation and deprotonation to happen at the surface of gate oxides[5]. The surface charge density of gate oxides will change as the concentration of hydrogen ions varies. The change in the surface charge density of gate oxides can be determined from the values of threshold voltage and the drain current of ISFET. The pH of the solution is measured in the form of threshold voltage that is induced electrochemically. According to Guoy-Chapmen-Stern theory, the double layer which is called the Guoy-Chapmen layer will form at the region near the surface of gate oxides and act like capacitor. The threshold voltage of ISFET, unlike MOSFET, is influenced by the electrolyte-insulator-semiconductor interface. Therefore, the expression for the threshold voltage for ISFET is [6]:

$$V_T = E_{Ref} - \psi_o + x_{sol} - (\phi_{si}/q) - (Q_{SS} + Q_{ox} + Q_B)/C_{ox} + 2\phi_F \qquad (2)$$

where reference electrode potential relative to vacuum is indicate by the E_{Ref} and ψ_o is the potential drop across the electrolyte-insulator interface. The x_{sol} and ϕ_{si} indicates the surface dipole potential of the solution and the work function of silicon respectively. The Q_{SS} is the surface state density while the Q_{ox} is the oxide charge of the ISFET. The Q_B is the depletion charge in the silicon and ϕ_F is the Fermi potential. All terms are constant in nature except for the ψ_o which is the surface potential that varies to make ISFET sensitive to the pH of the electrolyte. On the other hand, the drain current of ISFET is similar to MOSFET in unsaturated region which can be determined by the expression [8].

$$I_{DS} = (C_o\mu_n W)/L[(V_{gs} - V_{th})V_{ds} - 1/2 (V_{ds})^2] \qquad (3)$$

979-8-3503-2369-6/23 $31.00 © 2023 IEEE

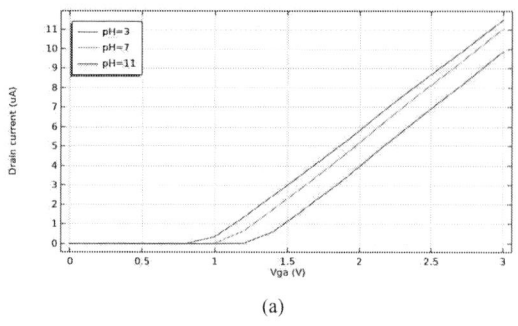
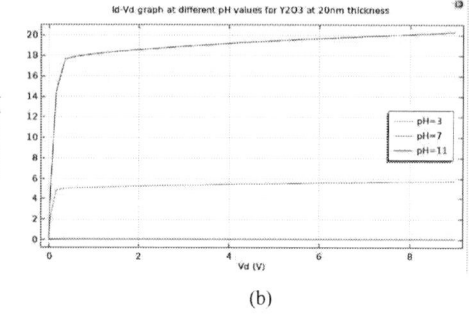

(a)   (b)

Fig. 3. (a) Id-Vg graph with 20 nm Y₂O₃ as gate dielectric materials at different pH (b) Id-Vd graph of 20 nm Y₂O₃ as ISFET gate dielectric materials at different pH

The pH sensing for the ISFET can be observed from the Id-Vg graph for the ISFET with 20 nm Y_2O_3 as gate dielectric materials at different pH in Fig. 3(a). The graph showed that the threshold voltage of ISFET will increase as the pH values of the electrolyte solution increases (become more alkaline).

Similar parameters are used to determine the relationship between the drain current and the pH. The drain current of ISFET decreases with increasing pH. The increase in pH indicates that the concentration of hydrogen ions is decreased and the potential drop across the electrolyte-insulator interface also decreases. According to (2), as the potential drop across the electrolyte-insulator interface decreases and becomes more negative, the value of the threshold voltage increases. Based on (3), the increase in threshold voltage will decrease the drain current in the ISFET. The Id-Vd graph of ISFET is shown in Fig. 3(b). The sensitivity of ISFET can be expressed with the change in threshold voltage per pH at fixed drain current. The expression for the voltage sensitivity can be extracted using the expressions [8]:

$$\text{Voltage sensitivity, } S = | \delta V_{ref} / \delta pH | \qquad (4)$$

The 20 nm gate dielectric Y_2O_3 is changed to other materials such as Ta_2O_5, HfO_2 and TiO_2 to investigate the sensitivity of ISFET with different gate dielectric materials. The results shown in Fig. 4 demonstrated that the TiO_2 has the highest voltage sensitivity at 56.25 mV/pH. The sensitivity in terms of change in threshold voltage in ISFET is close to the Nernst's limit of 59.2 mV/pH. The results showed that the high-k dielectric material can provide higher sensitivity for ISFET in pH sensing. The high-k gate dielectric can reduce leakage current and increase the gate capacitance [9]. More charge can be stored and more protonation and deprotonation can occur at the surface of the gate dielectric. The change in the concentration of hydrogen ions (protons) can be detected in a shorter time.

The effect of thickness of each gate dielectric materials on the performance of the ISFET is also analyzed. Fig. 5 shows that the voltage sensitivity of ISFET increases with decreasing gate dielectric thickness regardless of which gate dielectric materials are used. All the gate dielectric materials will have the highest voltage sensitivity at 20 nm. The high-k dielectric materials used can suppress the current leakage which commonly occurs when the gate dielectric materials used are too thin. Table II shows the overall tabulated data for the effects of varying gate dielectric materials and their thickness. The voltage

sensitivity for each case have been determined and recorded.

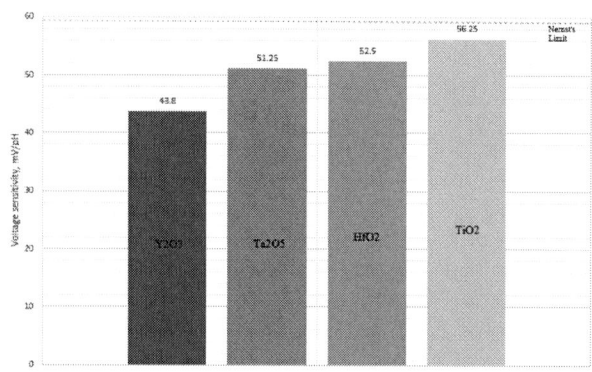

Fig. 4. Voltage sensitivity of ISFET using different 20 nm gate dielectric materials.

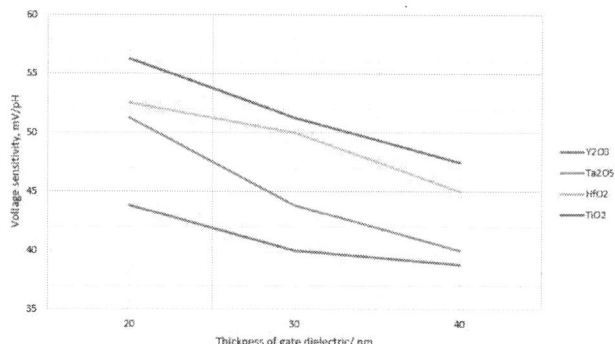

Fig. 5. Voltage sensitivity of different gate dielectric materials with different thickness

As compared to the previous study, the types of gate dielectric materials used in this work are different such as SiO_2, Si_3N_4, Ta_2O_5 and Al_2O_3 except for the Ta_2O_5, which is also included in the simulation [5]. Therefore, Ta_2O_5 is used as the benchmark to determine the reliability of the study. Table III shows the percentage difference from the previous study compared to this work. The current study showed that the 20 nm Ta_2O_5 can reach the voltage sensitivity of 51.25 mV/pH, which has 11.94% of percentage difference compared to the study done by Tiong Chi Ye et al[5]. The difference in the result is caused by the different parameters chose in the simulation. The previous study did not specify the parameters set in the simulation except for the types of the materials used. Therefore, several parameters involved in

979-8-3503-2369-6/23 $31.00 © 2023 IEEE

the simulation of this study are taken from other research studies.

TABLE II. VOLTAGE SENSITIVITY OF ISFET FOR DIFFERENT GATE DIELECTRIC MATERIALS AND THICKNESS.

Gate dielectric materials	Relative permittivity	Gate dielectric thickness, nm	Voltage sensitivity, mV/pH
Y_2O_3	15	20	≈ 43.8
		30	≈ 40
		40	≈ 38.8
Ta_2O_5	22	20	≈ 51.25
		30	≈ 43.8
		40	≈ 40
HfO_2	25	20	≈ 52.5
		30	≈ 50
		40	≈ 45
TiO_2	80	20	≈ 56.25
		30	≈ 51.25
		40	≈ 47.5

TABLE III. THE COMPARISON BETWEEN THE PREVIOUS STUDY AND THIS WORK.

Gate dielectric materials	Previous Study	This Work		Percentage difference
	Voltage sensitivity, mV/pH	Voltage sensitivity, mV/pH	Current sensitivity, $\mu A/pH$	
SiO_2	44.2 [5]	-	-	-
Si_3N_4	57 [5]	-	-	-
Al_2O_3	52.3 [5]	-	-	-
Ta_2O_5	58.2 [5]	51.25	5.19	11.94%
Y_2O_3	-	43.8	2.36	-
HfO_2	-	52.5	6.37	-
TiO_2	-	56.25	22.62	-

IV. CONCLUSION

The ISFET model has been successfully developed using COMSOL Multiphysics. The pH sensing using ISFET has been demonstrated. The ISFET with 20 nm TiO_2 as gate dielectric material has showed the highest voltage sensitivity at 56.25 mV/pH. The results have shown that the performance of the ISFET can be improved by using the thinner high-k dielectric material such as TiO_2.

ACKNOWLEDGMENT

The authors would like to acknowledge the support from the Faculty of Electrical Engineering, Universiti Teknologi Malaysia.

REFERENCES

[1] M. Baylav, "Ion-Sensitive Field Effect Transistor (ISFET) for MEMS Multisensory Chips at RIT." Order No. 1482950, Rochester Institute of Technology, United States -- New York, 2010.

[2] N. Choksi, D. Sewake, S. Sinha, R. Mukhiya, and R. Sharma, "Modeling and simulation of ion-sensitive field-effect transistor using TCAD methodology," IEEE Xplore, Apr. 01, 2017. https://ieeexplore.ieee.org/abstract/document/8076935 (accessed May 25, 2023).

[3] R. Bhardwaj et al., "Temperature compensation of ISFET based pH sensor using artificial neural networks," IEEE Xplore, Aug. 01, 2017. https://ieeexplore.ieee.org/abstract/document/8069141 (accessed May 25, 2023).

[4] J. Robertson, "High dielectric constant oxides," The European Physical Journal Applied Physics, vol. 28, no. 3, pp. 265–291, Dec. 2004, doi: 10.1051/epjap:2004206.

[5] T. C. Ye, S. F. Wan. M. Hatta, N. Soin, and A. A. Bakar Sajak, "Modeling and optimization of ISFET microsensor for the use of quality testing in food and pharmaceutical industry," IEEE Xplore, Aug. 01, 2021. https://ieeexplore.ieee.org/abstract/document/9511588 (accessed May 26, 2023).

[6] P. K. Chan and D. Y. Chen, "A CMOS ISFET Interface Circuit With Dynamic Current Temperature Compensation Technique," IEEE Transactions on Circuits and Systems I: Regular Papers, vol. 54, no. 1, pp. 119–129, Jan. 2007, doi: https://doi.org/10.1109/TCSI.2006.887977.

[7] C.-S. Lee, S. K. Kim, and M. Kim, "Ion-Sensitive Field-Effect Transistor for Biological Sensing," Sensors (Basel, Switzerland), vol. 9, no. 9, pp. 7111–7131, Sep. 2009, doi: https://doi.org/10.3390/s90907111.

[8] T. M. Abdolkader et al., "ISFET pH-Sensor Sensitivity Extraction Using Conventional MOSFET Simulation Tools," International Journal of Chemical Engineering and Applications, vol. 6, no. 5, pp. 346–351, 2015, Accessed: May 26, 2023. [Online]. Available: https://www.academia.edu/44366601/ISFET_pH_Sensor_Sensitivity_Extraction_Using_Conventional_MOSFET_Simulation_Tools

[9] T. M. Abdolkader and A. G. Alahdal, "Performance optimization of single-layer and double-layer high-k gate nanoscale ion-sensitive field-effect transistors," Sensors and Actuators B: Chemical, vol. 259, pp. 36–43, Apr. 2018, doi: https://doi.org/10.1016/j.snb.2017.12.0

Trade-offs and Optimization: Low Power Approaches for Area, Power Consumption, and Performance in Microprocessor Design

Nur Mahirah Sallehuddin,
School of Electrical
Engineering,
College of Engineering,
Universiti Teknologi MARA,
40450 Shah Alam
Selangor, Malaysia
nmahirahsalleh@gmail.com

Maizan Muhamad*,
Integrated Microelectronic
System and Application,
School of Electrical
Engineering,
College of Engineering,
Universiti Teknologi MARA,
40450 Shah Alam
Selangor, Malaysia
*maizan@uitm.edu.my

Hanim Hussin,
Integrated Microelectronic
System and Application,
School of Electrical
Engineering,
College of Engineering,
Universiti Teknologi MARA,
40450 Shah Alam
Selangor, Malaysia
hanimh@uitm.edu.my

Abdul Karimi Halim,
School of Electrical Engineering,
College of Engineering,
Universiti Teknologi MARA, 40450 Shah Alam
Selangor, Malaysia

Yasmin Abdul Wahab
Nanotechnology and Catalysis Research Centre,
Universiti Malaya,
50603 Kuala Lumpur, Malaysia
yasminaw@um.edu.my

Abstract— **This research paper presents the findings of implementing low-power techniques on RISC V microprocessors using 90 nm technology. The power consumption of smaller microprocessors is a concern, despite their advantages of reduced chip size and higher operating frequency. Low-power microprocessors often involve trade-offs with size and performance constraints. The objective of this project is to apply three low-power techniques - clock gating, multi-voltage, and multi-threshold approaches - and evaluate the resulting optimizations in timing, area, and power. The design is simulated and synthesized using Intel Quartus Prime Lite, Synopsys VCS, and Design Compiler tools. The synthesis with different compile options shows varied output values in terms of timing, area, and power consumption compared to the initial design without low-power techniques. The study concludes that setup timing slacks increase by 141%, while multi-voltage and multi-threshold approaches significantly reduce overall power consumption and area. Clock gating demonstrates the highest optimization in timing, area, and performance and is recommended as a standalone technique.**

Keywords— **RISC V microprocessor, PicoRV32, low power, clock gating, multi-voltage, multi-threshold, 90 nm technology**

I. INTRODUCTION

Smaller transistors provide advantages such as reduced area and increased chip operating frequency[1-2]. These advantages would help embedded systems that prioritize size in their design. However, in this scenario, increased transistor density raises supply currents and current density, resulting in increased power consumption [3]. This is an undesirable situation because the power budget is one of the most important aspects that microprocessor designers are concerned about, as it also represents the reliability and safety of the design[4].

Dynamic power, static power, and short circuit power are all components of overall power usage[5]. However, with significantly smaller circuits, leakage power dominates dynamic power and results in high power consumption of a CPU[6]. Aside from that, supply voltage and toggling frequency have an impact on dynamic power, which is why methods that minimize net capacitance and allow the system to function at lower voltage and frequency are required[7].

When efforts are made to minimize power use, the opposite occurs. Because it eventually increases the area and timing slacks, low power consumption raises the trade-offs between the other constraints of chip design, which are area and performance. As a result, low-power techniques are being deployed because they minimize these trade-offs while simultaneously decreasing power consumption to the required power standard. Approximate low-power techniques might be useful in this project to reduce area, power consumption, and timing slacks in microprocessor design, resulting in higher energy efficiency performance[8].

The goal is to reduce power by implementing three low-power techniques: clock gating, multi-voltage, and multi-threshold approaches and optimizing the area and performance. In this project, a 90 nm technology library is used to synthesize the design of PicoRV32 which is a size-optimized 32-bit RISCV CPU. The Verilog design and testbench files of the microprocessor are obtained through the GitHub website.

II. LITERATURE REVIEW

The rapid advancement of technology has brought forth challenges such as larger processors, complex designs, faster operating times, and increased power consumption, making power efficiency a top priority in VLSI design [9-10]. To address this, low-power ALU components are developed, and commonly used techniques like clock gating and multi-threshold are implemented in VLSI to save power [11]. Clock gating reduces dynamic power by disabling unused circuitry and minimizing unnecessary state changes, resulting in significant power reductions [12-13]. Meanwhile, static power from leakage current is addressed through sleep mode and wake-up techniques enabled by clock gating, preventing excessive power consumption [14-15]. The combination of clock gating and multi-threshold techniques has shown notable power savings in various processors, emphasizing their effectiveness [16]. Moreover, approaches like multi-voltage partitioning and dynamic voltage scaling further contribute to power reduction by optimizing power supply voltage and minimizing dynamic power [17]. Multi-threshold CMOS is widely employed, leveraging cells with different threshold voltages to optimize power consumption on critical and non-critical paths, significantly reducing leakage power and overall power consumption [18-22].

III. RESEARCH METHODOLOGY

The implementation of low-power techniques in this project follows an ASIC design flow, including design simulation, functional simulation, logic synthesis, and results analysis. Initially, knowledge acquisition is conducted by researching relevant topics such as RISC V and low-power techniques with 90 nm technology. Major issues in microprocessor designs, like high power consumption and large area, are identified, along with suggested solutions.

Next, the PicoRV32 design is verified through design simulation using ModelSim, where the basic timing functionality is validated. Simulation and synthesis are completed to verify the design's performance without a physical FPGA board.

Functional simulation is then performed using Synopsys VCS, ensuring the gate-level simulation of the design. Various test cases are employed to validate the functionality, and the results are compared with the previous design simulation in ModelSim.

The logic design synthesis of the PicoRV32 design takes place in Synopsys Design Compiler, utilizing a 90 nm technology library with low-power techniques. Techniques such as clock gating, multi-voltage, and multi-threshold are implemented, aiming for reduced power consumption, improved performance, and smaller chip area. Different compile options are explored to assess the effects of these techniques on the design.

The multi-threshold method is implemented using various DC setup files, while clock gating cells are included using specific clock gating technique implementation files. The Unified Power Format (UPF) is employed to divide the design into voltage zones or power domains, each with its power supply. Compile options like compile ultra and compile with high power effort are used in combination with multi-voltage and single library approaches, respectively. The UPF implementation allows for power domain definition and the expression of power purposes in the design.

Different commands and options are used for synthesis based on the specific compile option, ensuring the correct implementation of low-power approaches. The findings of various compile options are compared to avoid trade-offs while reducing power consumption.

After completing each of the different compilations, the timing slacks, total area, and power consumption are recorded. The graphs generated from the data that have been obtained and recorded are used for the analysis process. The results are examined whether they meet the theoretical concept of the low-power techniques implementation.

IV. RESULTS AND DISCUSSION

In the logic design synthesis stage, the synthesis of PicoRV32 using 90 nm technology with low-power implementations is performed. The effects of different compile options with low-power techniques are analyzed by comparing schematic diagrams, setup and hold timing slacks, total areas, and total power consumption values. The results of each compile option are compared to evaluate the impact of low-power techniques.

Regarding setup timing slacks, it is observed that the design generally meets the maximum timing slacks at a clock period of 40 ns for both multi-threshold and single library compilations. As the clock period increases, the timing slacks decrease. However, the compile ultra option with multi-voltage and multi-threshold only meets the maximum timing slacks at a clock period of 110 ns due to the optimized netlist area. The design's high transistor density at 90 nm technology contributes to the relatively high clock periods required to meet the maximum timing slacks.

Comparing compile ultra option with multi-voltage and single library compile in Fig 1, the maximum timing slacks are lower by approximately 16% to 67% compared to the same option under multi-threshold compile.

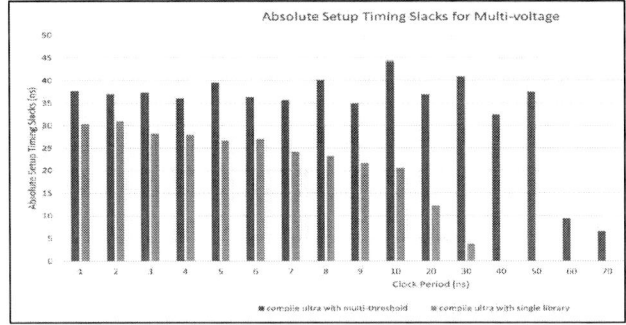

Fig. 1 Absolute Setup Timing Slacks for Multi-voltage

The hold timing slacks of the design are not affected by the different compile options as can be observed in Fig. 2. The timing values remain almost constant with small relative differences. All timing slacks within the clock period of 1 to

120 ns are successfully met without any violation. The maximum hold timing slack recorded is less than 2 ns, which is considered acceptable.

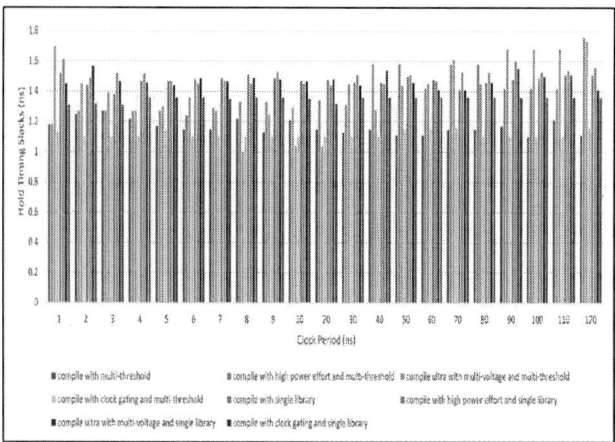

Fig. 2 Hold Timing Slacks for Multi-Threshold and Single Library Compile

Fig. 3 shows the overall area average of the design synthesis for multi-threshold and single library compile. The design synthesized under multi-threshold compile demonstrates higher area optimization, with a 16% reduction in total area when compiled with the compile ultra option compared to a normal compile synthesis. The total areas of the design compiled with clock gating are the lowest for both multi-threshold and single-library compile, showing area reductions of 26% and 20%, respectively, compared to other low-power techniques.

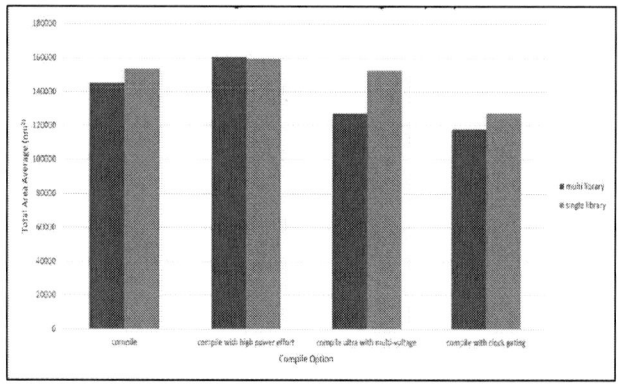

Fig. 3 Total Area Average for Multi-threshold and Single Library Compile

Fig. 4(a) indicates that around 80% to 95% of the power consumption is contributed by leakage power, attributed to the small size of the transistor. However, by implementing high power effort with multi-threshold compile, the leakage power can be minimized to as low as 2% of the total power. The effect of the clock gating approach is not as evident for total power outputs under multi-threshold compile, but Fig. 5 shows that clock gating manages to lower around 22% of dynamic power in the design under a single library compile. The comparison of setup timing slacks with leakage power under multi-threshold compile confirms the theoretical concept of the technique, as

an increase in setup timing slacks results in a decrease in leakage power.

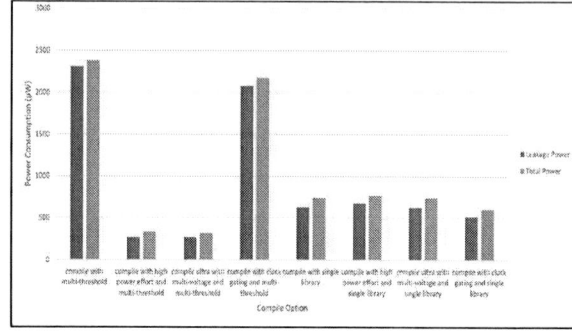

(a)

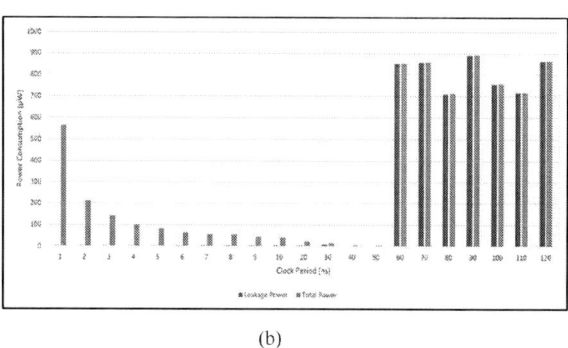

(b)

Fig. 4 (a) Leakage Power and Total Power Consumption of Multi-threshold and Single Library Compile, (b) Leakage Power and Total Power Consumption of Compile with High Power Effort and Multi-threshold

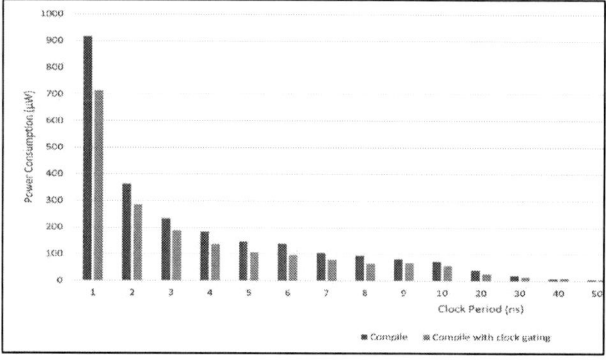

Fig. 5 Dynamic Power of Single Library for Compile and Compile with Clock Gating

Table 1 presents the average optimization percentages for setup and holds timing slacks, total area, and power consumption of a RISC V microprocessor implemented in 90 nm technology using various low-power strategies. The study highlights the consequences of low-power technology implementation on timing slacks, which are crucial for microprocessor performance but often overlooked in previous research. Although the multi-voltage with multi-threshold approach significantly reduces total power consumption, it results in a 141% increase in setup timing slacks. The clock gating approach stands out as the best option in this project, offering high optimization percentages in timing slacks, total area, and power consumption. The PicoRV32 implementation on 90 nm technology is more efficient than previous studies and achieves a higher percentage of power savings while maintaining similar power optimization levels. To further increase the optimization percentage for the area- and power-prioritized designs, the multi-voltage, and multi-approach can be used together, resulting in reduced power consumption and a 15.56% smaller area compared to other studies.

Table 1 Average Optimization Percentage on Timing Slacks, Total Area, and Power Consumption by Different Low Power Techniques

No.	Low Power Techniques	Average Optimization Percentage (%)					
		Setup Timing Slacks	Hold Timing Slacks	Total Area	Power Consumption		
					Dynamic Power	Leakage Power	Total Power Consumption
1	Clock gating	33.05	8.36	16.89	14.84	16.53	16.93
2	Multi-voltage	2.32	-0.05	0.49 *	-0.29	0.01	0.09
3	Multi-threshold	22.25	20.83	4.75	31.01	-300.76	-275.23
4	Clock gating and multi-threshold	-2.09	17.51	6.89	-27.94	-340.53	-317.75
5	Multi-voltage and multi-threshold	-141	4.53	15.56	48.40	61.59	63.15

V. CONCLUSIONS

Low-power methods are effective in addressing the challenges of chip design, specifically in optimizing area, power consumption, and performance. However, there are limitations in achieving simultaneous optimization of power, area, and speed. The decision on which constraint to prioritize depends on the specific application of the design. In this project, the focus was on reducing power dissipation while maintaining acceptable areas and speed performance. The findings indicate that clock gating can lead to improvements in timing, area, and power consumption. Combining multi-voltage and multi-threshold approaches may not fully optimize all constraints.

ACKNOWLEDGMENT

The author would like to thank the College of Engineering, UiTM for the funding for this paper and research work.

REFERENCES

[1] R. K. Cavin, P. Lugli, and V. v. Zhirnov, "Science and engineering beyond Moore's law," in *Proceedings of the IEEE*, May 2012, vol. 100, no. SPL CONTENT, pp. 1720–1749.

[2] D. Reusch, J. Strydom, and A. Lidow, "A new family of GaN transistors for highly efficient high-frequency DC-DC converters," in *IEEE Applied Power Electronics Conference and Exposition (APEC)*, 2015, pp. 1979–1985.

[3] D. Paramesh Kumar and D. Raja Ramesh, "IR Drop and Electro Migration Reduction Techniques in Deep Sub-Micron technologies," *International Journal of Engineering Research in Electronics and Communication Engineering (IJERECE)*, vol. 3, no. 8, pp. 318–327, 2016.

[4] S. Kumar and G. Hiremath, "Low power implementation of risk-v processor," *IOSR Journal of VLSI and Signal Processing (IOSR-JVSP)*, vol. 6, no. 3, pp. 59–64, 2016.

[5] M. Moradinezhad Maryan, M. Amini-Valashani, and S. J. Azhari, "A New Circuit-Level Technique for Leakage and Short-Circuit Power Reduction of Static Logic Gates in 22-nm CMOS Technology," *Circuits, Systems, and Signal Processing*, vol. 40, no. 7, pp. 3536–3560, Jul. 2021.

[6] S. Sreevidya, R. Holla, and R. Raghu, "Low Power Physical Design and Verification in 16nm FinFET Technology," in *2019 3rd International Conference on Electronics, Communication and Aerospace Technology (ICECA)*, 2019.

[7] R. Chadha and J.Bhasker, *An ASIC low power primer: analysis, techniques, and specification.* Springer Science & Business Media, 2012.

[8] C. Liu, J. Han, and F. Lombardi, "A low-power, high-performance approximate multiplier with configurable partial error recovery," in *2014 Design, Automation & Test in Europe Conference & Exhibition (DATE)*, 2014, pp. 1–4.

[9] V. Manuel and M. Jiménez, "Impact of physical low power techniques in a RISC-V processor," Master's thesis, Universitat Polit`ecnica de Catalunya, 2021.

[10] Y. Zhang et al., "Automatic Register Transfer level CAD tool design for advanced clock gating and low power schemes," in *2012 International SoC Design Conference (ISOCC)*, pp. 21–24.

[11] F. bin Muslim, A. Qamar, and L. Lavagno, "Low Power Methodology for an ASIC design flow based on High-Level Synthesis," in *2015 23rd International Conference on Software, Telecommunications and Computer Networks (SoftCOM)*, 2015, pp. 11–15.

[12] S. Sakthikumaran, S. Salivahanan, and V. S. K. Bhaaskaran, "16-Bit RISC processor design for convolution application," in *International Conference on Recent Trends in Information Technology, ICRTIT 2011*, 2011, pp. 394–397.

[13] J. Ravindra and T. Anuadha, "Design of Low Power RISC Processor by Applying Clock gating Technique," *Int. Journal of Engineering Research and Applications (IJERA)*, vol. 2, pp. 94–99, 2012.

[14] P. Bhattacharjee, A. Majumder, and T. D. Das, "A 90 nm leakage control transistor-based clock gating for low power flip flop applications," in *IEEE 59th International Midwest Symposium on Circuits and Systems (MWSCAS)*, Jul. 2016.

[15] G. Pouiklis and G. C. Sirakoulis, "Clock gating methodologies and tools: A survey," *International Journal of Circuit Theory and Applications*, vol. 44, no. 4, pp. 798–816, Apr. 2016.

[16] V. Melikyan, E. Babayan, A. Melikyan, D. Babayan, P. Petrosyan, and E. Mkrtchyan, "Clock gating and multi-VTH low power design methods based on 32/28 nm ORCA processor," in *Proceedings of 2015 IEEE East-West Design and Test Symposium, EWDTS*, 2016, pp. 1–4.

[17] H. K. Mondal, S. H. Gade, R. Kishore, and S. Deb, "Adaptive multi-voltage scaling in wireless NoC for high-performance low power applications," in *Design, Automation & Test in Europe Conference & Exhibition (DATE)*, 2016, pp. 1315 1320.

[18] P. Upadhyay, R. Kar, D. Mandal, and S. P. Ghoshal, "A design of low swing and multi-threshold voltage based low power 12T SRAM cell," *Computers and Electrical Engineering*, vol. 45, pp. 108–121, Jul. 2015.

[19] M. Gautam and S. Akashe, "Reduction of leakage current and power in full subtractor using MTCMOS technique," in *Inter. Conference on Computer Communication and Informatics*, 2013, pp. 1–4.

[20] G. Amuthavalli and R. Gunasundari, "Analysis and design of subthreshold leakage power-aware ripple carry adder at circuit-level using 90nm technology," *Proc. Computer Science*, vol. 48, pp. 660–665, 2015.

[21] B. Gupta and S. Nakhate, "Transistor gating: A technique for leakage power reduction in CMOS circuits," *International Journal of emerging technology and Advanced Engineering*, vol. 2, no. 4, pp. 321–326, 2012

[22] P. Kushwah, S. Khandelwal, and S. Akashe. "Multi-threshold voltage CMOS design for low-power half adder circuit," *International Journal of Nanoscience*, vol. 14, no. 5–6, Dec. 2015.

Design and Implementation of 32-bit SDRAM Memory Controller with Optimized Dynamic Power using ASIC

T. Zheng Hong
Faculty of Electrical
Engineering,
Universiti Teknologi Malaysia,
81310 Johor Bahru, Malaysia
zhenghong.5155@gmail.com

N. Ezaila Alias*
Faculty of Electrical
Engineering,
Universiti Teknologi Malaysia,
81310 Johor Bahru, Malaysia
ezaila@fke.utm.my

M.L Peng Tan
Faculty of Electrical
Engineering,
Universiti Teknologi Malaysia,
81310 Johor Bahru, Malaysia
michael@utm.my

Yasmin Abdul Wahab
Nanotechnology & Catalysis
Research Centre,
Universiti Malaya,
50603 Kuala Lumpur, Malaysia
yasminaw@um.edu.my

Abstract — **Dynamic random access memory (DRAM) is one of the four primary technologies used in the memory hierarchies of a computer system. To improve the operational speed of DRAM, synchronous DRAM (SDRAM) is introduced. Besides, a memory controller is required to manage the data flow between the selected application and the SDRAM. However, a high-speed memory controller that can cope with a high-performance processor will dissipate a lot of dynamic power. Hence, this work proposed a way to reduce the dynamic power dissipation of a 32-bit SDRAM controller through the implementation of clock gating. The design was implemented in Application Specific Integrated Circuit (ASIC) in which the clock gating cells were inserted in DC and further optimized in ICC. As compared to the case without clock gating, a 44.7 % reduction in the dynamic power of the memory controller was observed after implementing clock gating at the end of this work. Next, an average register gating efficiency of 61.6 % was achieved while the voltage drop in the power network is 57.1 mV or 2.54 %. Briefly, the results obtained show that clock gating is an effective way to optimize the dynamic power while maintaining the functionality and performance of the memory controller.**

Keywords — memory controller, clock gating, dynamic power, DRAM, SDRAM, ASIC

I. INTRODUCTION

The use of a clock enables SDRAM to eliminate the time for the memory and processor to synchronize. Besides, SDRAM has a speed advantage as it does not require specifying additional address bits to transfer the bits in a burst [1]. The data transfer performed by the SDRAM is considered a single data rate since only one word of data will be transmitted per clock cycle while double data rate (DDR) SDRAM can operate at a faster speed as data transfer can occur at both the rising edge and falling edge of the clock signal [2]. Apart from read and write operations, some general SDRAM commands include Activate (ACT), Pre Charge (PRE), Refresh (REF), and No Operation (NOP) [3]. A memory row should be activated for read and write operations while an inactive memory row is said to be idle. ACT command is used to access a particular row in a memory bank whereas the PRE command can be used to deactivate an opened row. At the end of the read or write operation, the auto pre-charge function can be enabled to deactivate the particular row. Meanwhile, the command inhibits command will prevent the SDRAM from executing any command regardless of the status of the clock signal while the NOP command can be used to prevent unwanted commands from being registered during an idle state. The REF command enables the original data will be written back to the memory cell.

A high-efficient SDRAM controller is required to manage the flow of data to/from SDRAM. Without the SDRAM controller, the processor of an application may have an extra workload. For instance, the processor's performance may be affected as it needs to refresh the SDRAM due to the absence of the SDRAM controller [4]. Because of the miniaturization of electronic devices especially portable and wearable devices address, the power dissipation of the VLSI design has become a major concern. It is truly a fact that a memory controller that can handle larger data bits may consume a larger area while a memory controller with high operational speed will dissipate a large amount of dynamic power. Hence, the dynamic power dissipation of the memory controller should be reduced by disabling unnecessary switching activities.

In this work, a 32-bit SDRAM controller with optimized dynamic power using the clock gating method was implemented by utilizing the Synopsys EDA tools and SAED 32 nm PDK.

II. METHODOLOGY

In most cases, multiplexers will be used to control register banks in a design. By replacing the multiplexers with clock gating cells and applying the clock enable signal, area consumption will be reduced while the clock signal will propagate only when it is required. Thus, clock gating can reduce dynamic power dissipation as it disables unnecessary switching activities.

Clock gating cells can be inserted into the intended design during logic synthesis. By utilizing the command, compile_ultra-gate_clock, the Power Compiler within Synopsys Design Compiler (DC) will implement the clock

979-8-3503-2369-6/23 $31.00 © 2023 IEEE

gating method to optimize power consumption without affecting the design's functionality. Fig .1 demonstrates the architecture of a single integrated clock gating (ICG) cell that can avoid the problem of glitches as compared to gate-based clock gating which consists of a single logic gate only [6]. The approach that uses a clock gating cell to drive another or a row of clock gating cells is known as multistage clock gating as shown in Fig. 2 [7]. Multistage clock gating allows many register banks to be combined whereas unnecessary clock pulses propagated to the clock gating logic can be reduced by cascaded gating circuits, leading to more power savings.

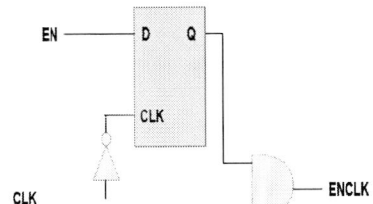

Fig. 1. Architecture of a latch-based ICG cell.

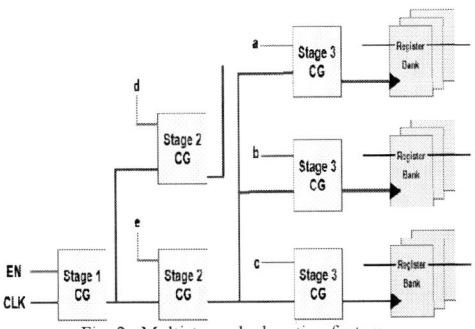

Fig. 2. Multistage clock gating feature.

Excessive implementation of clock gating logic may lead to undesired power and area consumption as well as timing problems. Thus, the efficiency of the clock gating should be governed. Next, the maximum allowable clock gating stage is determined by the set_clock_gating_style -num_stages command. After trial and error, the best dynamic power optimization can be achieved by setting the maximum allowable clock gating stage to 3 and enables reconfiguration of the number of clock gating stages.

III. SIMULATION SETUP

A. RTL Simulation

Verilog Compiler Simulator (VCS) was used to read and compile the open-source design files of the 32-bit SDRAM controller [8]. A testbench written in SystemVerilog will supply the stimulus to verify the functionality of the SDRAM controller during RTL simulation. Moreover, a backward Switching Activity Interchange Format (saif) file which contains static probability and toggle rates of the design was written at the end of the RTL simulation. If the design contains any functionality error, the RTL code must be modified before proceeding to the logic synthesis stage.

B. Logic Synthesis

Fig. 3 shows the general workflow of the logic synthesis. Initially, the relevant design files were loaded whereas the targeted process technology library was defined by specifying the paths to access the target library, synthetic library, and link library. Tool command language (Tcl) was used to specify the timing constraints, area constraints, and clock gating options. For better dynamic power optimization purposes, the generated saif file was read before performing logic synthesis.

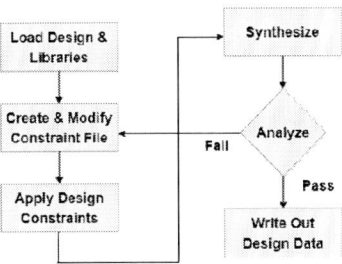

Fig. 3. Flowchart of logic synthesis [9].

After the RTL code inside the design files was converted into a technology-specific gate-level netlist, the EDA tool then performed optimization on the circuit to meet the user-defined constraints. Clock gating cells were inserted into the design at this stage whereas the generic cells were mapped into the technology-dependent cells.

C. Physical Design

Integrated Circuit Compiler (ICC) was used to implement the physical design as shown in Fig. 4. After setting up the environment for generating the layout, proper floorplanning is necessary to enhance area optimization and avoid congestion. IR drop analysis was performed to evaluate the effectiveness of the power network. Subsequently, the placement process will place the standard cells inside the core area. Clock tree synthesis (CTS) tends to achieve a balanced clock skew by inserting buffers or inverters along the clock path. Routing refers to the process of creating physical connections between signal pins using metal layers. To ensure the generated layout fulfills the fabrication-specific rules, Design Rule Check (DRC) was performed. Layout Versus Schematic (LVS) check will compare the connectivity of the layout netlist to the original Verilog netlist to determine if they match.

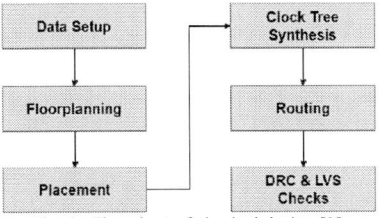

Fig. 4. Flowchart of physical design [9].

D. Post Layout Verification

Static timing analysis was performed by using PrimeTime (PT) to check all possible paths for timing violations. Setup time refers to the minimum time that the input must be stable before the next clock edge whereas hold time defines the minimum time that the input must be stable after the preceding clock edge. Timing requirements are met if both setup slack and hold slack of the critical path are greater than or equal to zero. The values of positive slacks will indicate the available margins to tolerate the possible timing violations.

IV. RESULTS AND DISCUSSION

A. VCS RTL Verification

Fig. 5 and Fig. 6 show that the 32-bit SDRAM controller can perform correct write operation and read operation respectively.

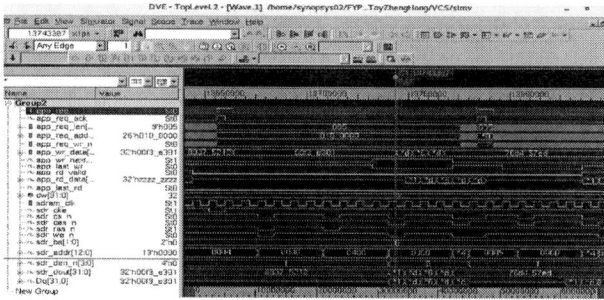

Fig. 5. Write operation.

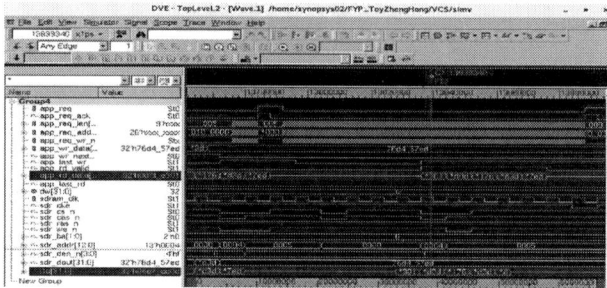

Fig. 6. Read operation.

B. DC Logic Synthesis

By referring to Fig. 7, 9 out of 28 clock-gating elements used were multistage clock gates.

Clock Gating Summary

Number of Clock gating elements	28
Number of Gated registers	341 (61.22%)
Number of Ungated registers	216 (38.78%)
Total number of registers	557
Number of multi-stage clock gates	9
Average multi-stage fanout	1.7
Number of gated cells	356
Maximum number of stages	3
Average number of stages	2.0

Fig. 7. Clock gating summary generated by DC.

61.22 % of the total registers were gated whereas the maximum number of clock-gating stages is 3. After applying the multistage clock gating approach, the total dynamic power dissipation was reduced by 55.07 %. A more accurate power analysis will be performed after the layout is generated.

C. ICC Physical Design

After constructing the power delivery network, IR drop analysis was performed to determine the voltage drop across the metal layers before reaching the standard cells. Fig. 8 shows that the highest IR drop of the design is 57.1 mV or 2.54 % out of the total voltage supply which is 2.25 V. The IR drop value is acceptable as it is less than 5 % of the total supply voltage.

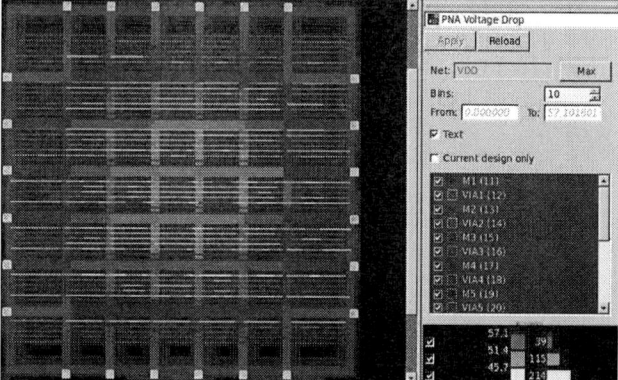

Fig. 8. IR drop based on power network analysis.

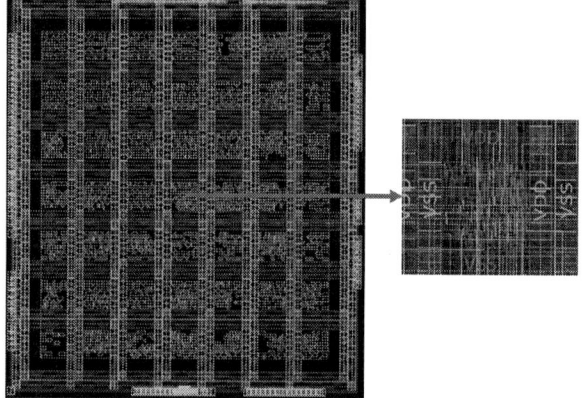

Fig. 9. Layout after routing.

Apart from checking the possible congestion, ICG cells were restructured during placement. Next, CTS constructed and optimized the clock tree to balance the clock skew among all sequential parts of the design. All the power pins and ground pins were connected correctly after the routing process was completed. Since no DRC violations and LVS violations were reported by the tool, it may be deduced that the design is free of any design rule violations, short nets, and open nets. Fig. 9 refers to the layout of the design after routing in which the expanded view shows the created routing area due to different metal layers.

D. PT Post Layout Verification

PT was used to analyze whether the design constraints were met at the post-layout stage. Fig. 10 shows the clock gating report which demonstrates the toggle savings on the clock tree from clock gating in the design. Due to the optimization of ICG cells during placement, the number of

979-8-3503-2369-6/23 $31.00 © 2023 IEEE

clock gating elements increased whereas an average register gating efficiency of 61.6 % was achieved.

```
Clock: clk
  + Clock Toggle Rate: 0.249985
  + Number of Registers: 525
    + Number of Un-gated Registers: 184 (35.0 %)
    + Number of Gated Registers: 341 (65.0 %)
  + Number of Clock Gates: 70
  + Max Number of Clock Gate Stage: 3
  + Average Clock Toggle Rate at Registers: 0.0959214
  + Average Register Gating Efficiency (savings with respect to root clock): 61.6%
-----------------------------------------------------------
Toggle Savings              Number of       % of
Distribution                Registers       Registers
-----------------------------------------------------------
100%                        87              16.6%
80% - 100%                  251             47.8%
60% - 80%                   3               0.6%
40% - 60%                   0               0.0%
20% - 40%                   0               0.0%
0% - 20%                    0               0.0%
0%                          184             35.0%
-----------------------------------------------------------
```

Fig. 10. Clock gating report generated by PT.

Table I summarizes the PT results of timing slacks, total dynamic power dissipation, and total area consumption for the cases with and without clock gating. Positive values of setup slack and hold slack indicate that the design is free of timing violations. Furthermore, the implementation of clock gating induced a 44.72 % reduction in the total dynamic power dissipation by disabling portions of the circuitry that are not in use. Since the multiplexers used to implement register banks were replaced by clock gating cells, the total area consumption was reduced by 7.80 %.

TABLE I. PT RESULTS FOR THE CASES WITH AND WITHOUT CLOCK GATING

	Without Clock Gating	With Clock Gating
Total Area (µm²)	10644.26	9814.17
Total Dynamic Power (µW)	490.10	270.95
Setup Slack	2.85	1.17
Hold Slack	0.29	0.14

However, the clock gating method only intends to optimize the dynamic power of the memory controller proposed in this work. To further optimize the power consumption of the design, future work may include ways to reduce the leakage power. For instance, the power gating method can be applied to shut down certain portions of the circuit when they are not in use. By applying the power gating with data retention and clock gating, a nearly 87 % reduction of the leakage power can be achieved for a 4x4 array multiplier circuit using a 32 nm technology node [10]. Thus, the power savings of the memory controller can be enhanced by optimizing both leakage power and dynamic power.

V. CONCLUSION

In this work, 61.6 % of the toggle savings on the clock tree were achieved while the total dynamic power dissipation was reduced by 44.72 % due to the implementation of clock gating. The results proved that the clock gating method is an effective way to minimize the dynamic power dissipation of a digital design by decreasing the unwanted switching activities. Since the design constraints were fulfilled whereas no DRC violations or LVS violations were reported at the post-layout stage, it may be assumed that the clock gating method was implemented successfully to optimize the design. To achieve better power savings, future work may include the implementation of power gating to optimize the leakage power of the SDRAM controller.

ACKNOWLEDGMENT

Authors would like to acknowledge the financial support under the UTM Fundamental Research Grant Project No. Q.J130000.3823.22H52. Also, thanks to the Research Management Center (RMC) of Universiti Teknologi Malaysia (UTM) for providing an excellent research environment in which to complete this work.

REFERENCES

[1] D. A. Patterson and J. L. Hennessy. (2014). *Computer Organization and Design: The Hardware/Software Interface, RISC-V Edition.* (5th ed.) [Online]. Available: https://ict.iitk.ac.in/wp-content/uploads/CS422-Computer-Architecture-ComputerOrganizationAndDesign5thEdition2014.pdf

[2] S. Gagan, M. Harshith, S. Savadatti, G. Surya and S. Tantry, "DDR Controller with Optimized Delay and Access Time," *2022 International Conference on Futuristic Technologies (INCOFT)*, Belgaum, India, 2022, pp. 1-5, doi: 10.1109/INCOFT55651.2022.10094548.

[3] S. R. Vudarapu and L. Lavanya, "Optimization of SDRAM memory controller for high-speed operation," *7th International Conference on Computing in Engineering & Technology (ICCET 2022)*, Online Conference, 2022, pp. 337-341, doi: 10.1049/icp.2022.0643.

[4] V. Vutukuri, V. B. Adusumilli, P. K. Uppu, S. Varsa and R. K. Thummala, "Verification of SDRAM controller using SystemVerilog," *2020 IEEE International Conference on Electronics, Computing and Communication Technologies (CONECCT)*, Bangalore, India, 2020, pp. 1-6, doi: 10.1109/CONECCT50063.2020.9198440.

[5] S. Iyengar and L. Shrinivasan, "Power, Performance and Area Optimization of I/O Design," *2018 International Conference on Inventive Research in Computing Applications (ICIRCA)*, Coimbatore, India, 2018, pp. 415-420, doi: 10.1109/ICIRCA.2018.8597347.

[6] T. Chindhu S. and N. Shanmugasundaram, "Clock Gating Techniques: An Overview," *2018 Conference on Emerging Devices and Smart Systems (ICEDSS)*, Tiruchengode, India, 2018, pp. 217-221, doi: 10.1109/ICEDSS.2018.8544281.

[7] Synopsys, Inc. *Power Compiler User Guide Version L-2016.03-SP4.* (2016). Accessed: May. 19, 2023. [Online]. Available: https://picture.iczhiku.com/resource/eetop/SHiEETiYeZaAwxMn.pdf

[8] D. Annayya, "SDRAM CONTROLLER Specification," opencores.org. https://opencores.org/projects/sdr_ctrl (accessed May. 19, 2023).

[9] N. E. Alias, S. Ishaak, K. J. Hong, M. L. P. Tan, A. Hamzah and Y. A. Wahab, "ASIC Implementation and Optimization of 16 Bit SDRAM Memory Controller," *2020 IEEE International Conference on Semiconductor Electronics (ICSE)*, Kuala Lumpur, Malaysia, 2020, pp. 81-84, doi: 10.1109/ICSE49846.2020.9166869.

[10] M. Saini, S. Shringi and A. Asati, "An Improved Power Gating Technique with Data Retention and Clock Gating," *2021 International Conference on Control, Automation, Power and Signal Processing (CAPS)*, Jabalpur, India, 2021, pp. 1-7, doi: 10.1109/CAPS52117.2021.9730489.

A Study of the Optimum Input Matching Simulation Networks for Integrated Differential Amplifier

Mohd Khier Alshamaileh
Dept. of Electrical Engineering
American University of Sharjah
Sharjah, UAE
b00098822@aus.edu

Nasir A. Quadir
Dept. of Electrical Engineering
American University of Sharjah
Sharjah, UAE
nquadir@aus.edu

Lutfi Albasha
Dept. of Electrical Engineering
American University of Sharjah
Sharjah, UAE
lalbasha@aus.edu

Abstract— **Differential amplifiers are the backbone of analog integrated circuit design and are used in various signal processing applications. This paper presents the design procedure of a differential amplifier with passive loads and degeneration resistance. The effect on the gain of a differential amplifier with different input signal matching topologies applied to it using two out-of-phase AC port power sources, sine wave AC sources, and an ideal transformer with a single AC input. It also presents the effect of shunting a resistor to the input signal in each design. It is found that the magnitude of the output signal is directly proportional to the shunted resistance across the input. This article explores the advantages and disadvantages of each topology, providing helpful insights for circuit designers. The results highlight the benefits of each topology and its choice based on the specific application requirements. The design simulation is carried out using Virtuoso Cadence GDBK 180nm technology.**

Keywords—Differential amplifiers, degeneration, input topologies, gain, NMOS, cascode design.

I. INTRODUCTION

Differential amplifier circuits are defined as operational amplifier circuits that amplify the difference between two input signals and reject common mode signals. A Differential Amplifier which is constructed using MOSFETs or bipolar transistors is one of the main building blocks in analog integrated circuit design, as it forms an input stage for other elements in the circuit such as operational amplifiers and is used in various applications such as signal processing. [1] This feature allows the amplifier to cancel out any common-mode noise that can be picked up by the two signals together. The primary purpose of the differential amplifier is to serve as an input stage for the operational amplifier, enabling the propagation of the signal. Thus, further operational amplifier gain stages do not get affected. [2]

Differential Amplifiers are designed with two types of loads which are passive loads and active loads. The selection between a differential amplifier with a passive or active load depends on the specific application and design requirements, such as input impedance, output impedance, gain, power consumption, and noise performance. [3]

II. DESIGN OF THE DIFFERENTIAL AMPLIFIER

A Design of a differential amplifier is presented in this section with passive load. The effect of source degeneration and drain resistance on gain are discussed.

A. Design Parameters

To attain optimal performance, the selection of design parameters such as R_d, R_s, and others should not be arbitrary.

Careful parameter value selection is necessary to achieve the desired gain and linearity of the amplifier.

1. Selecting R_d

Drain resistance Rd is considered the main factor in determining the gain of a passive load amplifier. The gain of the differential amplifier (Av) is expressed in (1).

$$Av = - g_m * R_d \quad (1)$$

Where *gm* is the transconductance which is the ratio of the change in drain current to the change in gate voltage. This value is governed by width over length ratio of MOSFET. In this design, the *gm* is approximately equal to 0.14 siemens by sitting width-to-length ratio to 4167. Fig. 1 shows different values of Rd that result in different output gain in dB. Rd is chosen to be 220 Ω, so the calculated gain using (1) is

$$Av = 0.14 * 220 = 30.8 \ dB \quad (2)$$

Which is almost equal to the simulated value (29db) shown in Fig. 1.

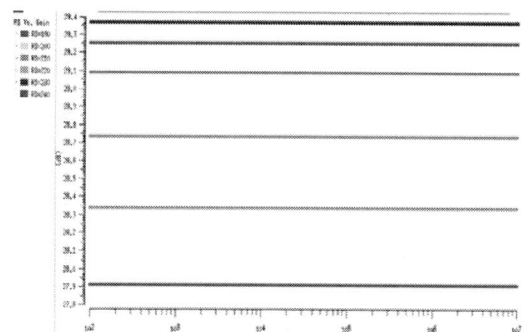

Fig. 1: Selecting Appropriate Rd value

2. Source Degeneration of Differential Amplifier

a) Overview

Source degeneration is a technique used to increase the linear region of operation for a differential amplifier and reduce the noise. This is done by inserting a resistance R_S at the source side of the MOS. [4] By increasing the value of R_S, the linear region of the differential amplifier becomes flatter. Hence, the differential amplifier becomes able to operate for a wider range of input voltage before reaching saturation.

Fig. 2 represents the schematic design and the DC operating points of the amplifier with the inclusion of R_S. A parametric sweep analysis is conducted to investigate the effect of using different values of R_S on the linear region of the amplifier. Fig. 3 describes the results obtained.

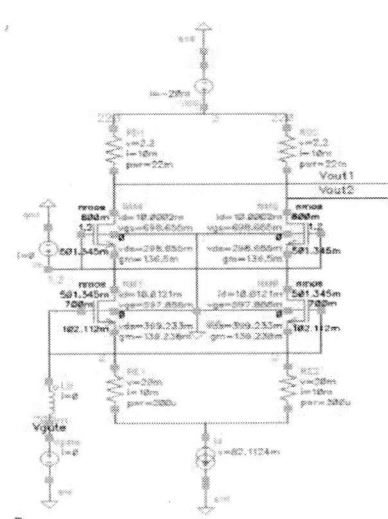

Fig. 2: Source Degenration

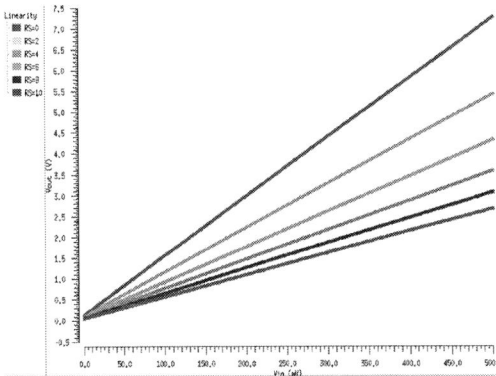

Fig. 3: Effect of degeneration on the linear region

b) Effect of R_S on the Gain of the Amplifier

R_S influences the gain of the amplifier in such a way that as the R_S value becomes higher the gain of the amplifier decreases. Fig. 4 explains the relationship between different values of R_S ($0\ \Omega - 10\ \Omega$ with $2\ \Omega$/ step). Hence, selecting the value of R_S is a trade-off between the desired gain and the range of linearity. In this design, R_S is chosen to be $2\ \Omega$ which corresponds to 26dB.

c) Listing of All Parameters

After investigating the dependence of the transconductance (gm) on the Width over Length (W/L) ratio and plotting the relationship between the drain resistor (RD), source resistor (RS), and gain, the gain of the amplifier was determined in part (b). The remaining parameter values used in the analysis are summarized in Table 1.

TABLE I.   TABLE 1: DESIGN SPECIFICATIONS

Specifications	Simulated Value
VDD	3 V
RD	220 Ω
RE	2 Ω
VG Stage 1 (Enable)	0.7 V
VG Stage 2 (Vb)	1.2 V
ID	10m A

Inductance (AC-Block)	20n H
Reference Voltage	GND (0 V)
NMOS	Multiplier = 1 Length = 180 nM Width = 750 um Fingers = 15
Gain	26 dB

III.   APPLYING DIFFERENT INPUT TOPOLOGIES

In this part of the design, the double output of the differential amplifier is converted into a single output by connecting the output wires to a current-controlled voltage source (CCVS) and grounding one of them. To ensure the same output gain as that of the differential amplifier, the internal resistance R_m of the CCVS must be chosen in a way that gives the same output gain of the differential amplifier. The output is taken across the terminals of the resistor connected after the CCVS. Table 2 shows the input signal and CCVS settings.

TABLE 2: INPUT AND OUTPUT SPECIFICATIONS

Type	Specifications
VIN 1	10mV (sin); Shift: 0
VIN2	10mV (sin); Shift: π
Input Frequency	1 MHz
Current Controlled Voltage Source Rm	200 Ω
R output	1 KΩ

A. Topology one: Direct Connection

The topology used for inserting the input voltages consists of two out-of-phase AC sources with an input frequency of 1 MHz, and this is considered the fundamental topology for this

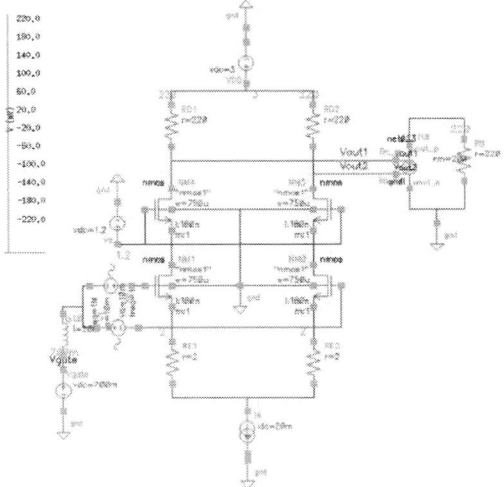

Fig. 5: Topology one schematic design

design. The schematic design for this topology is illustrated in Fig. 5, while Fig. 6 depicts the resulting transient response for topology one. The use of this topology enables the input signals to be effectively delivered to the differential amplifier, ensuring accurate and reliable results. In terms of efficiency, this topology has no input or output losses. Since the whole signal is injected in the amplifier and the output matches the expected gain.

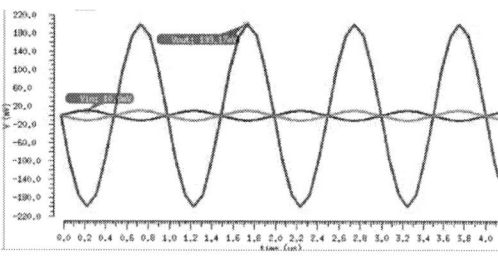

Fig. 6: Transient response of topology one

B. Topology 2: Adding Shunt Resistors to the Input Voltage

Shunt resistors are often added to the input of differential amplifiers to help improve their performance. These resistors are connected in parallel to the input terminals and are designed to reduce the common-mode voltage while maintaining the differential voltage. This section had been conducted to investigate if there is any effect of adding resistance to the input side of the amplifier on the output signal. The Vsin instrument does not identify the resistor branches. Hence, it is replaced by Psin in order to identify the shunt resistance. Fig. 7 and 8 illustrate the results obtained.

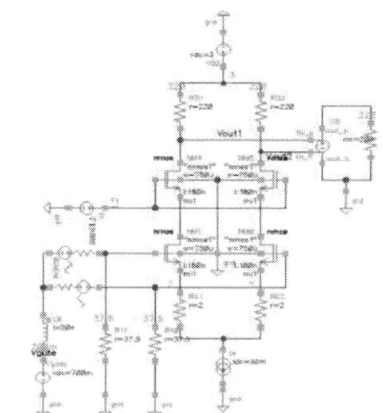

Fig. 7: Topology two schematic design

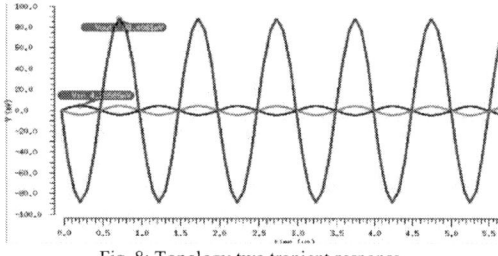

Fig. 8: Topology two tranient response

One of the disadvantages of shunting resistors to the input is that they can introduce additional noise into the amplifier circuit, and reduce the overall voltage gain, which may not be desirable in certain applications. In this topology, the addition of a parallel resistance ($r = 37.5 \ \Omega$) at the input terminal was observed to have a significant effect. Specifically, the added resistance caused a voltage drop of 5.72mV, which in turn resulted in only 4.2mV entering the amplifier. This voltage drop, in turn, led to a roughly halved output waveform when compared to topology one, thereby significantly impacting the overall performance of the differential amplifier.

By sweeping the input resistance magnitude, it is found that the value of Vin which is entering the amplifier is inversely proportional to the value of the resistance. In other words, the output voltage is directly proportional to the value of the shunt resistor.

C. Topology 3: Replacing the Vsin with Port

In this topology, the Psin AC voltage sources are replaced by Port in Cadence analogLib library The value of input voltage (Vpk) is set to half of the value used in the Vsin; this is because that the voltage on the internal source is set twice the value specified on the port [5]. Hence when it is set 5 mV, the wave produced is 10 mV peak as the one used in the other topologies.

The Port is another type of AC source that can be utilized within the Virtuoso Cadence software. The main advantage of using a Port instead of a Psin is that it allows for the generation of multiple types of signals, including DC, sine, pulse, and exponential signals. In contrast, the Psin source is limited to only producing sine waves. To illustrate the use of a Port, Fig. 9 and 10 show the schematic design and the corresponding transient response, respectively.

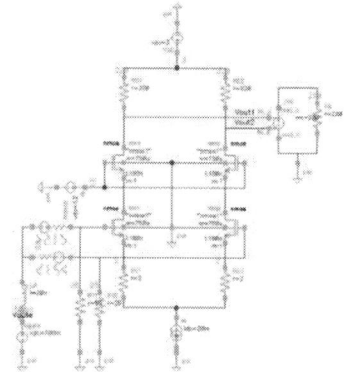

Fig. 9: Topology three schematic design

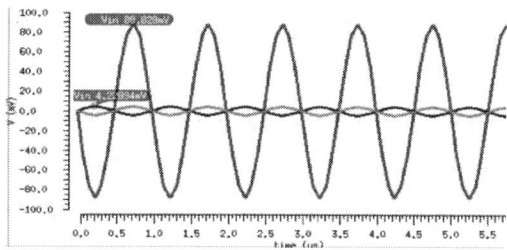

Fig. 10: Topology three tranient response

The purpose of this section was to investigate the potential effects of replacing the Psin source used in the second topology with a Port. The result shown in Fig. 10 is compared to this obtained using the Psin source in Fig. 8. The comparison revealed that both responses were identical, indicating that there is no significant difference in using a Port or Psin to generate the input voltage signal. Therefore, the use of either source can be chosen based on the type of signal needed, with the Port offering greater flexibility in generating various signal types.

D. Topology four: Single Vsin with an Ideal Transformer

Using a transformer with one voltage source as an input to the differential amplifier can be helpful in certain situations. One example is when there is a need to provide galvanic isolation between the input source and the amplifier. The transformer provides isolation, which means there is no electrical connection between the input and output sides of the transformer, and both circuits are magnetically coupled. This can be important in situations where there is a risk of high voltage or current on the input side, which could potentially damage the amplifier. Additionally, it can be used to provide different voltage levels by stepping the voltage up or down based on the desired output. Overall, using an ideal transformer with one voltage source as an input to the amplifier can provide isolation and noise reduction. [6]

In this simulation, the transformer used is named "ideal_balun," which is a component available in the analogLib library. The transformer has a transformation ratio of one, meaning that the input and output voltages have the same magnitude. Fig. 11 and 12 summarize the results obtained.

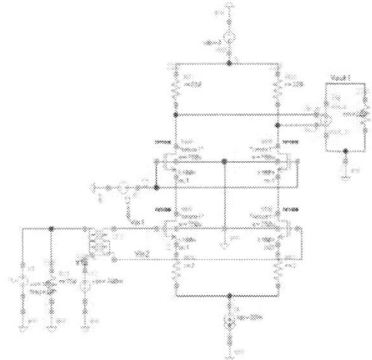

Fig. 11: Topology four schematic design

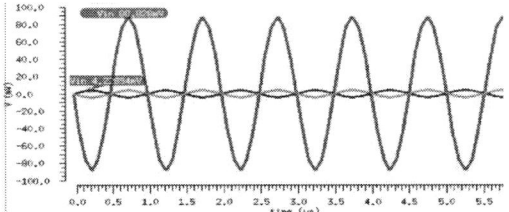

Fig. 12: Topology four tranient response

This topology is useful when using the differential amplifier as an operational amplifier to amplify a single input signal. By connecting the signal to an ideal transformer, the signal is split into two out-of-phase input signals with a peak voltage that is half of the original signal's peak voltage. The differential amplifier then amplifies the difference between the two inputs, which is precisely equal to the input signal that was applied to only one of the terminals of the differential amplifier. This approach effectively doubles the voltage swing, which can be beneficial in situations where the available power supply voltage is limited.

E. Results

The input voltage used for all topologies was 10 mV sin wave with a frequency of 1MHz. Table 3 summarizes the results obtained in all topologies.

The first topology involves directly connecting two out-of-phase AC sources to the amplifier for accurate and reliable results. The second topology adds shunt resistors to the input terminals to reduce common-mode voltage, but it can introduce additional noise and reduce voltage gain. The third topology replaces the Psin AC voltage sources with a Port in

Cadence analogLib library, which allows for the generation of multiple types of signals. The fourth topology uses an ideal transformer with one voltage source as an input to provide galvanic isolation, noise reduction, and effectively double the voltage swing. Table 3 summarizes the results obtained.

TABLE 3: RESULTS

T	Number of inputs used	Vin	Vout	Observations
1	2 (vsin)	10.0 mV	199.5 mV	Normal operation
2	2 (Psin)	4.3 mV	88.8 mV	Effect of shunt resistance is observed. ($R_{shunt} \propto$ Vout)
3	2 (port)	4.3 mV	88.8 mV	Effect of shunt resistance is observed. ($R_{shunt} \propto$ Vout)
4	1 (vsin) with Transformer	5.0 mV	99.2 mV	The single voltage wave is splitted into two out-of phase waves with half the magnitude

IV. CONCLUSION

Differential Amplifiers are one of the fundamental components in analog integrated circuit design which is mainly used in several signal processing applications. The principle of amplifying the signal relies on the difference between two input signals. Hence, differential amplifiers can be used in common noise-cancellation applications.

In this paper, a proposed cascode design of a differential amplifier with passive loads is discussed, tested, and simulated with different input signal topologies. The design is built using MOSFETs (NMOS) and simulated for DC, frequency, and transient response using Cadence ADE L simulator.

REFERENCES

[1] Lutfi Albasha (2010) Design and measurement of an integrated wideband radio frequency low-noise amplifier for terrestrial digital television applications, International Journal of Electronics, ,587-604.

[2] Q. M. Abubaker and L. Albasha, "Balun LNA Thermal Noise Analysis and Balancing With Common-Source Degeneration Resistor," in IEEE Access, vol. 8, pp. 64949-64958, 2020).

[3] A. N. Darwish, L. Albasha and H. Alrifai, "An overview of design techniques for high frequency wide-band low noise amplifiers," 2018 IEEE Symposium on Computer Applications & Industrial Electronics (ISCAIE), Penang, Malaysia, 2018, pp. 139-144.

[4] D. Malathi and M. Gomathi, "Design of inductively degenerated common source RF CMOS Low Noise Amplifier," Sādhanā, vol. 44, no. 1, Dec. 2018, doi: https://doi.org/10.1007/s12046-018-1017-5.

[5] Tawna (2018). How to set input power in transient simulation? - RF Design - Cadence Technology Forums - Cadence Community. community.cadence.com. Available at: https://community.cadence.com/cadence_technology_forums/f/rf-design/40218/how-to-set-input-power-in-transient-simulation#:~:text=Amplitude%201%20(Vpk)%20%2D%20When [Accessed 12 Apr. 2023].

[6] A. Pini, "The Basics of Isolation Transformers and How to Select and Use Them," www.digikey.com, May 20, 2020. https://www.digikey.com/en/articles/the-basics-of-isolation transformers-and-how-to-select-and-use-them

Study of Error Amplifiers for Low Power Capacitorless Low Dropout Voltage Regulator using 110 nm CMOS Technology

Wan Maziyah Ab. Halim
Electronic Technology Section
Universiti Kuala Lumpur
British Malaysia Institute
Selangor, Malaysia
maziyah.halim@s.unikl.edu.my

Julie Roslita Rusli
Electronic Technology Section
Universiti Kuala Lumpur
British Malaysia Institute
Selangor, Malaysia
julie@unikl.edu.my

Yuzman Yusoff
Analog Mixed Signal
Department
Symmid Corporation Sdn Bhd
Kuala Lumpur, Malaysia
yuzman.yusoff@symmid.com

Mohd Azraie Mohd Azmi
Electronics Technology Section
Universiti Kuala Lumpur
British Malaysia Institute
Selangor Malaysia
mazraie@unikl.edu.my

Abstract— **Capacitorless low dropout (CLDO) voltage regulators are becoming a key power management sub-block in contemporary electronic circuits. However, a high quiescent current in a CLDO regulator drains more power from the system and rapidly reduces the battery charge. Therefore, this study aims to identify the effect of different error amplifier topologies on the quiescent current of a low-power 110 nm CLDO regulator by simulating two CLDO regulators which utilize telescopic and two-stage error amplifier architectures operating at a nominal input voltage of 1.2 V and regulating an output voltage of 1.0 V with maximum load up to 50 mA. The obtained results show that both CLDO regulators are able to achieve single-digit micro-range quiescent current with the telescopic-base CLDO having better quiescent current of 4.81 μA compared to the two-stage-base CLDO quiescent current of 5.18 μA. However, the two-stage-base CLDO regulator demonstrates better overall performance in terms of better accuracy, stability and noise rejection although has a slightly higher quiescent current.**

Keywords— cldo, ldo, telescopic, amplifier, regulator.

I. INTRODUCTION

Rapid advancements in the Internet of Things (IoT), smart wearables, sensor-based mobile devices, and implantable electronic devices, along with the emergence of advanced complementary metal oxide semiconductor (CMOS) technologies have led to surge in the demands for extendable power supplies that prolong battery life and cut down on standby power consumption [1]. In order to provide longer battery life, low-power circuitry is necessary especially in battery-powered stand-alone devices and systems-on-chips (SoC). As a result, good power management has become critical in modern electronic circuit designs. One of the key components in a power management system of an electronic device is a low dropout (LDO) voltage regulator. It converts and manages an external voltage coming from a battery or other power source so that any voltage requirements in the system can be met. In the recent development of LDO regulators, capacitor-less LDO (CLDO) regulators have gained more interest because it is preferred that LDO regulators can operate in stand-alone mode and become system-on-chip (SoC) solutions to match the critical demands of mobile and low-power devices [2]. Although the elimination of the large external capacitor of the CLDO greatly reduces the cost and power consumption due to fewer I/O pins and smaller printed circuit board (PCB) area, it also suffers from the major disadvantage of degraded

stability, as the dominant pole is now shifted from the large external capacitor into the internal nodes of the CLDO regulator [3]. The power consumption of a CLDO regulator can be reduced either by lowering the supply voltage or optimizing the amount of current flowing in the CLDO regulator during its idle state. This current during idle mode is called quiescent current. It is current when the CLDO regulator has zero load and is described as the difference between the input and output currents [4]. Having a high current flow at zero load will draw more power from the system and discharge the battery faster, therefore reducing the quiescent current contributes greatly in prolonging the battery life. A CLDO regulator mainly consists of four main parts; error amplifier, pass transistor, feedback network and load [5], [6]. The contributing blocks to a CLDO regulator's quiescent current are the error amplifier and the feedback network. Minimizing the current in these two areas will effectively reduce the quiescent current of the CLDO regulator, thus reducing the power dissipated during the idle mode and therefore contributing to extending the battery life.

This study focuses on optimizing the quiescent current of a 110 nm CLDO regulator in the error amplifier block by implementing and simulating two CLDO regulators with different topologies of error amplifiers, telescopic and two-stage error amplifiers and the performances of the two CLDO regulators are analysed. The design utilized C11AL 110 nm CMOS technology with 1.2 V nominal input voltage supply, and the CLDO regulator is expected to regulate 1.0 V of output voltage with maximum load current of 50 mA. This paper is organised into four sections, with section II of this paper describes the methodology used to design and verify the respective CLDO regulators and section III discusses and analyses the results obtained. The overall study is concluded in Section IV.

II. CLDO REGULATOR DESIGN

Two topologies of CLDO regulators are selected for this study, a telescopic error amplifier-based CLDO regulator and two-stage error amplifier-based CLDO regulator. The telescopic error amplifier is selected mainly because it has better power consumption performance and relatively larger DC gain due to its high output impedance that comes from the cascoded transistor architecture without the need for second-stage amplification [7], [8]. This greatly contributes to the minimization of the total current due to the exclusion of the second-stage node. Maintaining a large DC gain is also important because it contributes to a better power supply rejection ratio (PSRR) which is also a key performance

979-8-3503-2369-6/23 $31.00 © 2023 IEEE

criterion of CLDO regulator [9]. The two-stage error amplifier, on the other hand, is a popular option due to its various advantages, among them are high DC gain and large voltage swing [10]. On top of that, two-stage error amplifier is included in this study due to bigger available overhead voltage it offers which is crucial to maintain the transistors in the saturation mode.

A. Telescopic Error Amplifier CLDO Regulator (TCLDO)

The design process of the TCLDO is described in the flow chart in Fig. 1. The design begins with defining the reference current for the telescopic amplifier. This step is crucial to obtain the minimum reference current which can properly bias the transistor at the reference node. Minimizing the reference current will contribute in minimizing the total current of the telescopic error amplifier since the current at other nodes will be derived based on a multiplication factor of the reference current. The next stage is to design the telescopic error amplifier. Since this telescopic amplifier is a PMOS-input amplifier, the current mirrors in the overall circuit will also be PMOS current mirrors. The tail current of the telescopic amplifier is the multiplication of the reference current by a determined factor, and it is achieved by manipulating the width of the PMOS. The next stage is the sizing of the PMOS input pairs of the telescopic error amplifier. It is advisable to avoid using the minimum length for input pair transistors to reduce mismatch possibility as the transistor mismatch is inversely proportional to the transistor gate dimension [11]. The size of the input PMOS pair is decided so that the voltage at the drain of the tail PMOS provides enough overhead voltage for it to operate in a saturation region.

The next stage is the sizing of the input cascode PMOS pair. Both PMOS share the same gate voltage VBP which is initially supplied by an ideal DC voltage source. VBP has decided so that it is sufficient to provide enough gate-to-source voltage V_{GS} to turn on both input cascode transistor pair, and the size is decided so that the output voltage of the telescopic amplifier is 0.6 V, which is half of V_{DD}. The next stage is to determine the size of the middle load pair NMOS which is biased by the voltage VBN. VBN is required to be able to supply the V_{GS} to both NMOS whose size is determined by leaving sufficient overhead voltage for bottom load NMOS pair to be in saturated region. Finally, the bottom load NMOS pair transistors size is determined by ensuring that all transistors in the telescopic error amplifier can operate in saturation region, which is indicated by region 2 in the simulator. It is important for all transistors in the telescopic error amplifier to be in saturation mode to ensure no changes in the amplifier characteristic for any voltage or current change which can affect the overall performance of the error amplifier if it occurs.

The design is continued with the completion of the bias circuit. Bias circuit is needed to generate the bias voltages VBN and VBP which initially come from ideal sources. The full bias circuit section is shown in the full schematic view in Fig. 2 with the bias section labeled. VBP is achieved by connecting five PMOS transistors in series with the NMOS current mirror while VBN is achieved by connecting five NMOS transistors in series with PMOS current mirror.

The resistance values in the feedback network are calculated based on the reference voltage V_{REF} and the decided current at the feedback network node. Since the output voltage V_{OUT} and the feedback voltage V_{FB} of the TCLDO regulator are known, the resistances are calculated

based on the current decided at the node. The dimensions of the resistors are decided based on the known resistance values and selected fixed parameter while considering proper physical layout factors such as parasitic and matching.

The size of the pass transistor is decided based on the targeted maximum load current I_L with minimum length applied. It should be large enough to drive the large maximum load current of the CLDO regulator. In this study the decided maximum load current I_L is 50 mA. The current at the pass transistor node needs to be relatively small to reduce the quiescent current, but at the same time should have an acceptable gap with the maximum load current I_L to avoid slow transient.

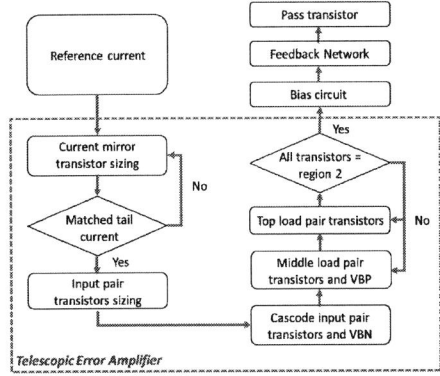

Fig.1 Design flow chart of TCLDO regulator

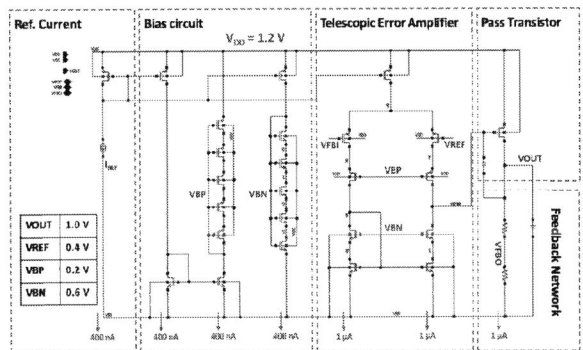

Fig.2 Full schematic diagram of TCLDO regulator

B. Two-Stage Error Amplifier CLDO Regulator (2SCLDO)

The design process of 2SCLDO is almost identical to PTCLDO except for slight difference in error amplifier design steps. The process begins with biasing the transistor in the reference node as shown in the flow chart in Fig. 3. The reference current is set to be 400 nA based on the minimum operable V_{GS} of the 110 nm CMOS technology PMOS with a length that is not too narrow to avoid parasitic effect due to mismatch which will affect the accuracy of the mirroring current. NMOS input type is not considered because the reference voltage of 0.4 V is insufficient to cover two levels of NMOS thresholds comprising the tail transistor and the input pair transistors. This makes the current mirrors utilized for 2SCLDO are also of PMOS type. The current at the mirroring node which is also the tail current of the first stage of the error amplifier is decided to be 2 µA. This means the mirroring PMOS will have a multiplication factor of 5 to amplify the 400 nA to 2 µA. At this stage the other yet to be sized transistors of the

979-8-3503-2369-6/23 $31.00 © 2023 IEEE   115

amplifier are represented by an ideal resistor with pre-calculated current and voltage so that the gate voltage *va* matches the voltage *vb* at the drain of the tail PMOS transistor as shown in Fig. 4 by running DC analysis.

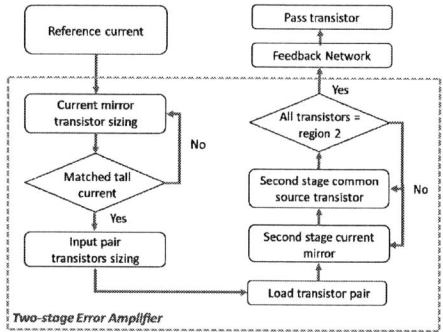

Fig. 3 Design flow chart of 2SCLDO regulator

Sufficient voltage at *vc* and *vd* is needed to cover the NMOS load pair threshold with acceptable overhead voltage, and the size of the PMOS input pair is determined with this consideration. This is again verified by running DC analysis on the circuit. The remaining NMOS load pair in the first stage amplifier are sized so that the output voltage of the first stage error amplifier *vd* is sufficient to drive the second stage common source NMOS, and this step completes the first stage of the error amplifier. The process is then continued with the second stage part of the two-stage error amplifier by adding the NMOS with its gate driven by *vd* and its drain connected to a PMOS current mirror. The size of the second-stage NMOS is decided so that the output voltage of the second-stage amplifier *vea* is able to drive the large pass transistor of the 2SCLDO. This concludes the error amplifier section of the 2SCLDO. The rest of the sections i.e., the feedback network and the pass transistor have the same process as the previous TCLDO since both CLDO regulators have the same specs and will not be elaborated again in this section.

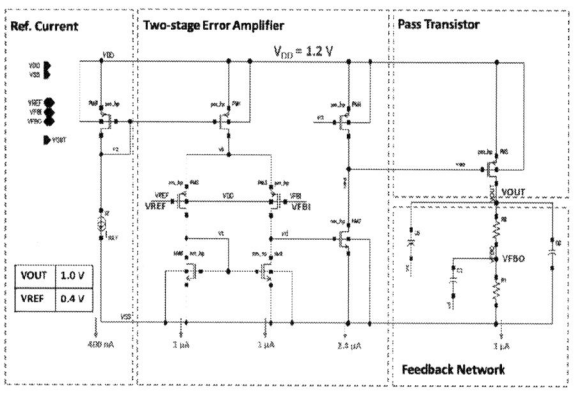

Fig. 4 Full schematic diagram of 2SCLDO regulator

In terms of verification, DC simulations are done to obtain the DC operating points of both CLDO regulators, while transient simulations are done to obtain the functionality and the expected voltages and currents. Closed-loop verification is also done to obtain the frequency responses of the CLDO regulators such as loop gain, stability and PSRR.

III. RESULTS AND ANALYSIS

Two CLDO regulators TCLDO and 2SCLDO, with nominal supply voltage of 1.2 V and regulate an output voltage of 1.0 V at maximum load current of 50 mA are simulated, and their respective results are discussed and analyzed in this section. The summary of results for both CLDO regulators is shown in Table I. Based on the transient response of TCLDO and 2SLDO as shown in Fig. 5, both CLDO regulators are able to regulate the output voltage of 1.0 V at maximum load of 50 mA with 2SCLDO having better accuracy than TCLDO. This can be observed in the differences of the obtained output voltage VOUT and feedback voltage VFB from the expected values. This means 2SCLDO is better in terms of maintaining the regulated voltage even during zero load conditions. As for TCLDO, it is quite risky for the output voltage to change level in ultra-low load conditions even though the CLDO is not expected to regulate at zero load condition.

In terms of frequency response, the 2SCLDO demonstrates better DC gain compared to TCLDO with 103 dB at maximum load, while the DC gain for TCLDO is only 47 dB at the same load current, shown in Table I. This explains the better accuracy of 2SCLDO since higher gain contributes to better accuracy of the amplifier. The unity gain bandwidth (UGBW) of the 2SLDO is also higher at 306 kHz compared to 251.2 kHz for TCLDO. In terms of phase margin, TCLDO shows better results with 81.65° compared to 2SLDO phase margin of 63°, which implies better stability although both CLDO regulators are still in the acceptable phase margin range of above 60°. The high loop gain of 2SCLDO also contributes to better overall PSRR performance which can be seen in Fig. 6. The PSRR of 2SCLDO is -86 dB at maximum load which is much larger than the -33 dB of TCLDO at DC.

TABLE I   SUMMARY OF RESULTS OF PROPOSED ERROR AMPLIFIER

Parameter	TCLDO	2SCLDO
Input Voltage (V_{IN})	1.2 V	1.2 V
Reference Voltage (V_{REF})	0.4 V	0.4 V
Reference Current (I_{REF})	400 nA	400 nA
Max Load Current	50 mA	50 mA
DC Gain (I_{LMAX}=50mA)	47 dB	103 dB
Phase Margin	81.65°	63°
PSRR (I_{LMAX}=50mA)	-33 dB	-86 dB
Unity Gain Bandwidth	251.2 kHz	306 kHz
Quiescent Current	4.8 µA	5.8 µA
Power	5.76 µW	6.96 µW

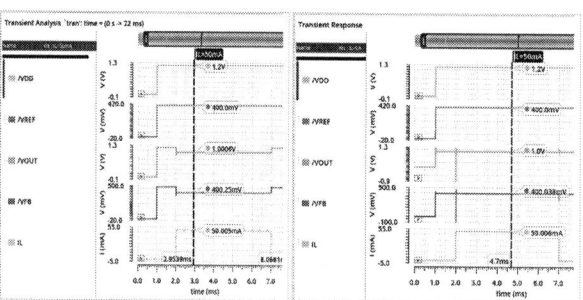

Fig.5 Transient simulation result of TCLDO (left) and 2SCLDO (right) regulators at maximum load current of 50 mA

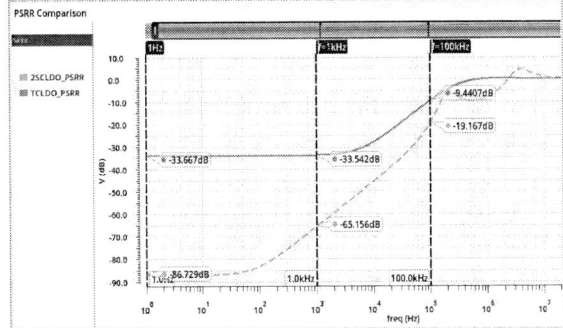

Fig. 6 Quiescent current of TCLDO and 2SCLDO regulators

In terms of quiescent current, TCLDO obtained a better quiescent current of 4.81 µA compared to the quiescent current of 2SCLDO which is 5.18 µA as shown in Fig. 7. The second stage node of the error amplifier in 2SCLDO regulator consumes more current than the other nodes, contributing to its higher quiescent current. Even though TCLDO introduces more current nodes than 2SCLDO due to the bias circuit needed for the extra bias voltages, it is still possible to keep the total current at a low level due to the low reference current. This emphasizes the importance of proper definition of the reference current to maintain low quiescent current even with extra current nodes introduced. However, both quiescent currents are still in the single-digit micro-range which contributes to micro-range power consumption, hence maintaining both CLDO regulators in the ultra-low power category.

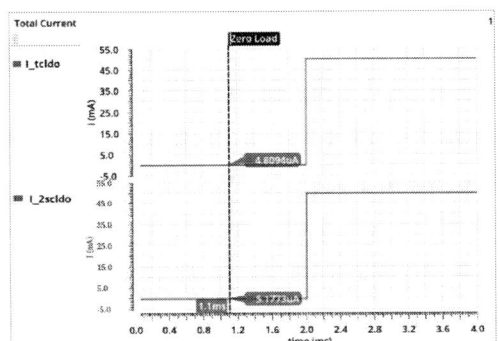

Fig. 7 Quiescent current of TCLDO and 2SCLDO regulators

IV. CONCLUSION

It can be concluded that TCLDO achieves better quiescent current of 4.81 µA than 2SCLDO which records 5.18 µA. However, looking at the overall performance of the CLDO regulator, 2SCLDO demonstrates better overall CLDO performance such as accuracy, stability and noise rejection while still being able to maintain the quiescent current at micro-range, and with proper optimization, it is possible for the 2SCLDO to achieve better performance while maintaining its low quiescent current. The absence of the second stage of the telescopic error amplifier contributes to its lower quiescent current, while the addition of the second stage of the two-stage error amplifier enhances the overall performance of the CLDO regulator.

ACKNOWLEDGMENT

The authors would like to thank Universiti Kuala Lumpur British Malaysia Institute (UniKL BMI) and Universiti Putra Malaysia (UPM) for supporting this research. This research is sponsored by STRG (STR19085) by UniKL BMI.

REFERENCES

[1] T. Guo, W. Kang, and J. Roh, "A 0.90-uA; A Quiescent Current High PSRR Low Dropout Regulator Using a Capacitive Feed-Forward Ripple Cancellation Technique," *IEEE J Solid-State Circuits*, 2022, doi: 10.1109/JSSC.2022.3161014.

[2] M. U. Abbasi, D. Bagnall, and V. Bn, "A high PSRR capacitor-less on -chip low dropout voltage regulator," in *SIISY 2010 - 8th IEEE International Symposium on Intelligent Systems and Informatics*, 2010. doi: 10.1109/SISY.2010.5647405.

[3] J. Zarate-Roldan, M. Wang, J. Torres, and E. Sanchez-Sinencio, "A Capacitor-Less LDO with High-Frequency PSR Suitable for a Wide Range of On-Chip Capacitive Loads," *IEEE Trans Very Large Scale Integr VLSI Syst*, vol. 24, no. 9, pp. 2970–2982, Sep. 2016, doi: 10.1109/TVLSI.2016.2527681.

[4] B. S. Lee, "Application Report Understanding the Terms and Definitions of LDO Voltage Regulators," *Texas Instruments*, no. October, 1999.

[5] M. A. S. Bhuiyan *et al.*, "CMOS Low-Dropout Voltage Regulator Design Trends: An Overview," *Electronics (Switzerland)*, vol. 11, no. 2. MDPI, Jan. 01, 2022. doi: 10.3390/electronics11020193.

[6] W. M. A. Halim, J. R. Rusli, S. Shafie, Y. Yusoff, and C. C. Yin, "Study on performance of capacitor-less LDO with different types of resistor," in *2019 4th IEEE International Circuits and Systems Symposium, ICSyS 2019*, doi: 10.1109/ICSyS47076.2019.8982395.

[7] M. Noormohammadi and K. Hajsadeghi, "A new class AB multipath telescopic-cascode operational amplifier," in *ICEE 2012 - 20th Iranian Conference on Electrical Engineering*, 2012. doi: 10.1109/IranianCEE.2012 6292326

[8] R. Todani and A. K. Mal, "A power efficient and digitally assisted CMOS complementary telescopic amplifier with wide input common mode range," in *Proceedings - International Symposium on Quality Electronic Design, ISQED*, 2013. doi: 10.1109/ISQED.2013.6523645.

[9] J. Torres *et al.*, "Low Drop-Out Voltage Regulators: Capacitor-less Architecture Comparison," *IEEE Circuits and Systems Magazine*, vol. 14, no. 2, pp. 6–26, 2014, doi: 10.1109/MCAS.2014.2314263.

[10] Y. Xin, X. Zhao, B. Wen, L. Dong, and X. Lv, "A high current efficiency two-stage amplifier with inner feedforward path compensation technique," *IEEE Access*, vol. 8, pp. 22664–22671, 2020, doi: 10.1109/ACCESS.2020.2967870.

[11] M. A. Lupercio and J. L. Del Valle, "Mismatch compensation in low power operational transconductance amplifiers using MIFGMOS," in *2017 IEEE International Autumn Meeting on Power, Electronics and Computing, ROPEC 2017*, 2017. doi: 10.1109/ROPEC.2017.8261578.

Chitosan as Natural Binder for Eco-Friendly Printable Conductive Ink

Nur Iffah Irdina Maizal Hairi
Department of Electrical and Computer Engineering,
Kuliyyah of Engineering, IIUM
Kuala Lumpur, Malaysia
iffah.irdina@live.iium.edu.my

Aliza Aini Md Ralib
Department of Electrical and Computer Engineering,
Kuliyyah of Engineering, IIUM
Kuala Lumpur, Malaysia
alizaaini@iium.edu.my

Anis Nurashikin Nordin
Department of Electrical and Computer Engineering,
Kuliyyah of Engineering, IIUM
Kuala Lumpur, Malaysia
anisnn@iium.edu.my

Rosminazuin Ab Rahim
Department of Electrical and Computer Engineering,
Kuliyyah of Engineering, IIUM
Kuala Lumpur, Malaysia
rosmi@iium.edu.my

Muhammad Farhan Affendi Mohamad Yunos
Jabil Circuit Sdn. Bhd, Malaysia
Department of Electrical and Computer Engineering, IIUM, Malaysia
farhanfendi93@gmail.com

Lim Lai Ming
Jabil Circuit Sdn. Bhd,
Bayan Lepas Industrial Park Phase 4,
Penang, Malaysia
laiming_lim@jabil.com

Abstract— Conductive inks have been extensively investigated in printed electronics for the development of wearable devices. Typical conductive inks consist of conductive filler, polymer binder and solvent. However, involvement of synthetic polymer binder in printable conductive ink emits volatile organic compounds (VOCs) that can impact human health. Chitosan (CS) biopolymer provides alternative solution as natural binder because it exhibits high tensile strength, non-toxic and environmental-friendly. Hence, the objective of this paper was to evaluate chitosan as a natural binder for eco-friendly printable conductive ink. The CS and multi-walled carbon nanotube (MWCNT) were synthesized and characterized using FESEM, Raman spectroscopy and FTIR analysis. The conductivity and bending test of CS/MWCNT printable conductive ink were evaluated. Rheological properties of CS/MWCNT printable conductive ink recorded viscosity of 1 Pa·s and behave as non-Newtonian fluid with shear-thinning characteristic. Homogenous dispersion and proper disentanglement of MWCNT fillers within CS polymer was depicted through surface morphology analysis. Raman and FTIR analysis illustrated that CS were successfully synthesized with MWCNT filler. The measured conductivity for CS/MWCNT printable conductive ink was 4.46×10^{-3} S/m which was comparable to the previous work. The bending test proved that higher weightage of CS will result to strong bond between CS and MCWNT and can prevent crack, resulting to flexible CS/MWCNT printable conductive ink. Therefore, integration of CS as natural binder for eco-friendly printable conductive ink provides promising solution for printed electronics applications.

Keywords— *chitosan/MWCNT, printable conductive ink, natural binder*

I. INTRODUCTION

Printed electronics are expanding rapidly as they are widely used in fabricating sensors and electronics. It is evolving as it promises to enable a huge number of applications due to its scalability, cost-effectiveness, and customizability. Generally, printed conductive ink requires three major elements which are the substrate, conductive ink and the printing setup [1]. Among these three elements, composition of conductive ink has been an interesting field to be explored as selection of material and ink synthesis will impact fabricated printable conductive ink. The composition of the conductive ink consists of filler, solvent, binder and additives [2]. Selection of binder is crucial to hold the ink component together upon solvent evaporation. It also helps to bind the filler and printed traces onto the substrate. However, the use of synthetic polymer binder in ink synthesis emits VOCs that may impact the environment and human health [3]. Recent studies are focusing on the needs of propylene glycol as co-solvent in synthesizing MWCNT/Triton X-300 printable conductive ink to improve the ink viscosity [4]. Singla et. al. described usage of polystyrene as binder required additional preparation upon synthesizing with MWCNT [5]. However, there were limited studies on integrating CS as binder with MWCNT and the impact of CS towards CS/MWCNT printable conductive ink. CS is a biopolymer, produced by alkali deacetylation of chitin obtained from exoskeletons of edible marine crustaceans such as shrimps and crabs. It is known in having high tensile strength [3], biocompatibility and non-toxicity [3] and environmental friendly [6]. The presence of chitosan produces non-breakable ink that improves the binding between conductive filler and polymer matrix. Hence, this work aims to evaluate chitosan as a natural binder for eco-friendly CS/MWCNT printable conductive ink. The materials, synthesis and characterization of CS/MWCNT printable conductive ink were discussed in Sect. II. Sect. III focused on analyzing the printable conductive ink fluid characteristic and materials characterization which includes FESEM, RAMAN and FTIR analysis. The conductivity and bending test of CS/MWCNT printable conductive ink were also discussed in this section. The conclusion was presented in Sect. IV.

II. SYNTHESIS AND CHARACTERIZATION

A. Materials

The conductive filler used in this work was MWCNTs, purchased from Sigma Aldrich. Chitosan powder with 85% degree of acetylation was purchased from Alfa Aesar. Sodium dodecyl sulphate (99.0%) was supplied by Sigma Aldrich. Acetic acid glacial was purchased from Systerm.

B. Synthesis of CS/MWCNT printable conductive ink

The CS/MWCNT printable conductive ink was prepared using simple ultrasonication and mechanical stirring process. A constant weight of MWCNT powder was dispersed with CS in deionized (DI) water. An adequate amount of acetic acid was required to improve dispersibility of CS as CS in insoluble in water. After uniform dispersion of MWCNT within CS polymer, sodium dodecyl sulphate (SDS) was added to facilitate better dispersion of the MWCNT and obtain uniform CS/MWCNT dispersion. The solution continued to be sonicated followed by mechanical stirring for 20 minutes.

C. Fabrication of CS/MWCNT printable conductive ink

The fabrication of CS/MWCNT printable conductive ink was initiated by depositing the ink onto thermoplastic polyurethane (TPU) substrate. The ink was then cured in the oven at 70°C for 30 minutes. Fig. 1. shows the synthesized printable conductive ink and CS/MWCNT deposited on TPU substrate.

Fig. 1. (a) Synthesized CS/MWCNT printable conductive ink (b) CS/MWCNT printable conductive ink on TPU substrate.

D. Characterization and Conductivity of CS/MWCNT Printable Conductive Ink

Rheology analysis was performed on CS/MWCNT printable conductive ink to analyze the fluid characteristic and measure the viscosity using Anton Paar MCR 72 rheometer with shear rate 0-1000 s^{-1}. Surface morphology analysis was conducted using FESEM JEOL JSM-7800F to observe the dispersion of MWCNT particles within CS polymer. Raman spectra was analyzed using Renishaw1000 micro-Raman spectrometer with 785nm excitation wavelength to study the structural information of CS/MWCNT printable conductive ink. The changes of chemical structure of CS/MWCNT printable conductive ink were analyzed using FTIR (NicoletTM iS50 FTIR, ThermoFisher Scientific, Massachusetts, United States). For electrical properties, the four-point probe method was used to measure the conductivity of CS/MWCNT printable conductive ink. A bending test was performed to analyze the presence of crack of CS/MWCNT printable conductive ink towards two different CS weightage.

III. RESULTS AND DISCUSSION

A. Rheological properties of CS/MWCNT printable conductive ink

A non-Newtonian fluid with the shear-thinning characteristic of CS/MWCNT composite ink was demonstrated in Fig. 2. It can be observed that viscosity decreases as shear rate increases. This is due to disentanglement of CS/MWCNT particles that occurred allowed them to slip past another while shear was applied, resulting to viscous structure. Plus, shear-thinning fluid behavior was required in this work as it can hinder the ink from smudging upon deposition onto TPU substrate during the fabrication of printed conductive ink. CS/MWCNT

printable conductive ink recorded the highest viscosity of 1 Pa·s, indicating that the amount of binder will affect the ink viscosity [7]. The presence of CS as binder was crucial as it can influence the viscosity of ink. In addition, CS/MWCNT printable conductive ink obey the ideal viscosity of printed ink as it reflects a non-Newtonian fluid characteristic, unlike water or solvent [7][8].

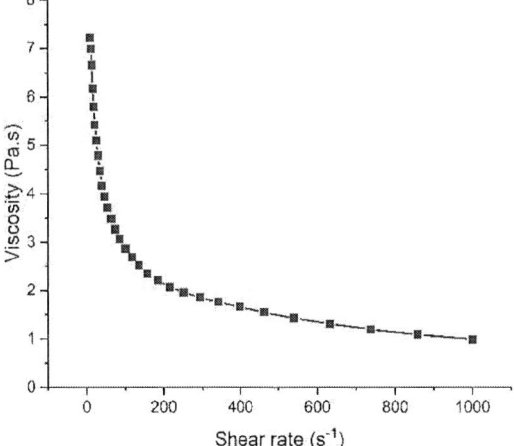

Fig. 2. Shear-thinning fluid behavior and recorded ink viscosity for CS/MWCNT printable conductive ink.

B. Characterization of CS/MWCNT printable conductive ink

The FESEM image of raw MWCNTs and CS/MWCNT printable conductive ink were shown in Fig. 3, respectively. Upon dispersion within CS polymer matrix, effective disentanglement of raw MWCNTs were observed as illustrated in Fig. 3(b). Thus, this indicates the homogenous dispersion of MWCNT fillers. Besides, the sonication upon MWCNT fillers within CS matrix also affects the MWCNT diameter [9]. A reduction in diameter can be observed for MWCNTs in CS/MWCNT where the average diameter of MWCNTs were reduced to 20-30 nm from 0.6-1.7 µm of rawMWCNTs. The structural information of the synthesized printable conductive ink was investigated using RAMAN spectra. Fig.4. illustrates the RAMAN spectra for MWCNT, CS and CS/MWCNT. Two characteristic peaks were observed in all samples: the D-band and the G-band. MWCNT powder depicted D-band centered at 1327 cm-1 and G-band centered at 1589 cm-1, representing two consistent graphitic CNTs peak [8][9].

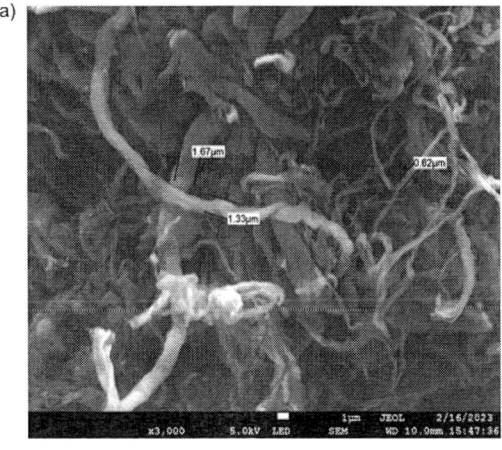

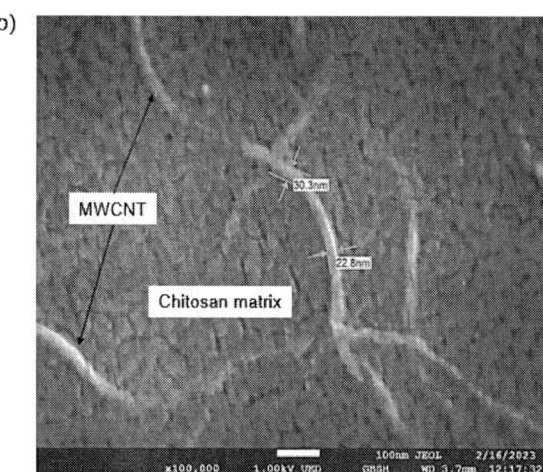

Fig. 3. FESEM image of (a) Agglomerated and entangled raw MWCNT (b) Disentangled and homogeneous dispersion of MWCNT particles within CS/MWCNT printable conductive ink.

The presence of D-band was related to the disordered structure from sp^2 carbon atoms. Meanwhile, the G-band was attributed to the tangential mode caused by in- plane stretching of the graphitic carbon. Upon addition of CS, the D-band shifted to 1329 cm^{-1} and the G-band shifted to 1617 cm^{-1}, indicating that the CS has interacted well with the MWCNT. This is due to the increased vibrational energy between the (-C=OCH3) of CS and (C=C) bands resulted to shifting of the Raman peaks. In addition, this result showed that the presence of SDS as surfactant did not significantly affect and disrupt the nanotubes structures. All peaks recorded were in line with previous literature [12][13].

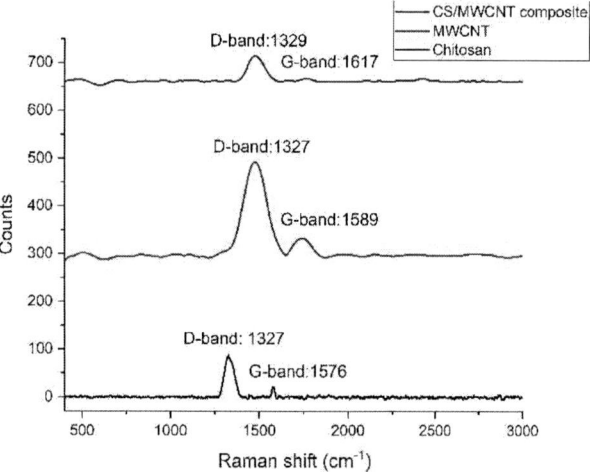

Fig. 4. The presence of D-band and G-band in RAMAN spectra of CS, MWCNT and CS/MWCNT printable conductive ink.

FTIR analysis was conducted to elucidate changes of chemical structure of CS/MWCNT printable conductive ink. As shown in Fig. 5, a broad absorption band appeared at 3200 - 3500 cm^{-1} corresponding to -OH stretching for CS and CS/MWCNT printable conductive ink. Raw MWCNT showed peak at 1512 cm^{-1} proved the presence of C-C stretching and peak at 1752 cm^{-1} attributed to C=O stretching. Meanwhile, spectrum recorded for pure CS possessed main peaks at 3357 cm^{-1} and 2881 cm^{-1} due to occurrence of -OH stretching and N-H group. Besides, the band at 1664 cm^{-1}

attributed to the C=O band implying presence of amide class I band and 1020 cm^{-1} peak of C-O stretching was observed indicating characteristic peak of CS. For CS/MWCNT printable conductive ink, peaks shifting was observed mainly corresponding to the -OH stretching peak and the N-H, while other peaks matched with peaks obtained for raw MWCNT and pure CS [14]. It can be observed that -OH peak for CS/MWCNT printable conductive ink shifted to 3226 cm^{-1} and N-H shifted to 2884 cm^{-1} compared to 2881 cm^{-1} and in pure CS, respectively. Plus, broader absorption peak was observed at peak 1545 cm^{-1} for CS/MWCNT printable conductive ink replacing peak at 1604 cm^{-1} for CS indicating the C=O stretching band converted from amino groups into amide bond. This implies successful attachment of MWCNT towards the CS. All these absorption peaks were in line with previous work [15][16].

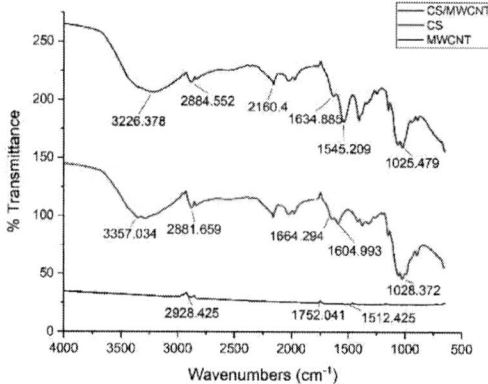

Fig. 5. FTIR profiler for raw MWCNT, CS and CS/MWCNT printable conductive ink.

C. Conductivity and bending test for CS/MWCNT printable conductive ink

The conductivity of CS/MWCNT printable conductive ink was evaluated using four-probe point device. The average conductivity of CS/MWCNT printable conductive ink recorded was 4.46×10^{-3} S/m. This conductivity is in line with previous work which used synthetic polymer binder for its MWCNT printable conductive ink synthesis, as tabulated in Table I. This indicates that integration of adequate CS will not impact or deteriorate electrical properties of MWCNT. This is supported by Menon et. al. where appropriate amount of binder will not affect the electrical properties of synthesized MWCNT printed conductive ink [8].

Two different CS/MWCNT printable conductive ink having different CS weightage were tested for bending test performance. The sensors were bent and relaxed six consecutive times in horizontal and vertical form. No crack was observed for higher CS weightage as shown in Fig. 6(a). The compact structure of CS/MWCNT printable conductive ink was illustrated through microscopic image as depicted in Fig. 6(b). This indicates adequate amount of CS holding the MWCNT particles within the ink component. Meanwhile, lower weightage of CS resulted in the presence of crack upon bending and several voids can be observed through the microscopic image as shown in Fig. 6(c). This is due to the inadequate amount of CS for MWCNT to bind. Hence, the presence of CS as natural binder proved the ability of CS to bind and hold MWCNT fillers and prevent ink breakage upon bending. In addition, high Young Modulus of MWCNT as

979-8-3503-2369-6/23 $31.00 © 2023 IEEE

filler could prevent complete cracking of CS/MWCNT printable conductive ink upon bending [16].

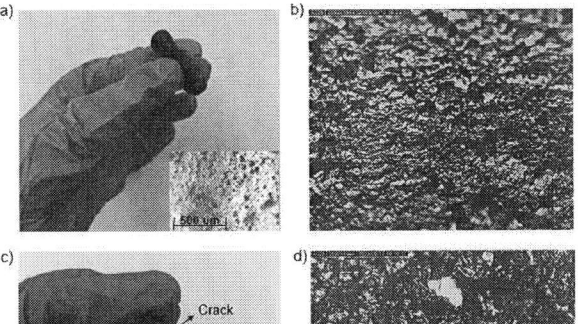

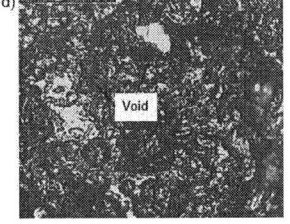

Fig. 6. Bending test for CS/MWCNT printable conductive ink with two different CS weightage (a) No crack observed upon bending for higher CS weightage (b) Severe crack observed upon bending for lower CS weightage. (c) Compact structure for higher CS weightage (d) Presence of crack (void) for lower CS weightage.

TABLE I.   TABLE I COMPARISON WITH PREVIOUS WORK

Conductive Filler	Binder	Conductivity (mS/m)	Ref
MWCNT	Polystyrene	0.002	[5]
	Triton X-100	81.1	[4]
	Polydimethylsilo-xane (PDMS)	0.03	[17]
	CS	4.46	This work

IV. CONCLUSION

In conclusion, the integration of CS as natural binder and MWCNT as conductive filler for printable conductive ink synthesis were evaluated and analyzed. Rheological property showed ink viscosity at 1 Pa·s. Two characteristic peaks were observed in all samples which were the D-band and the G- band for Raman analysis. FTIR analysis illustrated that MWCNT been successfully bind with CS. The conductivity of CS/MWCNT printable conductive ink was 4.46 mS/m. This indicates that the presence of CS does not affect electrical properties of MWCNT. The bending test showed that no crack for higher CS weightage implying that presence of CS able to bind and hold ink particles together, resulting to flexible CS/MWCNT printable conductive ink. Therefore, integration of CS as natural binder for eco-friendly printable conductive ink provides promising solution for printable electronics.

ACKNOWLEDGMENT

This work is fully supported by the Ministry of Higher Education (MOHE) Fundamental Research Grant Scheme (FRGS 21-249-0858) (Grant No: FRGS/ 1/2021/TK0/UIAM/02/14).

REFERENCES

[1] J. Khan and M. Mariatti, "Effect of natural surfactant on the performance of reduced graphene oxide conductive ink," J. Clean Prod., vol. 376, no. September, p. 134254, 2022, doi: 10.1016/j.jclepro.2022.134254.

[2] K. Senthil Kumar, P.-Y. Chen, and H. Ren, "A Review of Printable Flexible and Stretchable Tactile Sensors," Research, vol. 2019, pp. 1–32, 2019, doi: 10.34133/2019/3018568.

[3] N. L. Lukman Hekiem, A. A. Md Ralib, M. A. M. Hatta, F. B. Ahmad, R. A. Rahim, and N. F. Za'bah, "Performance Analysis of VOCs Detection using Polyisobutylene and Chitosan Overlayed on QCM Sensor," Proc. - 2021 IEEE Reg. Symp. Micro Nanoelectron. RSM 2021, pp. 157–160, 2021, doi: 10.1109/RSM52397.2021.9511614.

[4] J. O. Akindoyo, N. H. Ismail, and M. Mariatti, "Development of environmentally friendly inkjet printable carbon nanotube-based conductive ink for flexible sensors: effects of concentration and functionalization," J. Mater. Sci. Mater. Electron., vol. 32, no. 9, pp. 12648–12660, 2021, doi: 10.1007/s10854-021-05900-y.

[5] T. Singla, A. Pal Singh, S. Kumar, G. Singh, and N. Kumar, "Characterization of MWCNTs-polystyrene nanocomposite based strain sensor," Proc. Inst. Mech. Eng. Part E J. Process Mech. Eng., vol. 235, no. 2, pp. 463–469, 2021, doi: 10.1177/0954408920966301.

[6] S. Maliki et al., "Chitosan as a Tool for Sustainable Development: A Mini Review," Polymers (Basel)., vol. 14, no. 7, 2022, doi: 10.3390/polym14071475.

[7] Z. Gu and W. Mirihanage, "Development of carbon ink for wearable sensors," M.S. thesis, Faculty of Science and Engineering, Manchester 2021.

[8] H. Menon, R. Aiswarya, and K. P. Surendran, "Screen printable MWCNT inks for printed electronics," RSC Adv., vol. 7, no. 70, pp. 44076–44081, 2017, doi: 10.1039/c7ra06260e.

[9] I. Y. Q. and I. A. F. H. Ma'an F. Alkhatib, Mohamed E.S. Mirghani, "Immobilization of Chitosan onto Carbon Nanotubes for Lead Removal from Water." Journal of Applied Sciences, 2010, doi: https://doi.org/10.3923/jas.2010.2705.2708.

[10] S. Yang et al., "Design of chitosan-grafted carbon nanotubes: Evaluation of how the -OH functional group affects Cs+ adsorption," Mar. Drugs, vol. 13, no. 5, pp. 3116–3131, 2015, doi: 10.3390/md13053116.

[11] Q. Luo et al., "Carbon Nanotube/Chitosan-Based Elastic Carbon Aerogel for Pressure Sensing," Ind. Eng. Chem. Res., vol. 58, no. 38, pp. 17768–17775, 2019, doi: 10.1021/acs.iecr.9b02847.

[12] R. P. Shukla and H. Ben-Yoav, "A Chitosan–Carbon Nanotube- Modified Microelectrode for In Situ Detection of Blood Levels of the Antipsychotic Clozapine in a Finger-Pricked Sample Volume," Adv. Healthc. Mater., vol. 8, no. 15, 2019, doi: 10.1002/adhm.201900462.

[13] F. Paquin, J. Rivnay, A. Salleo, N. Stingelin, and C. Silva, "Multi- phase semicrystalline microstructures drive exciton dissociation in neat plastic semiconductors," J. Mater. Chem. C, vol. 3, no. 207890, pp. 10715–10722, 2015, doi: 10.1039/b000000x.

[14] E. A. K. Nivethaa, S. Dhanavel, V. Narayanan, and A. Stephen, "Fabrication of chitosan/MWCNT nanocomposite as a carrier for 5-fluorouracil and a study of the cytotoxicity of 5-fluorouracil encapsulated nanocomposite towards MCF-7," Polym. Bull., vol. 73, no. 11, pp. 3221–3236, 2016, doi: 10.1007/s00289-016-1651- 1.

[15] M. Guo et al., "Carbon nanotube-grafted chitosan and its adsorption capacity for phenol in aqueous solution," Sci. Total Environ., vol. 682, pp. 340–347, 2019, doi: 10.1016/j.scitotenv.2019.05.148.

[16] Y. Yang, C. Luo, J. Jia, Y. Sun, Q. Fu, and C. Pan, "A wrinkled Ag/CNTs-PDMS composite film for a high-performance flexible sensor and its applications in human-body single monitoring," Nanomaterials, vol. 9, no. 6, pp. 1–17, 2019, doi: 10.3390/nano9060850.

[17] J. Du et al., "Optimized CNT-PDMS Flexible Composite for Attachable Health-Care Device," Sensors, vol. 20, no. 16, p. 4523, Aug. 2020, doi: 10.3390/s20164523.

Achieving Compact Structure and Good Mechanical Properties of AlN Thin Film through Low Temperature HiPIMS

Zulkifli Azman
Faculty of Electrical and Electronic
Universiti Tun Hussein Onn Malaysia
Batu Pahat, Malaysia
zulkifli.bin.azman@gmail.com

Nafarizal Nayan
Microelectronics and Nanotechnology-Shamsuddin Research Centre
Universiti Tun Hussein Onn Malaysia
Batu Pahat, Malaysia
nafa@uthm.edu.my

Chin Fong Soon
Microelectronics and Nanotechnology-Shamsuddin Research Centre
Universiti Tun Hussein Onn Malaysia
Batu Pahat, Malaysia
soon@uthm.edu.my

Ahmad Shuhaimi Abu Bakar
Low Dimensional Material Research Centre
Universiti Malaya
Kuala Lumpur, Malaysia
shuhaimi@um.edu.my

Ahmad Nasrull Mohamed
Microelectronics and Nanotechnology-Shamsuddin Research Centre
Universiti Tun Hussein Onn Malaysia
Batu Pahat, Malaysia
nasrull@uthm.edu.my

Norain Sahari
Faculty of Engineering Technology
Universiti Tun Hussein Onn Malaysia
Pagoh, Malaysia
norains@uthm.edu.my

Yusmar Palapa Wijaya
Department of Electronic System Engineering
Politeknik Caltex Riau
Pekanbaru, Indonesia
yusmar@pcr.ac.id

Mohd Yazid Ahmad
Nanorian Technologies Sdn Bhd, Kajang, Malaysia
yazid@nanoriantech.com

Abstract— Thin films of aluminium nitride (AlN) were successfully deposited using the High Power Impulse Magnetron Sputtering (HiPIMS) technique at room temperature. X-ray diffraction (XRD) analysis confirmed a hexagonal close-packed (hcp) AlN crystal structure in the films. Notably, a highly pronounced orientation along the (002) plane was observed at the lowest sputtering pressure of 3 mTorr, while higher pressures led to the formation of a polycrystalline structure. The films deposited at 3 mTorr exhibited a dense and pebble-like morphology, resulting in impressive mechanical properties. The measured mechanical hardness was determined to be 17 GPa, and the Young's modulus was found to be 187 GPa. By reducing the sputtering pressure, the crystal quality, surface morphology, and mechanical properties of the films were observed to improve. This study demonstrates the potential of the HiPIMS technique to produce high-quality thin films at low temperatures, opening opportunities for various applications in fields such as electronics, optoelectronics, and materials science. The findings highlight the importance of process parameters in tailoring the properties of thin films for specific applications, paving the way for further advancements in thin film technology.

Keywords— AlN thin films, HiPIMS, compact structure, mechanical properties, room temperature deposition

I. Introduction

Due to its wide band gap at 6.2 eV, high thermal conductivity at 320 W/mK, aluminium nitride (AlN) is a very technologically advantageous material for the semiconductor industry [1]. Rocksalt, zinc blende, and wurtzite phase are the three main phases in which AlN can be found [2]. The wurtzite or hexagonal close-packed (hcp) phase is the most stable of the three phases. The native crystal structure of AlN is hcp-AlN, which has a higher thermodynamic stability than zinc blende. Due to its compatibility with a variety of materials and great thermal and chemical stability, it is frequently employed as a substrate for the growth of other materials such as Gallium Nitride (GaN) [3]. The hcp-AlN is desirable for high-temperature electronic applications, optoelectronic, and surface acoustic devices (SAW)[4].

A wide range of techniques have been used to produce hcp-AlN, including molecular beam epitaxy (MBE)[5] metal-organic chemical vapour deposition (MOCVD) [6] plasma enhanced chemical vapour deposition (PECVD)[7] and pulsed laser deposition (PLD)[8]. For the most part, these techniques needed high temperatures as the deposition process developed. Though, the high deposition temperature may increase the surface roughness and compressive stress which produce negative impact on the devices' performance [9]. Among them, reactive magnetron sputtering offers superior sputtering parameter control, a straightforward operation at a low cost, high coating material adherence, and deposition at a comparatively low temperature [1][4].

One of the current trends in thin film deposition by magnetron sputtering is high-power impulse magnetron sputtering (HiPIMS) [10]. HiPIMS is recognised for producing a high proportion of target material at a low frequency and short pulse length, which increases plasma density and ionisation rate [11]. Thus, it results in dense, higher hardness, smoother surfaces, and better adherence of thin layer deposition at lower process temperatures [12]. The ability of HiPIMS to produce dense AlN thin films at lower substrate temperatures is crucial in preserving the integrity of temperature-sensitive substrates, such as plastics or organic materials.

In a previous study, K. Ait Aissa et al. investigated the hcp-AlN thin film's deposition rate and residual stress analysis prepared by HiPIMS and DC magnetron sputtering at various sputtering pressures with very short target to

979-8-3503-2369-6/23 $31.00 © 2023 IEEE

substrate distance (TSD), at 30 mm [13]. For a larger chamber dimension that permits complex co-sputtering processes, the TSD is considerably longer. In magnetron sputtering, the TSD plays a crucial role in influencing the film density as it directly impacts the energy of the sputtered particles [14]. It became a challenge since the hcp-AlN with preferred orientation along (002) plane known require higher sputtering energy [15]. In this study, the authors demonstrate the impact of sputtering pressure on the crystal structure evolution of hcp-AlN thin film on Si (111) at a longer TSD without the aid of adding heat to the substrate, by utilizing the benefits of HiPIMS technique. Traditional sputtering techniques often require elevated substrate temperatures to facilitate film growth and adhesion, which can lead to substrate degradation or limited material choices. Therefore, the aim of this study is to investigate the potential of the HiPIMS technique to deposit hcp-AlN on Si (111) at room temperature with longer TSD by adjusting the common sputtering parameters, such as sputtering pressure. This study provides a detailed analysis of the structural, surface morphology, and mechanical properties of the deposited thin films.

II. MATERIALS AND METHODS

A. Materials and Sputtering Paramters Conditions

In this experiment, hcp-AlN thin films were produced on silicon substrates using HiPIMS techniques without heat assistance during deposition. Prior to deposition, Si (111) substrates were cleaned with a hydrofluoric acid (HF) solution mixed with deionized (DI) water (1:5 ratio) to eliminate any contaminants that could hinder the growth of the epitaxial layer of AlN on top of the silicon substrate and remove the native oxide layer present on the substrates. The cleaning process involved submerging the substrates in diluted HF solution for 2 minutes, followed by washing with DI water and drying using nitrogen flow. Subsequently, the substrates were loaded into the sputtering chamber using a load lock system attached to the setup (SNTEK PSP 5004, Korea).

In this instance, a single sputter gun was used to sputter a 3-inch-diameter 5N purity of aluminium target (ITASCO, Korea), which was positioned 120 mm above the substrate inside the chamber. The sputtering system is outfitted with a DC power source (PSPLASMA SDC1024A, Korea) and a High-Power Impulse power source (STARFIRE, USA), both of which have the ability to create DC pulses with maximum 2 kHz repetition rates and pulse widths between 1 and 400s.

At first, the chamber was inflated to base pressure at about $7 \times 10-6$ Torr. For each experiment, the system was filled with 5N purity of argon and nitrogen gas in a 2:1 ratio at fixed 100 sccm and 50 sccm flow rates, respectively, with varying working pressures of 3, 5, and 10 mTorr. At an average sputtering power of 300 W, the deposition runs for 2 hours. The voltage supply was kept at 350 V with an average duty cycle of 6% at maximum 2kHz repetition rate. Throughout the deposition process, the substrate holder was kept rotated to ensure uniform thin film deposition achieved.

B. Characterization Method

X-ray diffraction (XRD, Panalytical, UK) with a Cu-K radiation source (1.5406 nm) was used to evaluate the crystal structure of AlN thin films. The 1/2 divergence slit and 0.5°

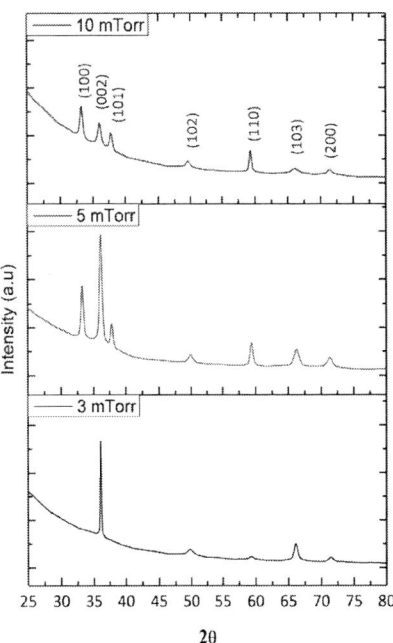

Fig.1   XRD Spectrum of hcp-AlN at different sputtering pressure

incidence angle were used to collect the diffraction pattern over the 20° to 80° range.   Field emission scanning electron microscopy, FESEM (JOEL-JSM1763, Japan) was use to examine the surface morphology. The influence of sputtering parameters to the mechanical properties evaluated by mechanical hardness (H) and modulus elasticity (Hf) calculated through nano-indentation measurement (Bruker – Hysitron TI PREMIER, USA) at peak force of 2000 µN.

III. RESULTS AND DISCUSSION

A. Crystal Structure of hcp-AlN Thin Films

The XRD spectrum from the Fig. 1 shows all samples exhibit the wurtzite crystal structure of AlN with reference ICSD 98-003-1169. It clearly observed that at higher sputtering pressure by this sputtering technique produces polycrystalline structure shown by various peaks. It slowly shows dominant (002) peak as the sputtering pressure reduced to 3 mTorr. There are two types of Al-N bonds in hexagonal AlN which are B1 bonds and B2 bonds. In these thin films, the B1 and B2 bonds are likely to be randomly oriented, resulting in a polycrystalline structure at higher pressure as shown by several XRD peaks with 10 mTorr sputtering pressure. The B1 bond is a polar covalent bond, in which electrons are shared between the Al and N atoms. This bond is formed between Al atoms in the octahedral coordination and N atoms in the trigonal planar coordination.

The B1 bond is stronger than the B2 bond and has a higher bond energy. The B2 bond is a weaker covalent bond, in which electrons are shared between two N atoms. While the (002) plane orientation have only B1 bond, higher kinetic energy required to form a highly (002) plane orientation of hcp-AlN . In addition, Fang, Liping, et al. claimed that in the case of a non-heated substrate, the adatoms sputtered onto the substrate lacked the necessary kinetic energy, resulting in the formation of lower formation energy planes such as (100), (101), and (110). Only with further temperature

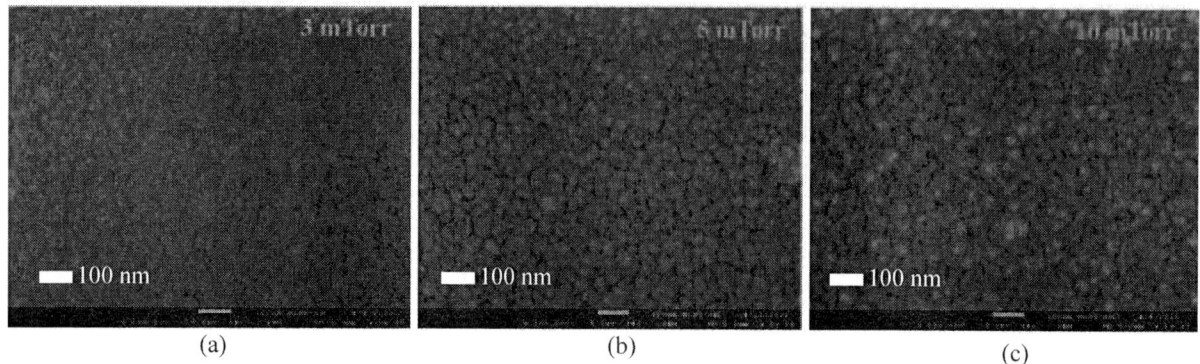

Fig. 2. FESEM images of hcp-AlN at (a) 3 mTorr, (b) 5 mTorr, and (c) 10 mTorr

assisted up to 200ºC will facilitate the formation of (002) plane hcp-AlN [15].

It is found that at lower sputtering pressures, both ions and neutral species in the sputtering chamber have larger kinetic energies. This is in accordance with the kinetic theory of molecular gases. These species condense and then become adatoms which are solid particles that fall to the substrate surface. Adatoms' higher kinetic energies enable them to move about more easily, which promotes the development of (002) plane hcp-AlN thin films [16]. It results the evolution from polycrystalline to highly (002) plane orientation of hcp-AlN shown in the XRD spectrum as the sputtering reduced. In this work, the energetic sputtering particles provided by HiPIMS with combination of low sputtering pressure has facilitate the growth of high crystalline hcp-AlN at no further heat assistance to the substrate.

B. Surface Morphology

The findings presented in Fig. 2 demonstrate the variations in the FESEM images (at 100x magnification) of hcp-AlN when deposited under different sputtering pressures. The analysis reveals a dense and compact with pebble-like structure for a sputtering pressure of 3 mTorr. The observed dense and pebble-like structure at a sputtering pressure of 3 mTorr suggests that HiPIMS can effectively produce compact AlN thin films with improved packing density. Dense and compact structure suggests that the AlN grains possess a preferred crystallographic orientation. This alignment indicates that a majority of the grains grow in a specific direction, resulting in a single dominant plane of crystal growth as shown in XRD spectrum. This is particularly beneficial for applications that require high-quality thin films with minimal defects

Fig. 2(c) illustrates that at 10 mTorr, there are several distinct shapes present. Less dense microstructure with prominent voids, indicating that the AlN grains lack preferred orientation and are randomly oriented. The presence of voids signifies a less uniform packing of atoms and a less compact structure. Consequently, the XRD pattern exhibits multiple peak planes, indicating the polycrystalline nature of the film. Each peak represents different crystallographic orientations of the AlN grains within the thin film.

C. Mechanical Properties

Fig. 3 shows indentation load-displacement curve for every sample including the AlN/Si template procured from

Kyma for comparison. The peak force of 2000 μN resulted in a penetration depth of approximately 75 nm. The hardness and Young's modulus of hcp-AlN thin films were determined using the load-displacement data, employing the analytical approach, established by Oliver and Pharr [17], as described by the following equation:

$$H = P_{max}/A_c \qquad (1)$$

where the H in (1) equal to the maximum indentation load, P_{max}, divided by projected contact area, A_c. The curve produced were observed almost identical for those samples. For a clear view, Table 1 shows the summarize calculated value of hardness and Young's modulus including the values form previous study. It can be seen that the calculated hardness and Young's modulus in this study were placed in the range while we reported that the deposition occurred at room temperature compared to other works. At lower sputtering pressure where dense pebble-like structure of hcp-AlN produce almost the same mechanical properties with commercially available AlN/Si template. Therefore, it suggests that the HiPIMS technique able to produce high quality AlN thin film at low temperature that meet the specification for mechanical coating application as well.

It was strongly correlate with the crystal structure and surface morphology analysis mention earlier. The highly oriented of (002) plane hcp-AlN with compact structure was observed having higher hardness and higher Young's modulus. For hcp-AlN with highly (002) oriented plane, the atoms are arranged more closely in the direction perpendicular to the film surface. This close packing and alignment result in stronger atomic bonds and enhanced interatomic forces.

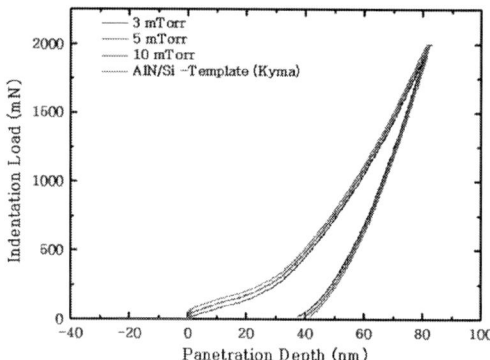

Fig. 3. Indentation load-displacement curve

TABLE I. MECHANICAL PROPERTIES OF HCP-ALN THIN FILM PRODUCED IN THIS STUDY AND COMPARISON WITH PREVIOUS REPORT

Reference	HiPIMS Duty Cycle (%)	Hardness, H (GPa)	Young's Modulus, E_f (GPa)	Deposition Temperature (°C)
C.-T. Chang et al.[18]	3.5	18.3	219	100
J.C. Ding et al.[19]	10.5	13	146	-
Y.-C. Yang et al.[20]	3.5	28	304	200
This work (3 mTorr) (5 mTorr) (10 mTorr)	6	17.48 14.93 10.34	187 132 120	Room Temperature
AlN/Si Template (Kyma)	-	19.13	187	-

The thin films tend to have higher bond strengths and greater resistance to deformation under applied stress. Therefore, it become the good reason why the highly (002) plane oriented of hcp-AlN chosen in application for piezoelectric devices [4]. As the thin film having increase in grain size at higher sputtering pressure, the number of grains decreases, resulting in a lower density of grain boundaries. With larger grains, there is a reduced number of interfaces between grains, and the relative proportion of grain boundary area decreases compared to the overall grain volume. In terms of mechanical properties, larger grain sizes generally lead to reduced strength and increased ductility. The fewer grain boundaries can act as preferred paths for dislocation movement, which can allow for more plastic deformation and strain accommodation. Thus, it shows higher value of Young's modulus at lower sputtering pressure where the grain boundary increased due to dense structure.

IV. CONCLUSION

The HiPIMS method is renowned for producing high-quality thin films with strong adherence. This study demonstrates the important role that HiPIMS's strong ionization ability plays in producing hcp-AlN thin films with a highly (002) plane orientation even at ambient temperature and longer TSD. Higher sputtering energy is needed to grow hcp-AlN along the (002) plane, hence the substrate needs additional energy help in the form of heat supply. This critical state can be overcome by HiPIMS by itself. Additionally, it was shown that the sputtering pressure was a key factor in shaping the characteristics of hcp-AlN, particularly the crystal structure and surface morphology. Its other qualities, such mechanical hardness and elasticity behaviour, are significantly impacted by the alteration in crystal structure. The possibility to study the deposition on specific substrate, notably heat-sensitive materials, has been made possible by the low temperature deposition of hcp-AlN using the HiPIMS technique.

ACNOWLEDGMENET

This research was supported by Malaysian Ministry of Higher Education (MOHE) through Fundamental Research Grant Scheme (FRGS) (FRGS/1/2020/STG05UTHM/023) and CREST industrial research grant (P28C1-17).

REFERENCES

[1] A. Iqbal and F. Mohd-Yasin, "Reactive sputtering of aluminum nitride (002) thin films for piezoelectric applications: A review," *Sensors (Switzerland)*, vol. 18, no. 6, pp. 1–21, 2018.

[2] B. Ahmed and B. I. Sharma, "Structural and electronic properties of aln in rocksalt, zinc blende and wurtzite phase: A dft study," *Dig. J. Nanomater. Biostructures*, vol. 16, no. 1, pp. 125–133, 2021.

[3] V. N. Bessolov, E. V Gushchina, E. V Konenkova, and T. V L, "Hexagonal AlN Layers Grown on Sulfided Si (100) Substrate," vol. 44, no. 1, pp. 81–83, 2018.

[4] K. A. Aissa *et al.*, "AlN films deposited by dc magnetron sputtering and high power impulse magnetron sputtering for SAW applications," *J. Phys. D. Appl. Phys.*, vol. 48, no. 14, p. 145307, 2015.

[5] P. Shao *et al.*, "Step-flow growth of Al droplet free AlN epilayers grown by plasma assisted molecular beam epitaxy," *J. Phys. D. Appl. Phys.*, vol. 55, no. 36, p. 364002, 2022.

[6] A. Kakanakova-Georgieva *et al.*, "MOCVD of AlN on epitaxial graphene at extreme temperatures," *CrystEngComm*, vol. 23, no. 2, pp. 385–390, 2021.

[7] N. Susilo *et al.*, "AlGaN-based deep UV LEDs grown on sputtered and high temperature annealed AlN/sapphire," *Appl. Phys. Lett.*, vol. 112, no. 4, 2018.

[8] S. Simeonov *et al.*, "Al/AlN/Si MIS Structures with Pulsed-Laser-Deposited AlN Films as Gate Dielectrics: Electrical Properties," *Rom. J. Inf. Sci. Technol.*, vol. 10, no. 3, pp. 251–259, 2007.

[9] J. W. Q. Zhang, G. F. Y. C. J. Yao, and Y. J. L. R. Sun, "Effect of substrate temperature and bias voltage on the properties in DC magnetron sputtered AlN films on glass substrates," 2015.

[10] J. Patidar *et al.*, "Improving the crystallinity and texture of oblique-angle-deposited AlN thin films using reactive synchronized HiPIMS," *Surf. Coatings Technol.*, vol. 468, p. 129719, 2023.

[11] J. M. Schneider, U. Helmersson, V. Kouznetsov, K. Maca, and I. Petrov, "A novel pulsed magnetron sputter technique utilizing very high target power densities," vol. 122, pp. 290–293, 1999.

[12] D. Lundin and K. Sarakinos, "An introduction to thin film processing using high-power impulse magnetron sputtering," *J. Mater. Res.*, vol. 27, no. 5, pp. 780–792, 2012.

[13] K. A. Aissa, A. Achour, J. Camus, L. Le Brizoual, P. Jouan, and M. Djouadi, "Comparison of the structural properties and residual stress of AlN fi lms deposited by dc magnetron sputtering and high power impulse magnetron sputtering at different working pressures," *Thin Solid Films*, vol. 550, pp. 264–267, 2014.

[14] H. U. Ha, H.-J. Seok, S. Yoon, D.-G. Lee, D.-W. Kang, and H.-K. Kim, "Plasma damage control via adjusting the target to substrate distance used to prepare semi-transparent perovskite solar cells," *Vacuum*, vol. 212, p. 112053, 2023.

[15] L. Fang *et al.*, "Substrate Temperature Dependent Properties of Sputtered AlN : Er Thin Film for In-Situ Luminescence," 2018.

[16] C.-M. Lin *et al.*, "Growth of highly c-axis oriented AlN films on 3C–SiC/Si substrate," in *Tech. Dig. Solid-State Sens. Actuators Microsystems Workshop*, 2010, pp. 324–327.

[17] S. V Kontomaris and A. Malamou, "Hertz model or Oliver & Pharr analysis? Tutorial regarding AFM nanoindentation experiments on biological samples," *Mater. Res. Express*, vol. 7, no. 3, p. 33001, 2020.

[18] C. Te Chang, Y. C. Yang, J. W. Lee, and B. S. Lou, "The influence of deposition parameters on the structure and properties of aluminum nitride coatings deposited by high power impulse magnetron sputtering," *Thin Solid Films*, vol. 572, pp. 161–168, 2014.

[19] J. Cheng, Q. Min, Z. Ren, S. Jeong, T. Fei, and K. Ho, "In fl uence of bias voltage on the microstructure , mechanical and corrosion properties of AlSiN fi lms deposited by HiPIMS technique," *J. Alloys Compd.*, vol. 772, pp. 112–121, 2019.

[20] Y. Yang, C. Chang, Y. Hsiao, J. Lee, and B. Lou, "Surface & Coatings Technology In fl uence of high power impulse magnetron sputtering pulse parameters on the properties of aluminum nitride coatings," *Surf. Coat. Technol.*, 2014.

FDTD Simulation for Optical Characteristics Study of Inverted Micro-pyramidal Surface Structure of Black Silicon

Md. Yasir Arafat
Faculty of Institute for Advanced Studies
University of Malaya
Kuala Lumpur, Malaysia
yasir@um.edu.my

Yasmin Binti Abdul Wahab
Faculty of Institute for Advanced Studies
University of Malaya
Kuala Lumpur, Malaysia
yasminaw@um.edu.my

Mohammad Aminul Islam
Faculty of Engineering
University of Malaya
Kuala Lumpur, Malaysia
aminul.islam@um.edu.my

Sharifah Fatmadiana Bt Wan Muhammad Hatta
Faculty of Engineering
University of Malaya
Kuala Lumpur, Malaysia
sh_fatmadiana@um.edu.my

Nurul Ezaila Alias
Faculty of Electrical Engineering
University of Technology Malaysia
Kuala Lumpur, Malaysia
ezaila@utm.my

Abstract— **This paper reports optical characteristics study of Black Silicon (BSi) containing inverted micro-pyramidal structure on p-type single-crystalline silicon (100) (sc-Si) surface. Finite-Difference Time-Domain (FDTD) simulation is conducted in three diverse modellings for analyzing the effects of light incident and reflectance based on surface morphology. The approach of each modelling is described by designing the microstructure and then acquiring the reflectance results, followed by observing the simulation results of the microstructure characteristics of vectors and electromagnetic fields as a function of position and wavelength. The final approach investigates the optical characteristics in the influence of irregularly distributed inverted micro-pyramidal structures of the real BSi structure surface, an analogous structure with exact dimensions. The characterized samples were obtained by metal-assisted chemical etching containing inverted micro-pyramidal surface structures. The findings revealed the optical characteristics of inverted pyramidal microstructure using a numerical 3D image reconstruction algorithm followed by FDTD simulation.**

Keywords— *Black Silicon, Inverted pyramid microstructure, Texturization, Single crystalline Silicon, FDTD*

I. INTRODUCTION

The implementation of micro-scale textured structures on the surface of silicon is a crucial step in minimizing light reflectance [1]. The enhancement of device performance, such as solar cells, requires a greater light absorption mechanism. This can be achieved by reducing the reflectance of the front surface. For minimizing this surface reflectance researcher have introduced the concept of black silicon (BSi) that applies surface modification on the c-Si wafer for creating a micro/nanotextured surface layer, which results in excellent light absorption abilities in terms of photon trapping mechanism [2].

Among the several methods for surface texturization metal-assisted chemical etching (MACE) is a highly effective method for surface texturization, which has gained significant traction due to its capacity to generate precise microstructures with homogeneity on silicon surfaces, by

utilizing metal nano- particles as a catalyst [3]. This study presents an analysis of the optical, physical, and morphological characteristics of black silicon (BSi) produced through the utilization of nickel-metal assisted chemical etching (MACE) employing innovative two-step methods. In the initial phase, a p-type single-crystalline silicon (100) wafer was treated with immersion in a concentrated aqueous solution comprising a mixture of ammonium fluoride and nickel (II) sulfate hexahydrate $(NH_4)_3FNi(SO_4)2(H_2O)_6$, with the aim of facilitating the deposition of nickel nanoparticles. During the second stage of the experiment, Si wafers that had been deposited with Ni were subjected to etching at a temperature of 25°C. This was achieved by immersing the wafers in an aqueous solution containing a mixture of hydrofluoric acid (HF) and hydrogen peroxide (H_2O_2).

The samples' morphological and optical changes were characterized using field emission scanning electron microscopy and UV-vis spectroscopy. The sample featuring an inverted micro- pyramidal structured surface that was to exhibit the minimum mean reflectance of approximately 4.80% within the wavelength range of 300-1200 nm. The study utilized the Finite-Difference Time-Domain (FDTD) method to examine the surface morphology of BSi. To investigate the effects of light incident and reflectance based on surface morphology through the FDTD simulation process the present work aims to propose and validate a simulated modelling method able to represent more accurately the textured surface structures obtained by the MACE fabrication.

II. METHOD

The FDTD simulation has been carried out using two different approaches. In this first approach, the inverted shapes were designed as a unit structure to represent the inverted micro-pyramidal shape structure of the real BSi sample surface. The synthesised capacitance approach [4] similarly uses single-structure unit cells for simulations. The design was based on the measurements of the top view and cross- sectional view of FESEM micrographs of the BSi sample. Based on the correlation in the height pattern

979-8-3503-2369-6/23 $31.00 © 2023 IEEE

between the analogous shape and the complete BSi sample, a much higher acceptable geometry for the corresponding BSi unit cell in this section was identified. Respective measurements depending on the topographical features were determined, namely height (h), structure spacing (d), and angles (θ), collected from the MACE fabricated sample as illustrated in Fig 1.

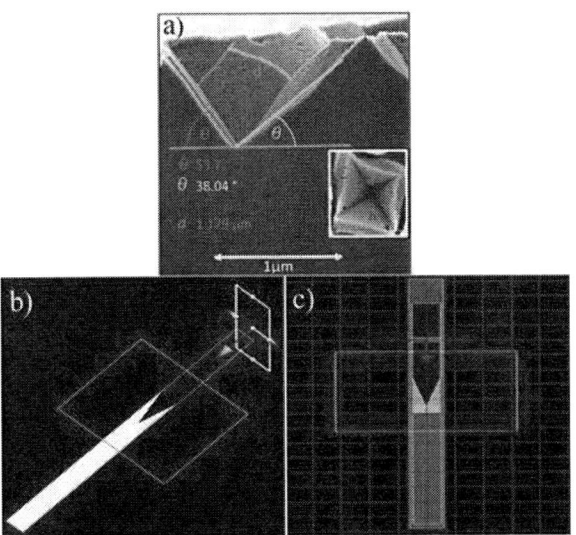

Fig. 1. (a) FESEM top and cross-sectional view images with mentioned structural height, space, and angles of the inverted pyramidal structure of sample 5, (b) Perspective view of FDTD simulated model structure for sample cell, (c). XY view of FDTD simulated model structure.

In the subsequent investigation, the surface data produced by BSi was reconstructed using Ansys Lumerical, HFSS, and MEEP technologies, run by finite element and finite-difference time-domain methods. This technique integrated small portions of the BSi surface morphology into the simulation for reflectance evaluations [5].

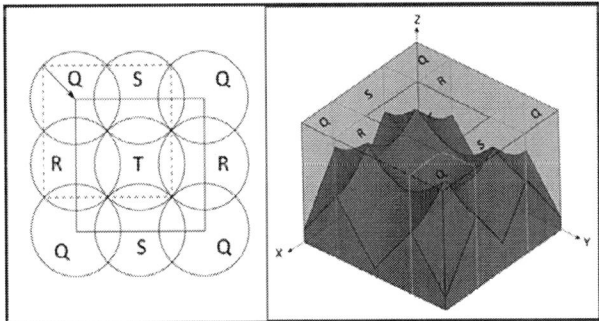

Fig. 2. a) Comparative visualization of the arrangement of the textured area from the selected region in terms of achieving a symmetrical structure regardless of the inverted micro-pyramidal structure's size. b) 3-D view of the HFSS-modeled unit cell structure.

The QRST region in Fig. 2, which is marked as the dotted square, is offset by 50% of the textured area from the selected region to achieve a symmetrical structure regardless of the size of the inverted micro-pyramidal structure. The rectangles indicated as QRST represent the part of the sample designed using HFSS. In the y-axis, each section contains peripheral proportions, approximately half of the average structural gap. In the x-axis, each section contains horizontal proportions equal to that same average structural gap. The final model proposed in Lumerical FDTD

illustrated in Fig. 3 provides a cross-sectional view of FESEM and a top view of the selected region. It was simulated by replicating the structure based on asymmetric boundary constraints, allowing it to perform simulations with the quasi-incident angular position..

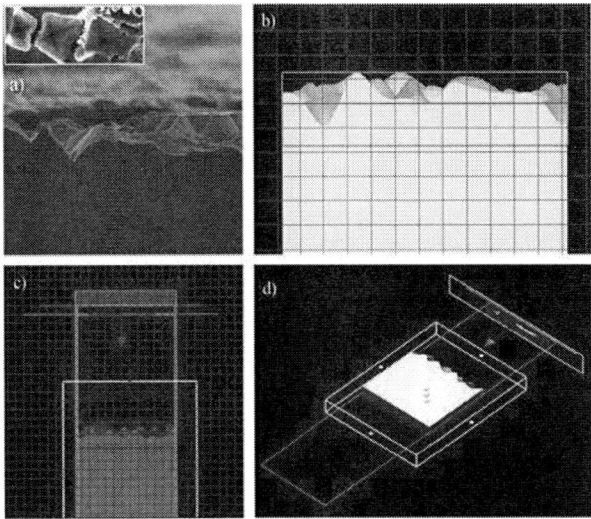

Fig. 3. ESEM cross-sectioal view along with the top view of the selected region, b) Simulated area, c) XY view of the FDTD simulated modelling construction, d) Perspective view of the simulated model.

In the single-structure cell model simulation section, although the modelled contours derived using inverted cones have a similar trend to the respective data, significant inconsistencies or variations are existing at real surface topography section (illustrated in Fig.3). The overall variation in the height propagation between the equivalent modelled form and the BSi sample is reflected in these inconsistencies. Thus, to address this inconsistency an equivalent unit cell constructed of numerous structures with changes in the inverted pyramids' height and distance is included to resemble the BSi non-periodic nature more accurately.

III. RESULTS AND DISCUSSION

A. Single-structured cell model

The first simulation has proceeded to investigate the height distribution of different geometrical shapes as it would give a significant in-depth idea of optical confinement behaviour of micro- inverted pyramidal structural shapes. This proceeds to compare the height distribution of different geometrical shapes with the ones obtained from BSi sample. Since the unit region is considered in a periodic-squared structure, the inverted pyramid structure illustrated in Fig.4 (a) is condensed in a cuboid. The expansion of the inverted pyramidal area, denoted as distance (d), and the groove depth denoted as height (h). This allowed for the proper representation of the single-structure cell model structure on the BSi sample's surface. The Seven different heights and distances were defined for the simulated structure, whereas sample-I represents the equivalent structure from sample cell.

979-8-3503-2369-6/23 $31.00 © 2023 IEEE

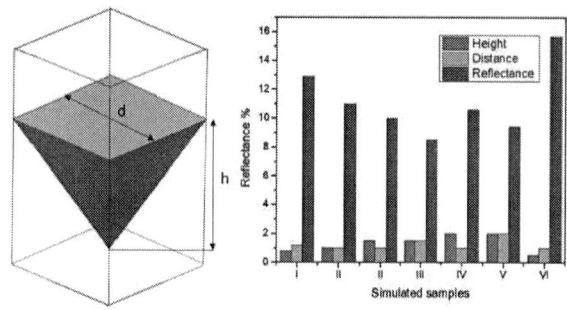

Fig. 4. (a) Measurement lines for the inverted pyramid structures (b) Impact of inverted micro-pyramidal structure's height and distance on reflectance.

TABLE I. COMPARISON BETWEEN THE TOPOGRAPHIC PARAMETERS OF EACH MICRO-PYRAMIDAL STRUCTURE REGION WITH THEIR CORRESPONDING REFLECTANCE

Simulated Sample No.	Height μm	Width μm	Reflectance %
I	0.79	1.179	12.9
II	1	1	11.01
III	1.5	1	9.97
IV	1.5	1.5	8.5
V	2	1	10.6
VI	2	2	9.41
VII	0.5	1	15.7

The above simulation results disclosed that an inverted pyramidal shape with the same height and distance results in better light trapping. The original structural shape with a height of 0.79 μm and a width of 1.179 μm produced a reflectance of 12.9%. Increasing the height and width of the structure results in a more efficient reflectance of approximately 11%. The reflectance was reduced by increasing the height and width. However, when the height was around 2 μm and the width was half of the height, the reflectance increased. A similar outcome occurred when the length of the height was decreased.

An advanced FDTD analysis was executed to comprehend the evolution of the microstructure's shape by pointing vector, electric field, and magnetic field distributions as a function of position and frequency/wavelength. Since the wavelength distribution throughout the changing frequency irradiation significantly changes the electron distribution in the first femtoseconds, a new optimal dielectric function for the excited states of silicon must be considered.

Accordingly, a simulation model was created solely for optical measurements to fulfil these criteria, where the inverted pyramidal structure was built around 3 μm in length and 1 μm in width. A thin layer of Si_3N_4 was used as an anti-reflection coating, and SiO_2 was considered for the recombination mechanism in getting a more accurate idea about photonic absorbance and reflectance incidents. The amplitude and direction of energy flow in electromagnetic waves are described by the Poynting vector. The Poynting vector, as a function of position and wavelength, represents the energy flux vector to indicate the direction where the energy is transported. Fig. 5 shows that the electromagnetic field absorption is significantly more substantial at the bottom of the inverted pyramid structure than at the top of the ripples. Absorption is maximum at the bottom rather than

the topside, with the presence of ripples and beads. This phenomenon explains the evidence that a strong electromagnetic distribution occurs inside the inverted pyramidal structure. The simulation results substantiate the reason for the least amount of reflectivity for BSi sample cell by confirming the higher photon harvesting capability of the inverted pyramidal microstructure surface.

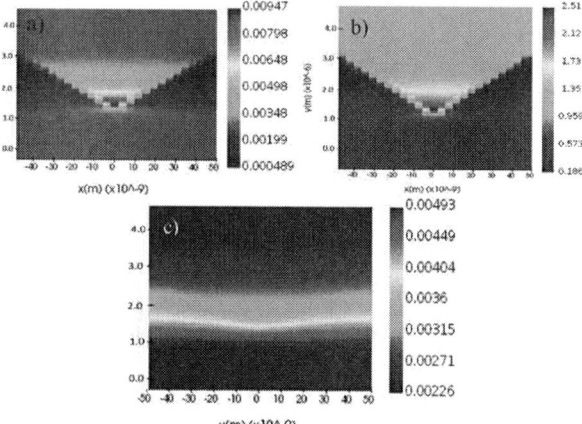

Fig. 5. a) Poynting vector as a function of position and frequency/wavelength, b) Electric field data as a function of position and frequency/wavelength, c) Magnetic field data as a function of position and frequency/wavelength.

B. Real surface topography

In the second approach, a comparison was made between the simulated reflectance of the area within the equivalent BSi sample and other model structures, as shown in Fig. 6 (a). The equivalent unit cell produced 6.35% reflectance, while the simulated structures of other models 1, 2, and 3 resulted in 6.94%, 6.69%, and 5.99%, respectively.

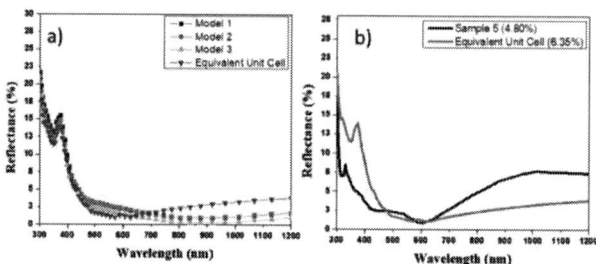

Fig. 6. a) Simulated reflectance results based on real morphological data of BSi compared with corresponding model structures, b) Sample 5 UV-vis vs FDTD simulated equivalent unit cell's comparative reflectance measurements.

The statistical model properties of each region and the equivalent unit cells are tabulated in Table II. The minimum, maximum, average, and standard deviation of heights are denoted as h_{min}, h_{max}, h_{μ}, and h_{σ}, respectively. The average depth, d_{μ}, is calculated by subtracting the average height from the maximum height of the structures.

For each model, additionally, compute the average surface common angle,

$$\tan \theta = \frac{h_n}{P_{u/2}} \qquad (1)$$

where, h_N depends on the average surface normal angle and P_u is the mean distance of the structures.

TABLE II.   MORPHOLOGICAL PARAMETERS OF SIMULATED MODELS FOR COMPARING THE RESULTED OPTICAL CHARACTERISTICS DIFFERENCES

Units	Parameters			
	Equivalent Unit Cell	Model 1	Model 2	Model 3
h_{min} (μm)	0.00	2.50	3.31	4.41
h_{max} (μm)	3.91	5.50	6.00	6.21
h_μ (μm)	1.28	2.00	2.50	3.0
d_μ (μm)	2.63	3.5	3.75	3.21
h_σ (μm)	0.99	0.84	0.86	0.90
θ (°)	65.49	75.82	60.63	70.67
Reflectance	6.35%	6.94%	5.99%	6.69%

Different reflectivity results indicate the effect of structural heterogeneity on the total reflectance of a broader area that consists of structures of various sizes and forms. Furthermore, a decrease in the average reflectance response can be observed as the height of each section from the equivalent unit cell to Model 3 increases. When the inverted pyramid structure height has a high distribution, the average reflectance decreases. In Fig. 6 (b), the reflectance results of BSi sample are compared with the equivalent unit cell model's reflectance. It is apparent some dissimilarities in terms of peaks in the reflectance curves that drives the contrast between the two results.

The absorption and reflection power in the BSi material against wavelengths spanning from 300 to 1200 nm is shown in Fig.7 The Fabry–Perot phenomenon refers to the appearance of reflected waves once they strike the structures and cause ripples.

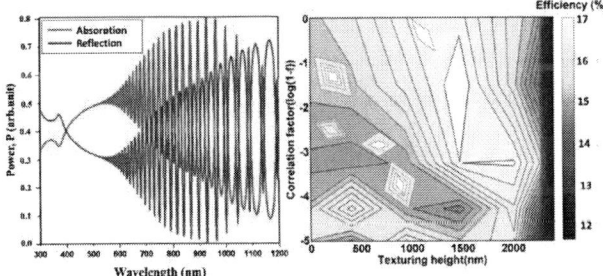

Fig. 7. a) Absorption and reflection power as a function of wavelength, b) Contour figure for 2D solar cells employing TM-polarized light encountered at standard projection, demonstrating computed short- circuit current density as a function of optimum texturing elevation as well as the correlation coefficient.

The absorbed power and the reflected power have an opposite correlation at first, meaning that for a moderate increment in the power received, the power rebounded decreases across the wavelength region around 300–700 nm. The lines increased in parallel after the cross- sectional point. A 2D solar cell simulation was conducted using the FDTD program (MEEP), as illustrated in Fig. 7 (b), to determine the ideal texturing height and correlated coefficient. A dual correlation theory was used to generate the arbitrary surface texturing method. A TM-polarised wave was encountered towards the vertical position within that 2D simulation.

Subsequently, the absorbance was used to calculate the short- circuit current density (Jsc), which is also a predictor for PV generation [6]. The electron-hole pair gives the photon current, so the J_{SC} is defined as,

$$J_{sc} = e \int \left(\frac{\lambda}{hc}(\lambda) \, I_{AM\,1.5}(\lambda) \right) d(\lambda) \qquad (2)$$

where, h is Plank's constant, c is the speed of light in the free space, and $I_{AM\,1.5}$ is the solar spectrum. The simulated J_{SC} for this condition was found around 17 mA/cm².

IV.   CONCLUSION

The findings in this paper suggest that computational models of optical outcome proportions associated with very multidimensional micro-nanostructured surfaces might be used to increase the accuracy of optical-electrical compiling results as solar cell technology. The difference in reflectance between the two projects is relatively low, which is approximately 1.55%. It can be observed from our findings that the tendency of the curves is similar but the reflectance rates at specific wavelengths are different. The reason might be stated as the regional dimension is insufficient to represent the behaviour of the whole BSi sample. If the size of the evaluated zone is increased, the local geometrical parameters approach can improve the consistency of the reflectance. The presented technique explored a highly possible proficiency to develop the accuracy of the complicated textured Si surface simulated design by evaluating acceptable comparative performance analysis between experimental observations and FDTD simulations, as well as the area to develop its accuracy improved by expanding the surface sampling.

ACKNOWLEDGMENT

The work is financially supported by the Ministry of Higher Education Malaysia (MOHE) via Fundamental Research Grant Scheme (FRGS/1/2022/TK09/UM/02/27). The authors also extend their appreciation to the Universiti Malaya (Grant No. ST055-2022).

REFERENCES

[1]   M. Y. Arafat et al., "Fabrication of black silicon via metal- assisted Chemical Etching—a review," Sustainability, vol. 13, no. 19, p. 10766, 2021.

[2]   W. Duan, High performance nanostructured black silicon biosensors. doi:10.17077/etd.005729

[3]   X. Liu, P. Coxon, M. Peters, B. Hoex, J Cole, D. Fray, "Black silicon: Fabrication methods, properties and solar energy applications," Energy Environ. Sci, pp.3223–3263, July 2015.

[4]   K. Peng, Y. Wu, H. Fang, X. Zhong, Y. Xu, and J. Zhu, "Uniform, axial-orientation alignment of one-dimensional single-crystal silicon nanostructure arrays," Angewandte Chemie International Edition, vol. 44, pp. 2737-2742, 2005.

[5]   D. Pera, J. Cardoso, D. Vilhena, G. Gaspar, K. Lobato, I. Costa, M. Serra, J.M. and J.A. Silva, computational optical analysis of 3d modeled crystalline silicon substrates randomly textured, September 2020.

[6]   H. Chung, K.Y. Jung, X.T Tee, and P. Bermel, "Time domain simulation of tandem silicon solar cells with optimal textured light trapping enabled by the quadratic complex rational function", Optics Express, 22(103), pp.A818-A832, 2014.

Investigation of The Performance Impact of Active Layer Parameter Variations on Inverted Perovskite Solar Cells Using GPVDM

A. M. A. Aziz
Faculty of Electronic and Computer Engineering
Universiti Teknikal Malaysia Melaka (UTeM)
Malaysia
m022210016@student.utem.edu.my

S. Muniandy
Faculty of Electronic and Computer Engineering
Universiti Teknikal Malaysia Melaka (UTeM)
Malaysia
m022020016@student.utem.edu.my

M. I. Idris
Micro and Nano Electronic Research Group (MiNE)
Faculty of Electronic and Computer Engineering
Universiti Teknikal Malaysia Melaka (UTeM), Malaysia.
idzdihar@utem.edu.my

Z. A. F. M. Napiah
Micro and Nano Electronic Research Group (MiNE)
Faculty of Electronic and Computer Engineering
Universiti Teknikal Malaysia Melaka (UTeM), Malaysia.
zulatfyi@utem.edu.my

Z. B. Zamani
Micro and Nano Electronic Research Group (MiNE)
Faculty of Electronic and Computer Engineering
Universiti Teknikal Malaysia Melaka (UTeM), Malaysia.
zarina@utem.edu.my

N. B. Norddin
Faculty of Electrical & Electronic Engineering Technology
Universiti Teknikal Malaysia Melaka (UTeM) Malaysia.
nurbahirah@utem.edu.my

M. Rashid
School of Physics
Universiti Sains Malaysia (USM) Malaysia.
marzaini@usm.my

Abstract— This research explores the performance of inverted perovskite solar cells (IPSC) using the General-purpose Photovoltaic Device Model (GPVDM) software. Alternatively, inorganic p-type semiconductors, especially NiOx which is the most widely used HTL, can provide intrinsically higher stability and exhibit lower cost than organic polymer-based HTL. The device structure in the simulation comprises ITO/NiOx/MAPbI3/C60/BCP/Ag. Various factors, including layer thickness, electrical parameters, absorption coefficient and refractive index of each layer, can influence the simulated IPSC's performance. GPVDM provides a comprehensive simulation platform to investigate the impact of these factors on the power conversion efficiency (PCE) of IPSCs. The simulation results from GPVDM exhibit a remarkable match and good agreement with achieving efficiencies of 17.35% and 17.57%. To optimize the results, two cases are analyzed and compared. Notably in Case 2, which employs experimental data for α and n from earlier research, outperforms the preceding journal with an efficiency of 18.23% compared to 17.57%. These simulation findings serve as a valuable guide for the fabrication of IPSCs utilizing NiOx, BCP, and C60 as active layers, offering insights into enhancing their performance.

Keywords— *GPVDM Software, Perovskite Solar cells, NiOx, Inverted Perovskite Solar Cells, Power Conversion Efficiency*

I. INTRODUCTION

Perovskite materials have emerged as an alternative as silicon-based solar cell technology approaches its limit [1]. Hybrid organic-inorganic perovskite solar cells have gotten much attention recently due to the PCE gains ranging from an initial 3.8% for the first prototype to 25.2% [2]. Besides, perovskite solar cells provide an enticing mix of low cost, simplicity of manufacturing, and good device performance [3][4]. NiOx, also known as nickel oxide, shows great

promise as a Hole Transport Layer (HTL) in Inverted Perovskite Solar Cells [5]. Its excellent conductivity allows efficient movement of positive charge carriers in these specific solar cell designs. NiOx can be doped with elements to enhance its properties and improve the performance of inverted perovskite solar cells. Its compatibility with different perovskite materials and stability makes it an attractive option for these types of solar cells. Overall, NiOx-based HTLs have the potential to play a significant role in advancing the technology of inverted perovskite solar cells.

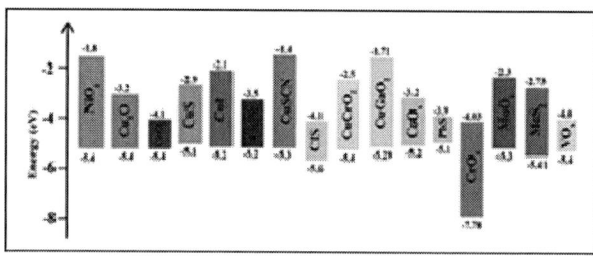

Fig. 1. The energy level diagram for generally used inorganic HTL [6].

This research aims to discover how the performance of active layer parameter variations can affect the power conversion efficiency (PCE) of perovskite solar cells such as layer thickness, electrical characteristics, the absorption coefficient (α) and the refractive index (n) of the materials used in each layer are some of these elements. The General-purpose solar Device Model (GPVDM) was used to analyze the consequences. This research aims on assessing the efficiency of inverted perovskite solar cells with MAPbI3 acting as the absorber layer. The research intends to acquire important insights into the influence of these parameters on the PCE of inverted perovskite solar cells by utilizing the capabilities of GPVDM, hence improving our understanding of their performance characteristics. These findings will help researchers produce more effective and sustainable solar energy technologies by enhancing the design and optimization of inverted perovskite solar cells.

II. METHODOLOGY

A. Device Structure

Figure 2 shows the layer of IPSC, which is composed of 6 layers, ITO/NiOx/MAPbI3/C60/BCP/Ag. This device structure was referred to from the previous simulation journal [7]. The absorber layer in this structure is MAPbI3, which is between the HTL (NiOx) and ETL (C60). In this study, NiOx was chosen as HTL because it is a promising HTL with its intrinsic p-type semiconductor characteristics, high transparency, wide bandgap and suitable work function with valence band (5.4 eV), which matched with the valence band of MAPbI3 [8][9].

Following that, C60 was employed as an ETL due to its high electron mobility and good band alignment. BCP was utilized as a blocking layer that acts as an exciton-blocking barrier, inhibiting exciton diffusion toward the Ag electrode, which would otherwise be quenched. PCE, Voc, Jsc, FF of this device structure are 17.57%, 1.042V, 21.72 mA cm-2 and 0.776 respectively.

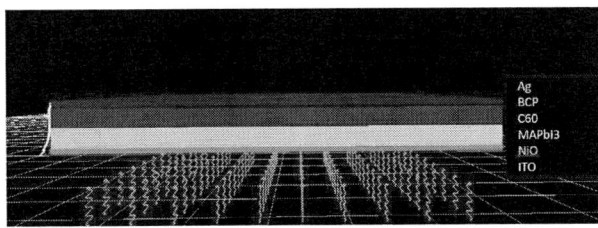

Fig. 2. The simulated IPSC in GPVDM software based on the reference thicknesses.

B. Simulation Methodology

GPVDM is a drift–diffusion base/Shockley-Read-Hall model [10]. It is a general-purpose simulation tool for optoelectronic and photovoltaic devices such as OLEDs, OFETs, and several types of first, second, and third-generation solar cells. ITO, perovskite and Ag thickness values were held constant. While the active layer such as NiOx was varied in between 50-550 nm, BCP and C_{60} layer properties were varied in between 5-50 nm to examine the impact of different variation on PCE. The effect of electrical parameters, α, n and layer thickness of the active layers were examined. Table 1 shows the reference electrical parameters of each active layer that were employed in simulation.

TABLE I.   REFERENCE OF ELECTRICAL PARAMETER OF ACTIVE LAYER

Parameters	NiOx [10]	MAPbI₃ [11][12]	BCP [13]	C₆₀ [14]
Electron mobility ($m^2v^{-1}s^{-1}$)	0.0028	6.86×10^{-7}	1×10^{-7}	8×10^{-5}
Hole mobility ($m^2v^{-1}s^{-1}$)	0.0028	0.0375	2×10^{-7}	3.5×10^{-7}
Effective density of free electron states @300K (m^{-3})	1×10^{26}	5×10^{26}	2.2×10^{9}	8×10^{13}
Effective density of hole electron states (@300K)	1×10^{26}	5×10^{26}	1.8×10^{12}	8×10^{13}

Parameters	NiOx [10]	MAPbI₃ [11][12]	BCP [13]	C₆₀ [14]
(m^{-3})				
Electron affinity, Xi (eV)	1.46	1.6	3.9	3.9
Band gap, Eg (eV)	1.3	1.5	3.5	1.7
Relative permittivity	5	5	5	5

III. RESULT AND DISCUSSION

A. The Analysis Of The Thickness Of Active Layers

TABLE II.   COMPARISON OF RESULT BASED ON THE THICKNESS THAT REFERRED FROM [16].

	Fabricated IPSC	a) With default electrical parameters	b) With reference electrical parameter (material from database)
NiO thickness (nm)	300	300	300
MAPbI₃ thickness (nm)	450	450	450
BCP thickness (nm)	6.5	6.5	6.5
C₆₀ thickness (nm)	26	26	26
PCE (%)	17.57	22.87	17.35
Voc (V)	1.042	0.924	0.727
Jsc (mA cm⁻²)	21.72	29.9	29.91
FF	0.776	0.827	0.798

From Table 2, the PCE in the simulation result of (a) was higher than the result from the journal. However, the result (a) was not accurate as the default electrical parameters were not specified for the materials used in the structure. The simulation result of (b) was slightly lesser than the result from the referred journal but was more accurate as compared to the result of (a). This was because the electrical parameter used was referred and specified for each of the material. Therefore, the result from GPVDM matched and showed good agreement with the fabricated IPSC device structure in the previous experiment, which was 17.34% and 17.57% [16]. Then, optimizing the IPSC by varying the thickness of the active layers such as NiOx, MAPbI3, BCP, and C60 was needed so that the performance would be greater than the experimental result in [8].

B. Optimization Of The Device Performance

This section was dedicated to varying the thickness of HTL, perovskite, and ETL layers to obtain the highest efficiency of the IPSC. Then, the α and n were replaced to obtain an accurate simulation result. The results were analyzed and grouped into two cases. In Case 1, the electrical parameters referred to in previous experiments were used, but α and n were obtained from the GPVDM database. The thickness of NiOx, BCP, and C60 was varied. In Case 2, the referred electrical parameters were retained, but the α and n were extracted from the experiment data. The thickness was varied again to achieve better efficiency compared to the fabricated IPSC device structure [15].

979-8-3503-2369-6/23 $31.00 © 2023 IEEE

C. Case 1: With referred electrical parameter for NiOx, MAPbI₃, BCP and C₆₀ layer.

The thickness of NiOx ranged from 50 nm to 550 nm. However, the thicknesses of BCP and C60 remained unchanged at 6.5 nm and 26 nm, respectively, according to the reference journal. Figure 3 showed that when the thickness of NiOx was 200 nm, the PCE was the greatest, at 24.89%. At this thickness, the Voc, Jsc, and FF were 0.923 V, 32.66 mA/cm-2, and 0.826, respectively. The ideal thickness for NiOx was determined to be 200 nm.

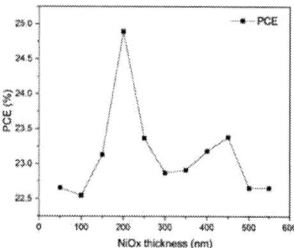

Fig. 3. Changes of PCE with different NiOx thickness starting from 50 to 550 nm.

The thickness of BCP was varied from 5 nm to 50 nm, whereas NiOx and C60 remained constant at 200 nm and 26 nm, respectively. Figure 4 showed that the highest PCE was 25.19% when the thickness of BCP was 30 nm. At this thickness, the Voc, Jsc, and FF were 0.924V, 32.99 mA/cm-2, and 0.826, respectively. The ideal thickness for BCP was determined to be 30 nm.

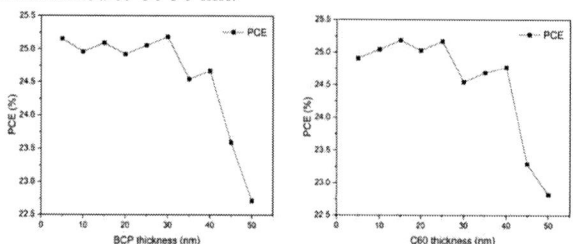

Fig. 4 Changes of PCE with different BCP thickness in between 5 to 50 nm, NiOx thickness now is 200 nm, C₆₀ thickness in between 5 to 50 nm.

The thickness of C60 was varied from 5 nm to 50 nm, while NiOx and BCP were kept constant at 200 nm and 30 nm, respectively. The highest PCE, as shown in Figure 4, was 25.18% when C60 was 15 nm. At this thickness, the Voc, Jsc, and FF were 0.924V, 33.02 mA/cm-2, and 0.826, respectively. The optimal thickness for C60 was determined to be 15 nm. The ideal thicknesses for NiOx, BCP, and C60, based on GPVDM software, were now 200 nm, 30 nm, and 15 nm, respectively, as they produced a PCE of 18.925% at these thicknesses.

D. Case 2: Replace the materials in database with manually inserted the α and n for the active layers with the same thickness obtained in Case 1

The thickness of the active layers from Case 1 was placed in the simulation that replaced α and n, which were extracted from the experiment data. The resulting table displayed the outcomes after the replacement of α and n.

TABLE III. RESULT AFTER MANUALLY INSERTED THE ABSORPTION COEFFICIENT AND REFRACTIVE INDEX

NiOx thickness (nm)	MAPbI₃ thickness (nm)	BCP thickness (nm)	C₆₀ thickness (nm)	PCE (%)	Voc (V)	J Jsc(mA cm⁻²)	FF
200	450	30	15	13.052	0.716	22.92	0.794

The optimum thickness from Case 1 was no longer suitable after manually inserting the absorption coefficient and refractive index for the active layers. As a result, the PCE decreased to 13.052%. Therefore, a new analysis needed to be conducted to construct the layers again.

The thickness of NiOx ranged from 50 nm to 550 nm. However, the thicknesses of BCP and C60 remained unchanged at 6.5 nm and 26 nm, respectively. Figure 5 showed that the PCE was greatest when NiOx was 500 nm, at 17.86%. The Voc, Jsc, and FF values at this thickness were 0.733 V, 30.49 mA/cm2, and 0.799, respectively. The optimal thickness for NiOx was determined to be 500 nm.

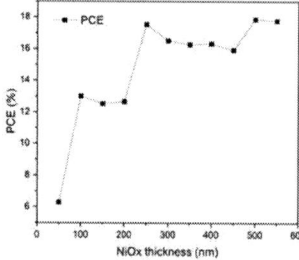

Fig. 5 Changes of PCE with different NiOX thickness starting from 50 to 550 nm.

The thickness of BCP was changed from 5 nm to 50 nm, while NiOx and C60 were fixed at 500 nm and 26 nm, respectively. According to Figure 6, the greatest PCE was 18.06% when BCP was 20 nm. The Voc, Jsc, and FF values at this thickness were 0.734 V, 30.83 mA/cm-2, and 0.798, respectively. The optimal thickness for BCP was now determined to be 20 nm.

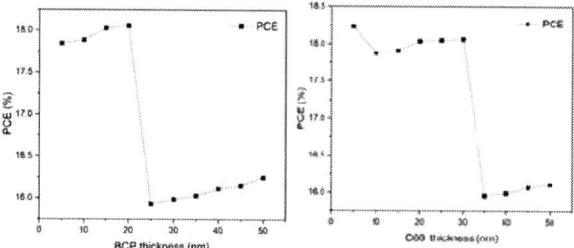

Fig. 6 Changes of PCE with different BCP thickness in between 5-50 nm, NiO thickness now is 500 nm, C₆₀ thickness in between 5-50 nm.

The thickness of C60 was changed between 5 nm and 50 nm, while NiOx and BCP remained constant at 500 nm and 20 nm, respectively. The maximum PCE exhibited in Figure 6 was 18.227% when C60 was 5 nm. At this thickness, the Voc, Jsc, and FF were 0.734 V, 31.15 mA/cm-2, and 0.798, respectively. The ideal thickness for C60 was determined to be 5 nm.

TABLE IV.   SUMMARY ON THE DIFFERENT CASES AFTER OPTIMIZATION [16].

Case	Fabricated IPSC result	Simulated IPSC result (Case 1)	Simulated IPSC result (Case 2)
NiOx thickness (nm)	300	200	500
MAPbI₃ thickness (nm)	450	450	450
BCP thickness (nm)	6.5	30	20
C₆₀ thickness (nm)	26	15	5
PCE (%)	17.57	18.925	18.227
Voc (V)	1.042	0.726	0.734
Jsc (mA cm^{-2})	21.72	33.012	31.15
FF	0.776	0.798	0.798

The preferable case to choose was the thickness of each layer from Case 2. The reason behind this decision was that the materials' α, n, and electrical parameters were taken from real experiment data obtained in previous research. This ensured that the simulation results were more accurate and yielded better efficiency after the optimization, compared to the reference journal [15]. Thus, the thickness of NiOx, MAPbI3, BCP, and C60 was based on the analysis conducted in Case 2.

Characteristic curves, as displayed in Figure 8, quantitatively indicated the performance of a solar cell. They showed Voc, Jsc, FF, and PCE. To calculate the PCE, the formula was used:

$$PCE = \frac{Jsc * Voc * FF}{100}$$

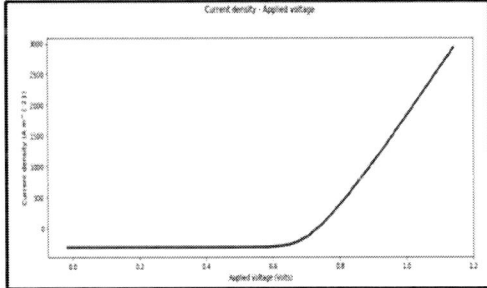

Fig. 8   J-V characteristic curve of ITO/NiOx/MAPbI₃/C₆₀/BCP/Ag

IV. CONCLUSION

In this study, we examined inverted perovskite solar cells (IPSCs), analyzing parameters like electrical parameter, absorption, refractivity, and active layer thickness. GPVDM results showed 17.34% (With reference electrical parameter (material from database)) and 17.57% (Fabricated IPSC) efficiencies. Case 2 proved optimal for active layer thickness, achieving 18.227% efficiency with real experimental data for α and n values. NiOx, MAPbI3, BCP, and C60 layer thicknesses were optimized at 500 nm, 450 nm, 20 nm, and 5 nm. These findings guide future IPSC fabrication using NiOx, BCP, C60, and MAPbI3 with Ag

back contact, aiming for higher PCE and improved performance.

ACKNOWLEDGMENT

We would like to thank the Ministry of Education and University Teknikal Malaysia Melaka for supporting this work under project No. of FRGS/1/2022/FTKEE/F00521.

REFERENCES

[1] A. Ghosh, S. Safat, and M. S. Islam, "Stable and Efficient Perovskite Solar Cell with Metal Oxide Transport Layers," *2nd International Conference on Electrical, Computer and Communication Engineering, ECCE 2019*, no. March 2020, 2019, doi: 10.1109/ECACE.2019.8679418.

[2] I. M. Maafa, "All-Inorganic Perovskite Solar Cells: Recent Advancements and Challenges," *Nanomaterials*, vol. 12, no. 10, p. 1651, May 2022, doi: 10.3390/nano12101651.

[3] H. S. Kim, S. H. Im, and N. G. Park, "Organolead halide perovskite: New horizons in solar cell research," *Journal of Physical Chemistry C*, vol. 118, no. 11, pp. 5615–5625, 2014, doi: 10.1021/jp409025w.

[4] Y. B. Martynov *et al.*, "On the efficiency limit of ZnO/CH3NH3PbI3/CuI perovskite solar cells," *Physical Chemistry Chemical Physics*, vol. 19, no. 30, pp. 19916–19921, 2017, doi: 10.1039/c7cp03892e.

[5] S. Muniandy, M. I. Bin Idris, Z. A. F. Bin Mohammed Napiah, and M. Rashid, "The effect of pH level and annealing temperature on NiO thin films as Hole Transport Material in Inverted Perovskite Solar Cells," in *2022 IEEE International Conference on Semiconductor Electronics (ICSE)*, IEEE, Aug. 2022, pp. 13–16. doi: 10.1109/ICSE56004.2022.9863126.

[6] G. M. Arumugam *et al.*, "Inorganic hole transport layers in inverted perovskite solar cells: A review," *Nano Select*, vol. 2, no. 6, pp. 1081–1116, Jun. 2021, doi: 10.1002/nano.202000200.

[7] G. Shen, Q. Cai, H. Dong, X. Wen, X. Xu, and C. Mu, "Using Interfacial Contact Engineering to Solve Nickel Oxide/Perovskite Interface Contact Issues in Inverted Perovskite Solar Cells," *ACS Sustainable Chemistry and Engineering*, vol. 9, no. 9, pp. 3580–3589, 2021, doi: 10.1021/acssuschemeng.0c09056.

[8] R. Singh, P. K. Singh, B. Bhattacharya, and H. W. Rhee, "Review of current progress in inorganic hole-transport materials for perovskite solar cells," *Applied Materials Today*, vol. 14, pp. 175–200, 2019, doi: 10.1016/j.apmt.2018.12.011.

[9] A. K. Mishra and R. K. Shukla, "Electrical and optical simulation of typical perovskite solar cell by GPVDM software," *Materials Today: Proceedings*, vol. 49, pp. 3181–3186, 2020, doi: 10.1016/j.matpr.2020.11.376.

[10] M. S. Shamna, K. S. Nithya, and K. S. Sudheer, "Simulation and optimization of CH3NH3SnI3 based inverted perovskite solar cell with NiO as Hole transport material," *Materials Today: Proceedings*, vol. 33, no. October 2021, pp. 1246–1251, 2019, doi: 10.1016/j.matpr.2020.03.488.

[11] A. bdelkader Hima, "GPVDM simulation of layer thickness effect on power conversion efficiency of CH3NH3PbI3 based planar heterojunction solar cell," *International Journal of Energetica*, vol. 3, no. 1, p. 37, 2018, doi: 10.47238/ijeca.v3i1.64.

[12] Y. Raoui, H. Ez-zahraouy, N. Tahiri, O. El, S. Ahmad, and S. Kazim, "Performance analysis of MAPbI 3 based perovskite solar cells employing diverse charge selective contacts : Simulation study," *Solar Energy*, vol. 193, no. February, pp. 948–955, 2019, doi: 10.1016/j.solener.2019.10.009.

[13] N. Touafek, C. Dridi, and R. Mahamdi, "Bathocuproine Buffer Layer Effect on the Performance of Inverted Perovskite Solar Cells," pp. 1–6, 2020.

[14] X. B. Yongjin Gan, P. M. and Q. J. and Yucheng Liu, e.t, *Numerical Investigation Energy Conversion Performance of Tin-Based Perovskite Solar Cells Using Cell Capacitance Simulator*, 2020.

[15] G. Shen, Q. Cai, H. et. al "Using Interfacial Contact Engineering to Solve Nickel Oxide/Perovskite Interface Contact Issues in Inverted Perovskite Solar Cells," *ACS Sustain Chem Eng*, vol. 9, no. 9, pp. 3580–3589, Mar. 2021, doi: 10.1021/acssuschemeng.0c090

Advanced Solar-Powered Seed Sowing Machine with Precision Seeding and Smart Control Features

Sadiq Ur Rehman
Department of Electrical Engineering
Hamdard University
Karachi, Pakistan
sadiq.rehman@hamdard.edu.pk

Asad A. Zaidi
Department of Mechanical Engineering
Hamdard University
Karachi, Pakistan
asad.zaidi@hamdard.edu.pk

Yasmin Abdul Wahab
Nanotechnology & Catalysis Research Centre
University of Malaya,
Kuala Lumpur, Malaysia
yasminaw@um.edu.my

Md. Yasir Arafat
Nanotechnology & Catalysis Research Centre
Universiti Malaya,
Kuala Lumpur, Malaysia
yasirisnow@gmail.com

Sharifah Fatmadiana Wan Muhamad Hatta
Department of Electrical Engineering, Faculty of Engineering
Universiti Malaya
Kuala Lumpur
sh_fatmadiana@um.edu.my

Abstract— This article proposes a system model to automate the labor-intensive task of digging and seed sowing in response to Pakistan's increasing demand for food production. The proposed model utilizes solar energy and a DC Stepper Motor to automate the process of digging, sowing seeds, and leveling the soil, resulting in reduced labor costs and improved crop productivity while maintaining soil quality. The Seed Hopper is designed to discard seeds at a specific distance, ensuring uniform spacing between seeds and rows. This system model has the potential to significantly reduce seed costs while improving the accuracy of seed count and spacing, which are often not achieved when farmers manually sprinkle seeds. Future research could optimize the system model for different crop types and evaluate its scalability for larger-scale implementation.

Keywords— Arduino, Agriculture, Solar Panel, NEMA, Hopper, Seed Sowing.

I. INTRODUCTION

The agriculture industry is a critical component of Pakistan's economy, contributing 19.2% to the GDP and employing 38.5% of the workforce [1]–[3]. The government is focused on supporting small and marginalized farmers and promoting innovative small-scale technology to foster growth in this sector [4], [5]. However, using manual, ox, and tractor techniques for seed sowing and fertilizer placement is prevalent in Pakistan, with manual and ox techniques taking a long time and producing little output [6]. Tractors run on fossil fuels, which emit harmful pollutants, resulting in air, water, and noise pollution and a potential energy crisis [7], [8].

To address these challenges, this article proposes a new model that automates digging and sowing crops, reducing the need for manual labor and speeding up cultivation. The prototype model uses solar energy, which is abundant in nature, and requires less energy than traditional agricultural instruments like tractors. The proposed model could also address the pollution issue in Pakistan by employing a solar plate. The proposed model saves seeds from waste, and the soil can be leveled and dug simultaneously using the machine. The prototype includes IR sensors [9] to alert

The work is financially supported by the Ministry of Higher Education Malaysia (MOHE) via Fundamental Research Grant Scheme (FRGS/1/2022/TK09/UM/02/27). The authors also extend their appreciation to the Universiti Malaya (Grant No. ST055-2022).

farmers when the hopper is empty. The objectives of the proposed model are to sow seeds, dig the soil, and level the surface with a movable blade. The model is user-friendly, low-cost, require less staffing, and has the potential to revolutionize the agricultural sector, which currently relies on conventional methods and heavy machinery that emit pollutants [10]. Therefore, the proposed model has the potential to be a game-changer and a significant step towards sustainable agriculture in Pakistan.

II. DESIGN OF THE PROPOSED MODEL

The design of the proposed machine is developed in AutoCAD software; using this software, the dimension of the machine structure was finalized, as can be seen in Fig. 1.

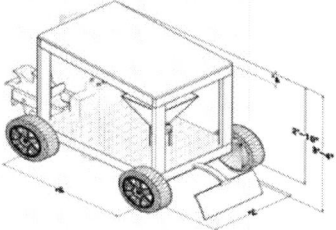

Fig. 1. Dimensions of the proposed machine.

Moreover, a visualization of the machine's 3-dimension along with the placement of solar panels, different sensors, batteries, motors, and seed hopper digging pulley, can be seen in Fig.2. To achieve maximum solar panel rating; a panel needs to be placed at the top surface of the structure.

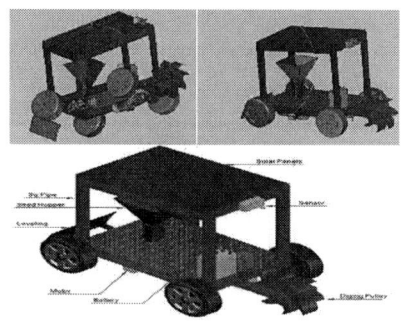

Fig. 2. Equipment placement and 3D view

979-8-3503-2369-6/23 $31.00 © 2023 IEEE

The mechanical structure begins with a 2ft 10 inches tall frame, 2ft 9 inches long, and 2ft 5 inches wide. The frame is made out of 1-inch angle iron. The solar panel frame is similarly made of 1-inch angle iron and has a width of 20.5 inches and a length of 32.5 inches.

Two front and back wheels shafts are located on the frame's underside. With one wheel on each side, the front wheel shaft is 20mm in diameter and 3ft long. The rear wheel shaft has a diameter of 20mm and a length of 3ft, with one wheel on each side. The machine is moved by four wheels, two in front and two in behind. The frames were designed according to the parameters of the motors and fixed on the below surface of the frame.

To obtain the selection and rating of solar panels and using a 12V battery and 60 watts of solar plates, the following calculation was performed,

A. Solar panels calculation

To find out the current that solar panels can drive.

$$\text{Power} = \text{Volt*current}, I= P/V \tag{1}$$

I= 60/12, I=5A

5A is the current from drive to load.

B. Battery Capacity

$$\text{Battery capacity in Ah} = \frac{KV \times h}{V} \tag{2}$$

$$= \frac{60 \times 8}{12}$$

$$= 40 \text{ Ah}$$

C. Battery Charging Current

$$\text{Battery charging current} = \frac{1}{10} \times \text{battery capacity} \tag{3}$$

$$\text{Battery charging current} = \frac{40}{10}$$

Battery charging current= 4A

D. Solar Plates

$$\text{Solar Plates Current} = \text{load current} + \text{charging current} \tag{4}$$

Solar Plates Current = 5+4

Solar Plates Current = 9A

E. Solar Plates Power

$$\text{Solar plates Power} = V \times I \tag{5}$$

$$= 12 \times 9$$

$$= 108W$$

Solar plates watts= 60 W

Solar Plates required = 108/60

$$= 1.38$$

Hence one solar panel of **60 W** is required to drive all the load of the proposed machine model.

Through this proposed system;

- The sowing rate can be controlled.
- Achieve seed spacing.
- Less staffing is required.
- The machine runs through solar no pollution is caused.

- Economical to sow a variety of seeds through this machine.

III. METHODOLOGY

This project's electrical setup begins with a 12V 60W Solar Panel at an angle on top of the frame to capture solar energy. The output voltages from the solar panel provided to the 12V 40Ah Battery are regulated and controlled by the 12V 10A Solar Controller. The 23/76 2-core cable connects the solar panel to the solar controller, as shown in Fig.3.

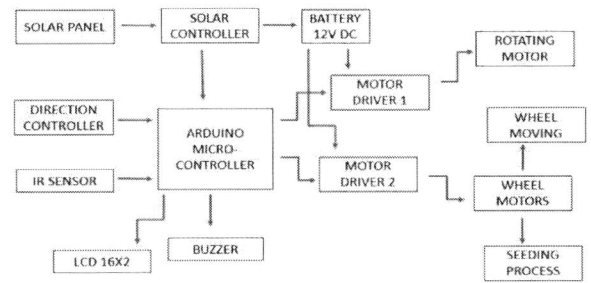

Fig. 3. System block diagram.

The solar controller (12V 10A) is fitted here to prevent the battery from overcharging. The controller's second purpose is to provide two 5V USB outputs that can be utilized to power the microcontroller. The controller's display shows the voltages originating from the solar panel being monitored. It is also able to keep track of how much the battery is charged. A 40/76 2-core wire connects the solar controller to the battery.

The 12V 40Ah Battery is installed to store energy from solar panels to power the two motors of the machine. The battery of 40Ah can charge for a total of 8 hours during daylight between 8 am- 4 pm. The battery can operate the machine for 3 hours, which means it can perform all its operations for at least 3 hours. The battery will power the two motors, i.e., the NEMA23[11] stepper motor for right/left and the NEMA34 stepper motor for forward/backward and hopper mechanism through the TB6600 drivers. The battery will be connected to TB6600 drivers with 40/76 2-core wires.

The Arduino [12] is used to control the directions of the stepper motors using TB6600 drivers for both stepper motors. The TB6600 will get electricity from the battery that powers the Stepper Motors and the control code from the Arduino. It will download the Arduino code and control the motor speed, bi-direction, and switching.

To shift the orientation of the mechanism right/left, a NEMA23 12V 2.7A Stepper Motor is utilized, which is controlled by the TB6600 motor driver. The rpm and rotations can be entered into the coding, which will rotate the motor and change the number of rotations. The 12V supply from the 12V battery will be routed through the TB6600 driver via the 23/76 2-core wire. The hopper mechanism is driven by a NEMA34 12V 3A stepper motor controlled by the TB6600 motor driver. The rpm and rotations can be entered into the coding, which will rotate the motor and change the number of rotations. The 12V supply from the 12V battery will be routed through the TB6600 driver via the 23/76 2-core wire.

Using an Arduino Uno, an IR sensor [13] is connected to a buzzer and a 16x2 LCD. An infrared sensor is put inside the hopper container to indicate whether or not the hopper is full of seeds. The IR sensor is connected to the Arduino by a 1mm control wire, which sends a signal to the microcontroller, which shows on LCD16x2 whether the seeds are filled or not, as well as turning on the buzzer, which indicates that the hopper is empty and seeds must be replenished.

The entire process of the proposed system can be seen in the flowchart presented in Fig.4.

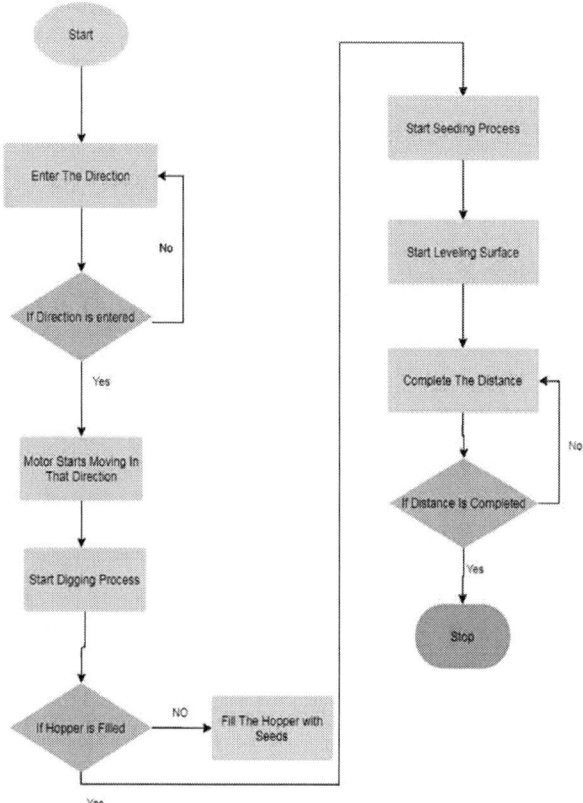

Fig. 4. Process flowchart.

The flowchart shows that when the power button is pressed, the motor driver energizes the motors, making them ready to accept directional signals and move in the user's selected direction. The user must enter the direction by pressing the button for any direction on the direction controller. The machine will only move once and when it is given instructions. The machine will begin going in that direction as soon as the direction is input through the controller. The digging mechanism, which you may manually adjust for depth, will begin digging the dirt for the seeds to be sown. As the machine moves, the hopper will drop the seed at a precise distance as the digging gear prepares the soil for seed sowing.

The hopper gear can be manually changed to fit the size of the seeds. If the hopper is empty, the IR sensor will signal to the microcontroller, activating the buzzer and showing the message on the LCD [14].

The seed-sowing operation will resume after the hopper is loaded with seeds. The leveler, placed at the back of the machine, will level the soil's surface when the seed is dropped from the hopper into the soil, covering the seed beneath the earth. After the entire field has been dug, seeds have been planted, and the surface has been leveled, the machine will travel the remaining distance until it reaches its starting location.

IV. RESULT

In Fig. 5, the LCD display shows the hopper status. In case when the hopper is empty, the IR sensor gets signals that seeds are not covering the sensor. Hence the display is showing an 'empty' status on LCD.

Fig. 5. Status of the hopper on LCD

The finalized structure of the seed-sowing machine can be seen in Fig. 6. All the electrical and mechanical components are mounted on the frame to provide the results. It can also be noticed that the proposed machine model maintains spacing between seeds and spacing between the rows. This prototype model has been tested on solid marble floor and small in-home garden.

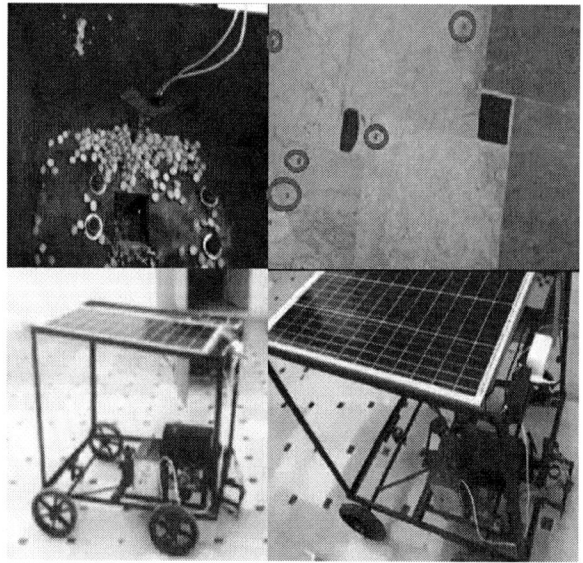

Fig. 6. Hoper and seed dropping by machine

It can be seen from Fig. 7 that the charge controller is receiving the voltages from the solar panel and giving the regulated output voltages to the lead-acid battery to get charged.

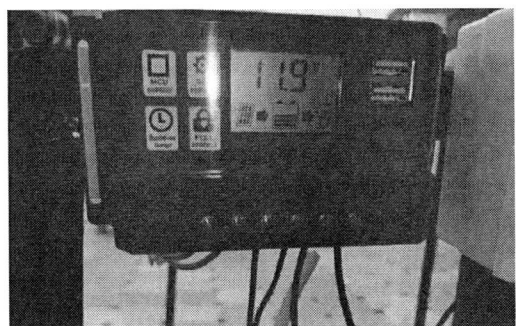

Fig. 7. Status of the solar battery charge controller

Table. 1 shows the performance and efficiency of the prototype machine compared with manual seed sowing when machine was tested in a small garden of approximately 1.8 meters with a solar panel of 60W rating.

TABLE I. COMPARISON OF PERFORMANCE AND EFFICIENCY

Metric	Machine	Manual
Time Taken to Sow (min)	22	30
Seed Utilization Efficiency (%)	65	60
Seed Spacing Accuracy (%)	75	60
Seed Placement Accuracy (%)	60	65
Soil Leveling Efficiency (%)	60	40
Energy Source	Solar	Manual
Energy Consumption (kWh)	0.18	N/A
Environmental Impact	Clean energy	Labor Intensive

The data presented in the table demonstrates that the proposed machine outperforms manual practices with higher seed utilization efficiency, precise spacing, and accurate seed placement, improving crop productivity and reducing labor costs. It saves time and uses sustainable solar energy. These findings suggest that the prototype has the potential to revolutionize seed sowing and promote sustainable agriculture.

V. CONCLUSION

Because most of Pakistan's population hails from villages, the country's economy is centered on agriculture. So, using non-conventional energy resources, this prototype machine can excavate and seed sow. It will benefit farmers who need more resources to cover gasoline costs. The use of solar energy can also help to lower the population of the planet. This can also save money for the government and fossil fuels. Some of the future works and enhancements that can be done with this machine to optimize its potential can be in a form of increasing the number of rows of digging mechanisms and seed hoppers that dig the soil at a set distance apart and installing numerous hoppers that can drop seeds in multiple rows at once at a set distance apart. This strategy can increase the amount of work completed in a single round.

Including a water spraying device can be considered as future work. This will result in a reduction in the workforce. Because this machine already does digging, seed sowing, and surface leveling, adding water spraying to the system will allow it to achieve its total labor output.

Currently, the solar panel on the frame of this proposed machine is fixed. Include a rotatory mechanism with a solar panel in the future scope, which will operate on the LDR sensor and indicate the maximum sun rays and spin the solar panel in that direction. We can get the most significant output voltage from the solar panels to charge the battery for our mechanism using this integration.

ACKNOWLEDGEMENT

The work is financially supported by the Ministry of Higher Education Malaysia (MOHE) via Fundamental Research Grant Scheme (FRGS/1/2022/TK09/UM/02/27). The authors also extend their appreciation to the Universiti Malaya (Grant No. ST055-2022).

REFERENCES

[1]. A. Malik, S. R. Qureshi, N. Abbas, and A. A. Zaidi, "Energy and exergy analyses of a solar desalination plant for Karachi Pakistan," Sustain. Energy Technol. Assessments, vol. 37, p. 100596, 2020.

[2]. A. Raza, M. B. Khan, W. Ali, M. J. Memon, and R. Daudpota, "An IoT based Smart Agriculture Monitoring and Control," Int. J. Electr. Eng. Emerg. Technol., vol. 4, no. SI 1, pp. 8–14, 2021.

[3]. R. Rai, A. R. Larik, K. Ahmed, S. Kumaramasy, and A. A. Zaidi, "Comparative Analysis of finned absorber plate with and without black paint in Solar Air Heater," in 2023 4th International Conference on Computing, Mathematics and Engineering Technologies (iCoMET), 2023, pp. 1–4.

[4]. G. Mehdi, N. Ali, S. Hussain, A. A. Zaidi, A. H. Shah, and M. M. Azeem, "Design and Fabrication of Automatic Single Axis Solar Tracker for Solar Panel," in 2019 2nd International Conference on Computing, Mathematics and Engineering Technologies (iCoMET), 2019, pp. 1–4.

[5]. H. Ahmed, A. Najib, A. A. Zaidi, M. N. Naseer, and B. Kim, "Modeling, design optimization and field testing of a solar still with corrugated absorber plate and phase change material for Karachi weather conditions," Energy Reports, vol. 8, pp. 11530–11546, 2022.

[6]. J. Lelieveld, K. Klingmüller, A. Pozzer, R. T. Burnett, A. Haines, and V. Ramanathan, "Effects of fossil fuel and total anthropogenic emission removal on public health and climate," Proc. Natl. Acad. Sci., vol. 116, no. 15, pp. 7192–7197, 2019.

[7]. M. N. Naseer et al., "Past, present and future of materials' applications for CO2 capture: A bibliometric analysis," Energy Reports, vol. 8, pp. 4252–4264, 2022.

[8]. M. Uzair, S. U. Hasan Kazmi, M. Uzair Yousuf, and A. Ali Zaidi, "Optimized performance of PV panels and site selection for a solar park in Pakistan," Trans. Can. Soc. Mech. Eng., vol. 46, no. 2, pp. 412–426, Jan. 2022.

[9]. S. U. Rehman, S. A. Khan, I. U. Khan, H. Mustafa, and M. A. Shaikh, "Interactive Smart Writing Technology (ISWT)," Univ. Sindh J. Inf. Commun. Technol., vol. 4, no. 1, pp. 1–8, 2020.

[10]. M. N. Naseer, Y. Noorollahi, A. A. Zaidi, Y. A. Wahab, M. R. Johan, and I. A. Badruddin, "Abandoned wells multigeneration system: promising zero CO2 emission geothermal energy system," Int. J. Energy Environ. Eng., vol. 13, no. 4, pp. 1237–1246, 2022.

[11]. M. Khairudin, R. Asnawi, and A. Shah, "The characteristics of TB6600 motor driver in producing optimal movement for the Nema23 stepper motor on CNC machine," Telkomnika, vol. 18, no. 1, pp. 343–350, 2020.

[12]. S. U. Rehman, H. Mustafa, and A. R. Larik, "Iot based substation monitoring & control system using Arduino with data logging," in 2021 4th International Conference on Computing & Information Sciences (ICCIS), 2021, pp. 1–6.

[13]. S. UrRehman, F. MohsinZakai, and M. Adeel, "Inspection on infrared-based image processing," in 2018 IEEE 21st International Multi-Topic Conference (INMIC), 2018, pp. 1–6.

[14]. S. U. Rehman, I. Khan, N. U. Rehman, and A. Hussain, "Low-Cost Smart Home Automation System with Advanced Features," Quaid-E-Awam Univ. Res. J. Eng. Sci. Technol. Nawabshah., vol. 20, no. 01, pp. 74–82, 2022.

Finite Element Simulation of Single Zinc Oxide Nanorod for Piezoelectric Nanogenerator

Muhammad Adhwa Fathullah
bin Nor Asmadi
*ECE Department, Kulliyyah of
Engineering, International
Islamic University Malaysia*
adhwa1729131@gmail.com

Aliza Aini Md Ralib
*ECE Department, Kulliyyah of
Engineering, International
Islamic University Malaysia*
alizaaini@iium.edu.my

Anis Nurashikin Nordin
*ECE Department, Kulliyyah of
Engineering, International
Islamic University Malaysia*
anisnn@iium.edu.my

Norazlina Bte Saidin
*ECE Department, Kulliyyah of
Engineering, International
Islamic University Malaysia*
norazlina@iium.edu.my

Abstract— The growing demand for sustainable and clean energy sources has motivated the development of wearable energy harvesters for portable and wearable electronic devices. However, the use of bulky and hazardous batteries poses challenges in terms of size, flexibility, and environmental impact. This paper addresses these challenges by presenting a 3D finite element simulation of single Zinc Oxide (ZnO) nanorod that has potential application as a wearable energy harvester. The effect of varying the aspect ratio (diameter/length) of ZnO nanorods toward the generated output voltage was investigated. The relationship between the variation of applied force to the output voltage and displacement of the vibration was also presented. The analysis results revealed that increasing the aspect ratio of the single ZnO nanorod led to higher generated output voltages. Similarly, applying higher forces resulted in increased voltage output. The optimum design of the single ZnO nanorod that has the highest output voltage is D=30nm L=9000nm force=500nN. The simulation results also demonstrated that the length and diameter of the nanorods influenced the generated piezoelectric potential.

Keywords— *Zinc Oxide nanorod, Nanogenerator, finite element simulation, aspect ratio, output voltage, energy harvester*

I. INTRODUCTION

Energy harvesting has emerged as a critical area of research, offering solutions to power various electronic devices without the need for external power sources. However, the use of bulky and hazardous batteries poses challenges in terms of size, flexibility, and environmental impact. By integrating piezoelectric energy harvesting systems into wearable electronics, it becomes possible to capture the mechanical energy generated by human motion[1]. Piezoelectric materials such as lead zirconate titanate (PZT), barium titanate (BaTiO$_3$), and ZnO shared the same unique crystal and composition that allows them to generate electrical charges in response to mechanical stress or strain [2]. However, PZT and BaTiO$_3$ materials are rigid and have limited options of growth techniques for nanogenerators [3]. ZnO exhibits a high piezoelectric coefficient, good mechanical strength, and good compatibility with flexible substrates. This makes it suitable for applications that require flexibility and conformability. This biocompatibility of ZnO also makes it a favorable choice for applications involving direct contact with the skin, such as wearable devices or biomedical applications [4]–[6].

Piezoelectric nanogenerators (PENGs) offer numerous benefits, including their high efficiency and facile fabrication process [7]. The growth of the nanostructure of ZnO will directly affect its energy harvesting performance. Compared with other nanostructures such as ZnO nanowires [8] and ZnO nanoneedles [9], ZnO nanorods [10] stand out due to their advantages such as simpler synthesis methods and no post-treatment needed [11] on most substrates. Previous work reported on the synthesis, fabrication, and characterization of ZnO nanorods for energy harvester applications [13]. However, limited findings have been explored on the prediction of geometry optimization through simulation. A recent study reported on finite element simulation of single nanorods using a different type of piezoelectric material such as Aluminium Nitride (AlN), ZnO, and Lead Zirconate Titanate (PZT) [12]. However, there are limited findings on the effect of the aspect ratio on the displacement of the vibration and the output voltage of the single nanorod.

Hence, this paper focuses on evaluating the performance of a single ZnO nanorod under varying conditions such as aspect ratio of the single ZnO nanorod and applied input force. The paper begins by explaining the design concept of a 3D model of a single ZnO nanorod and the concept of piezoelectricity. Finite element modeling was presented in Section II. Section III explains the simulation setup. Section IV presents the effect of variation of aspect ratio and applied input force towards the generated output voltage. The paper concludes the finding in Section V.

II. DESIGN CONCEPT

The single ZnO nanorod has been modeled as 3D cylinders with lengths from 500nm to 9000nm and radius from 15nm until 400nm as shown in Fig.1. A lateral force is applied to the upper part of the rod while the bottom is held fixed as illustrated in Fig. 2. To approximate the electrical boundary, the bottom of the rod is considered as a ground with no freely moving charges. The dimension of the nanorod used in the simulation is tabulated in Table I.

$$Di = d_{iq}T_q s + \varepsilon_{ik}^T E_k \qquad (1)$$

Equation (1) shows the equation for the direct piezoelectric effect. As shown in (1), the electric displacement (D$_i$), represents the net electric field generated within the ZnO nanorods. It shows the electric flux per unit area and is directly connected to the dielectric constant tensor, ε_{ik}^T. The dielectric constant tensor characterizes the ZnO nanorods' ability to store electrical energy when subjected to an electric field.[14]

979-8-3503-2369-6/23 $31.00 © 2023 IEEE

At the initial stage, the overlapping of the centers of Zn^{2+} cation and O^{2-} anion in ZnO occurs. However, when a strain is applied along the c-axis, the centers of Zn^{2+} cation and O^{2-} anion become misaligned. Consequently, positive and negative piezoelectric potentials are generated on the two sides of the strained ZnO, respectively. Furthermore, the application of strain along the c-axis induces an electric dipole in ZnO, resulting in the generation of a piezoelectric potential. This potential drives the flow of free electrons in an external circuit. [15]

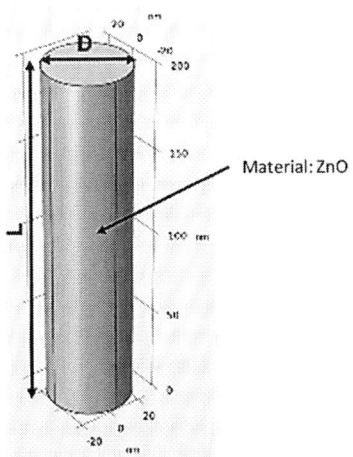

Fig. 1. ZnO nanorod structure design

TABLE I.   ZINC OXIDE NANOROD PROPOSED DESIGN

Design	Diameter, D (nm)	Length, L (nm)	Force (nN)
A (vary length)	30	500-900	100
B (vary diameter)	200-800	5000	100
C (vary applied input force)	300	9000	100-500

In Table I, the proposed design for the simulation of ZnO nanorod is presented. Design A focuses on the variation in the length of the nanorod, Design B explores the variation in the diameter of the nanorod, and Design C examines the effects of applied input force on the nanorod. Table II provides information on the ZnO material properties, including the piezoelectric constant, relative permittivity, and Young's modulus. The piezoelectric constant of ZnO nanorod is a measure of the piezoelectric effect in a material along a specific crystallographic direction. It also quantifies the induced polarization (charge separation) per unit of applied stress in that direction. The subscript indicates the direction of measurement within the crystal lattice.[16]

TABLE II.   ZINC OXIDE MATERIAL PROPERTIES

Property	ZnO
Crystal Structure	Wurtzite
Piezoelectric Constant (d_{33})	10 - 18 pC/N
Piezoelectric Constant (d_{31})	-30 - -110 pC/N
Piezoelectric Constant (d_{15})	-11 - -20 pC/N
Relative Permittivity (ε_r)	8.5 - 9.5
Density	5.606 g/cm³
Young's Modulus	100 - 200 GPa

III. 3D FINITE ELEMENT SIMULATION

To simulate the performance of different nanorod structures at nanometres length scales, the MEMS module within COMSOL Multiphysics® was specifically selected for analyzing a single ZnO nanorod.

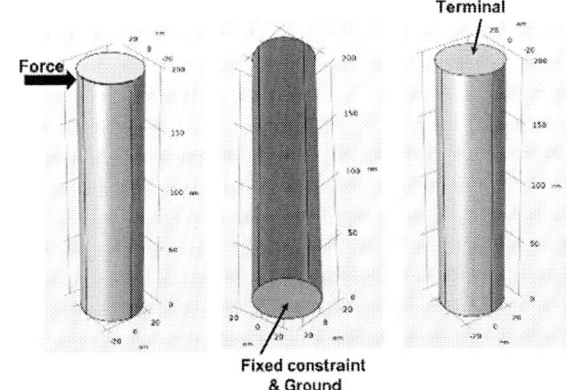

Fig. 2. 3D simulation setup of a single ZnO nanorod

A horizontal force is applied to the nanorod at the upper left tip along the x-axis. The terminal is positioned on top of the nanorod, while the ground is located at the bottom. Additionally, the bottom nanorod is kept fixed during the simulation as shown in Fig. 2. The finer mesh was used in this simulation. The simulation results can be analyzed to study the relationship between displacement and voltage output. Different factors, such as the magnitude of the force and nanorod dimensions were studied to investigate the effect on displacement and output voltage.

IV. RESULT AND DISCUSSION

A. Effect of applied input force on the output voltage of a single ZnO nanorod

The effect of input force on the output voltage was investigated. The single nanorod has a dimension of 9000 nm length and 300 nm diameter. The value of input force applied to the single ZnO nanorod is varied from 100nN to 500 nN. Fig. 3 shows the displacement distribution of a single ZnO nanorod for different applied input forces. The effect of force on a piezoelectric nanogenerator based on vertical ZnO nanorod is typically proportional to both the displacement and the electric potential. This relationship can be explained by the fundamental principles of piezoelectricity. Applying an external force to the piezoelectric nanogenerator results in mechanical deformation or displacement of the vertical ZnO nanorod as shown in Fig. 3. The degree of deformation is directly linked to the force's magnitude. As per the direct piezoelectric effect, this mechanical deformation generates an electric potential throughout the material. The displacement of the nanorod is directly proportional to the applied force, indicating that higher forces lead to larger displacements as shown in Fig. 3.

$$V = \frac{dFt}{\varepsilon_r \varepsilon_o A} \qquad (2)$$

As shown in equation (2), the direct piezoelectric effect refers to the phenomenon where an applied mechanical stress or strain induces an electric polarization in a piezoelectric material. This effect can be mathematically described by a set of equations V is voltage, d is the piezoelectric coefficient, F is force, t is thickness, ε_r is relative permittivity, ε_o is vacuum permittivity, and finally A is for the area. From Equation (2), it can be inferred that when the applied input force increases, there is a higher potential for the piezoelectric material to generate a higher voltage in response to the applied stress or strain. The relationship between the applied input force and voltage is likely to be linear, indicating that higher input forces have a greater potential to produce higher voltages through the direct piezoelectric effect, as proved in Fig. 4(b).

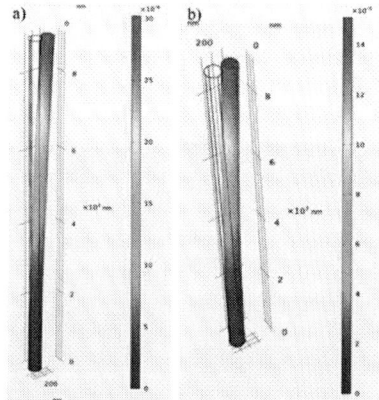

Fig. 3. Displacement along the length of the nanorod with force (a) 100nN and (b) 500nN

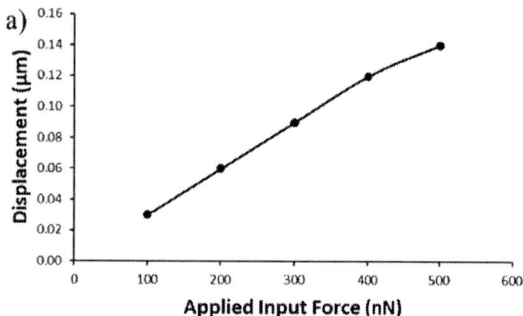

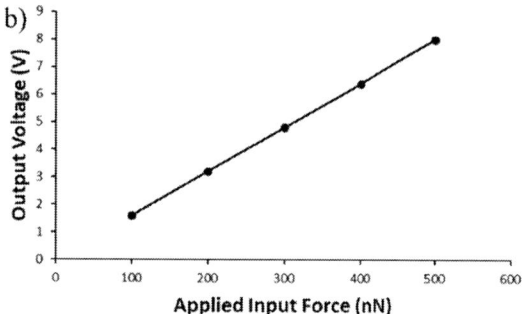

Fig. 4. (a) Linear relationship of displacement vs applied input force (100nN - 500nN) (b) Linear relationship of output voltage vs applied input force (100nN- 500nN)

B. Effect of aspect ratio (varies length) on the output voltage of a single ZnO nanorod

The aspect ratio, which is the ratio of length to diameter, can have a significant effect on the piezoelectric potential of vertical ZnO nanorods (NRs). The displacement distribution of a single ZnO nanorod for different nanorod lengths was investigated. The single nanorod has a dimension of 30 nm

diameter and the value of the input force applied is kept constant at 100nN. The values of the nanorod length were varied from 500nm to 900nm. When the length or aspect ratio of vertical ZnO nanorod increased, it generally leads to an increase in displacement, which is associated with the piezoelectric potential or output voltage as shown in Fig. 5. This is because longer nanorod will have higher mechanical deformation of bending when the same input forces were applied to the nanorod. As a result, when an external force or strain was applied, the longer nanorod experienced a greater amount of mechanical deformation compared to the shorter nanorod.

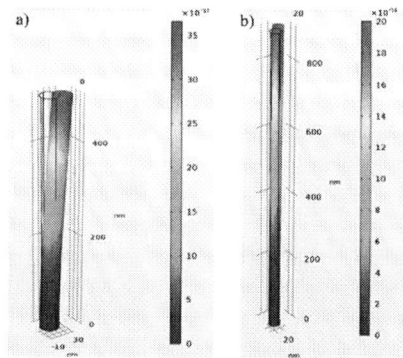

Fig. 5. Displacement along the length of the nanorod with lengths (a) 500nm and (b) 900nm

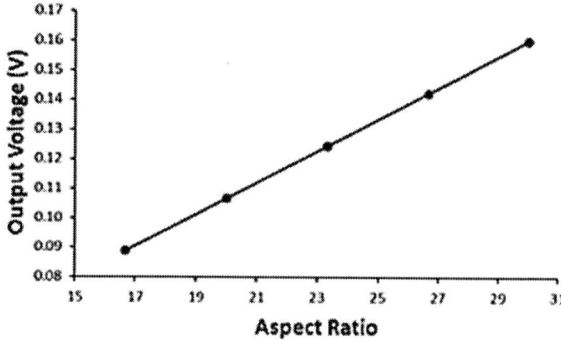

Fig. 6. Linear graph relationship of voltage vs aspect ratio when the length is varied from 500nm to 900nm

An increase in the length or aspect ratio of vertical ZnO nanorod generally results in a higher voltage generation, which is associated with the piezoelectric potential. This can be attributed to the larger surface area of longer NRs compared to shorter nanorods. The increased surface area allows for a greater accumulation and separation of charges. As the nanorod experienced mechanical deformation due to applied forces or strains, the charge separation became more pronounced, leading to an elevated electric potential. Consequently, longer nanorods with larger surface areas have the capability to generate a higher electric potential as shown in Fig. 6.

C. Effect of aspect ratio (varies diameter) on the output voltage of vertical ZnO nanorod

The effect of aspect ratio (varies diameter) was investigated. The single nanorod has a dimension of 5000 nm length and the value of the input force applied is 100nN. The values of the nanorod diameter of the single ZnO nanorod were varied from 200nm to 800 nm. The displacement distribution of a single ZnO nanorod for different nanorod lengths was illustrated in Fig. 7. When the diameter of the vertical ZnO nanorod increased or the aspect ratio decreased, it generally resulted in a reduction in displacement, which is associated with the piezoelectric potential. This can be attributed to the fact that increasing the diameter or decreasing the aspect ratio of the nanorod limited their ability to undergo significant mechanical deformation. With a larger diameter or lower aspect ratio, the nanorod became less flexible and less prone to substantial deformations.

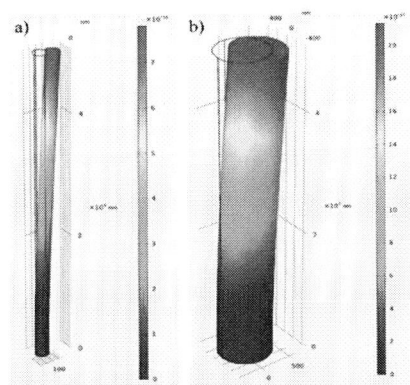

Fig. 7. Displacement along the length of the nanorod with diameter (a)200 nm and (b) 800nm

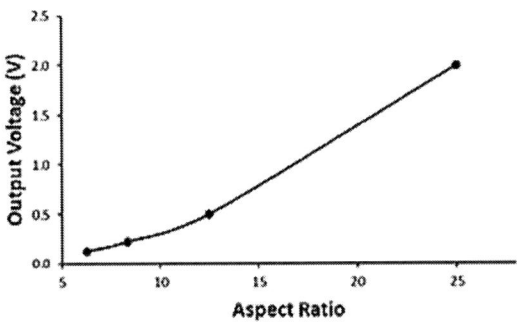

Fig. 8. Linear graph relationship of voltage vs aspect ratio when the diameter is varied from 200nm to 800nm

Thus, it can be concluded that when the area or diameter of the piezoelectric material increases, there is a decrease in voltage. This means that a larger area or diameter of the material leads to a lower voltage generated in response to applied stress or strain. The relationship between area/diameter and voltage is likely to be inversely proportional, indicating that increasing the size of the material results in a reduction in the voltage produced through the direct piezoelectric effect as shown in Fig. 8.

V. CONCLUSION

We presented in this paper detailed FEM simulation and performance analysis of three different analyses of a single ZnO nanorod. The displacement and voltage output of a single ZnO nanorod were recorded with the variation of aspect ratio and input force applied. The results demonstrated a proportional increase in displacement with the applied input force. Similarly, the voltage output exhibits a proportional increase with the aspect ratio as the length or diameter of the

nanorods varies. Output voltage also has a linear relationship with the input force applied. The optimum design of the single ZnO nanorod that has the highest output voltage is D=30nm L=9000nm force=500nN.

ACKNOWLEDGMENT

This work was supported by the Asian Office of Aerospace Research and Development (AOARD) Sponsored Project (International) (SPI) Research Project SPI21-082-0082

REFERENCES

[1] S. Chandrasekaran *et al.*, "Micro-scale to nano-scale generators for energy harvesting: Self powered piezoelectric, triboelectric and hybrid devices," *Phys Rep*, vol. 792, pp. 1–33, 2019.

[2] F. R. Fan, W. Tang, and Z. L. Wang, "Flexible nanogenerators for energy harvesting and self-powered electronics," *Advanced Materials*, vol. 28, no. 22, pp. 4283–4305, 2016.

[3] J. Briscoe, S. Dunn, J. Briscoe, and S. Dunn, "Nanostructured Materials," *Nanostructured Piezoelectric Energy Harvesters*, pp. 19–55, 2014.

[4] Z. L. Wang, "Zinc oxide nanostructures: growth, properties and applications," *Journal of physics: condensed matter*, vol. 16, no. 25, p. R829, 2004.

[5] S. Bahl, H. Nagar, I. Singh, and S. Sehgal, "Smart materials types, properties and applications: A review," *Mater Today Proc*, vol. 28, pp. 1302–1306, 2020.

[6] N. Izyumskaya, Y.-I. Alivov, S.-J. Cho, H. Morkoç, H. Lee, and Y.-S. Kang, "Processing, structure, properties, and applications of PZT thin films," *Critical reviews in solid state and materials sciences*, vol. 32, no. 3–4, pp. 111–202, 2007.

[7] O. Y. Pawar, S. L. Patil, R. S. Redekar, S. B. Patil, S. Lim, and N. L. Tarwal, "Strategic Development of Piezoelectric Nanogenerator and Biomedical Applications," *Applied Sciences*, vol. 13, no. 5, p. 2891, 2023.

[8] A. Jarjour, J. W. Cox, W. T. Ruane, H. Von Wenckstern, M. Grundmann, and L. J. Brillson, "Single metal ohmic and rectifying contacts to ZnO nanowires: A defect based approach," *Ann Phys*, vol. 530, no. 2, p. 1700335, 2018.

[9] X. Cha *et al.*, "Superhydrophilic ZnO nanoneedle array: Controllable in situ growth on QCM transducer and enhanced humidity sensing properties and mechanism," *Sens Actuators B Chem*, vol. 263, pp. 436–444, 2018.

[10] [W. Tie *et al.*, "Facile synthesis of carbon nanotubes covalently modified with ZnO nanorods for enhanced photodecomposition of dyes," *J Colloid Interface Sci*, vol. 537, pp. 652–660, 2019.

[11] M. Xie, D. Zhang, Y. Wang, and Y. Zhao, "Facile fabrication of ZnO nanorods modified with RGO for enhanced photodecomposition of dyes," *Colloids Surf A Physicochem Eng Asp*, vol. 603, p. 125247, 2020.

[12] R. Ahmed and P. Kumar, "Determining the most efficient geometry through simulation study of ZnO nanorods for the development of high-performance tactile sensors and energy harvesting devices," *arXiv preprint arXiv:2301.09370*, 2023.

[13] S. Abubakar *et al.*, "Controlled Growth of Semiconducting ZnO Nanorods for Piezoelectric Energy Harvesting-Based Nanogenerators," *Nanomaterials*, vol. 13, no. 6, p. 1025, Mar. 2023, doi: 10.3390/nano13061025.

[14] H. Sekimoto, T. Tamura, S. Goka, and Y. WATANABE, "IEEE Standard on Piezoelectricity, ANSI/IEEE Standard 176 IEEE Standard on Piezoelectricity, ANSI/IEEE Standard 176, 1987".

[15] M. Hyland, H. Hunter, J. Liu, E. Veety, and D. Vashaee, "Wearable thermoelectric generators for human body heat harvesting," *Appl Energy*, vol. 182, pp. 518–524, 2016.

[16] A. L. Kholkin, N. A. Pertsev, and A. V Goltsev, "Piezoelectricity and crystal symmetry," *Piezoelectric and acoustic materials for transducer applications*, pp. 17–38, 2

Acoustic Streaming in Microchannel as Micromixing

Anjam Waheed
Institute of Microengineering and Nanoelectronics
Universiti Kebangsaan Malaysia
Bangi Malaysia
anjamwaheed90@gmail.com

Muhamad Ramdzan Buyong
Institute of Microengineering and Nanoelectronics
Universiti Kebangsaan Malaysia
Bangi Malaysia
muhdramdzan@ukm.edu.my

MF Mohd Razip Wee
Institute of Microengineering and Nanoelectronics
Universiti Kebangsaan Malaysia
Bangi Malaysia
m.farhanulhakim@ukm.edu.my

Abstract—. **This paper proposes a micromixer made from poly(methyl methacrylate) (PMMA) actuated by a lead zirconate titanate (PZT) transducer for efficient fluid mixing The utilization of active microfluidic mixers in the microfluidic and lab on chip (LOC) platform for rapid and precise fluid mixing has generated significant interest. We proposed a micromixer made from PMMA actuated by PZT transducer. We fabricated our microfluidic channels using Computer Numerical Control (CNC) machining and the PMMA layers are bonded with a simple bonding consist of mixtures of IPA and acetone. We excited our PZT at 240KHz operating at 60Vpp in a straight microchannel and compared to the passive mode condition. To assess its functionality, our proposed micromixer was employed to mix two solution one is solution of DI water with organic dye and other one is phosphate buffered saline with DI water. Results show that this device has 91% of the mixing efficiency within microfluidic devices.**

Keywords—Microfluidics, Acoustophoresis, Mixing, Chaotic, Advection

I. INTRODUCTION

In recent decades, small-scale analysis systems have revolutionized fields such as forensic sciences, pharmaceuticals, medicine, and environmental monitoring. These systems offer automated platforms that handle fluid volumes from microliters to attoliters, enabling faster and more precise analyses for improved performance. Their compact size and portability make them ideal for point-of-care applications, especially in situations where access to traditional laboratories is limited or rapid testing is required. Additionally, the low cost of mass production allows for the creation of disposable devices, eliminating contamination risks and reducing waste generation for environmental sustainability [1].

Microfluidic structures, due to their small size and low flow rates, present challenges in precise fluid control as they primarily operate under laminar flow conditions with low Reynolds numbers. To overcome these limitations, there is a growing emphasis on developing mixing and pumping systems that can be integrated into microfluidic devices. Passive mixers, initially reported, rely on fluid pumping energy and unique microstructure geometries for flow management [2]. However, relying solely on diffusion and convection for mixing can result in complex channel configurations and prolonged transit and mixing times, particularly for fluids with low diffusion coefficients. In contrast, active mixers employ external forces to expedite the mixing process, demonstrating better efficiency and controllability for reconfigurable microfluidic systems [3]. However, some active mixers such as acoustic streaming [4], which utilizes ultrasonic acoustic waves to induce pumping and mixing, offer a promising approach. This technique can be implemented using silicon, glass, or polymeric-based chips in microfluidic devices [5].

Polymeric microfluidic devices offer numerous advantages over other materials. They are cost-effective and can be easily fabricated using well-established manufacturing processes like hot embossing and micro-injection molding. Additional steps such as polymer-polymer bonding and micro-milling can be employed to refine the device design. These methods also provide flexibility in material selection, allowing the use of various materials such as cyclic olefin copolymer(COC), polycarbonate(PC), polymethyl methacrylate (PMMA) in microfluidic chips [6]. This opens up opportunities for the application of acoustophoresis devices in medical settings beyond academia, where there is a growing demand for single-use devices to prevent cross-contamination in processes like blood plasma separation. The integration of acoustophoresis in lab-on-chip technology, widely used in various scientific applications, requires compatibility with polymer-based microfluidic platforms.

Although polymers are widely utilized in various attributes of microfluidics [7], research in the field of acoustofluidics involving polymers has been limited to a few researchers. The literature contains studies on acoustofluidic devices that employ polymer materials such as PMMA or PS, which have been utilized for bacterial and blood cell separation [8], purification of lymphocytes [9], platelet separation [10], focusing [11] and particles flow separation [12]. However, single channel devices often suffer from a low throughput when compared to silicon and glass devices. This is likely due to the fact that they are typically designed for an acoustic resonance between the channel walls similar to the glass or silicon-based devices. This assumption may not hold true in the case of polymer-based devices since the difference in acoustic impedance the liquid and chip material, which causes acoustic reflection, may be

979-8-3503-2369-6/23 $31.00 © 2023 IEEE

considerably low. This is demonstrated by the occasionally unexpected optimal operating frequency [13].

II. MICROFLUIDIC DEVICE

The 3D microfluidic design was created using AutoCAD and then we fabricated a device with Roland MDX-40A Desktop Machine i.e CNC milling process. One of the advantages of utilizing this process was the ability to achieve precise design control, resulting in a durable and robust device. The schematic diagram is shown in figure 1.

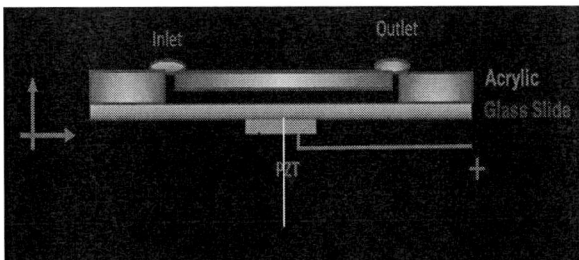

Fig.1. Schematic Diagram of Device

The microfluidic device used in the channel has two inlets and one out outlet. The inlet and outlet connectors are joined with glue. The acrylic sheet with dimensions 3cm x 1.5cm x 0.15cm is placed as a cover for the microchannel by utilizing IPA and acetone in the ratio 20:80 as glue and then placed in the oven at 70 °C for 15minute. The width of the channel is taken as 3mm and depth is 500um. The PZT of frequency 240kHz is bonded at the middle of the channel. This frequency is calculated by using formula f=v/2l where v is the speed of the sound in DI-water and 'l' is the width of the channel of device.

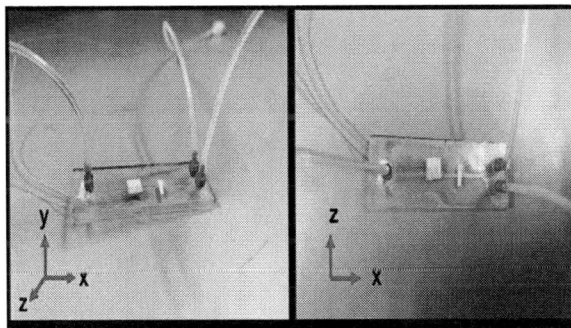

Fig.2. Microfluidic device

The generation of flexural waves initiated from the piezoelectric actuator situated at the apex of the device. These waves travel through the chip material and stimulate acoustic waves within the microchannel containing liquid. It is important to highlight that our microchannel structure allows streaming effect of particles across the width of the channel which is the ultimate goal of this study. The microparticles (polystyrene beads) can be loaded from the inlet and collected from the outlet of the device. The piezoelectric actuator has dimensions of 7x8x0.2mm with the thickness mode resonance estimated to be approximately 240kHz manufactured by STEMIC. The one side of the actuator has been divided into two sections to ensure that there is no conductive linkage between them. One section is utilized as a positive electrode and other section is used as a negative electrode. These two electrodes provide an out of phase excitation pattern when connected with power supply. The resulting vibration pattern generates an acoustic pressure that directs the movement of particles inside the channel.

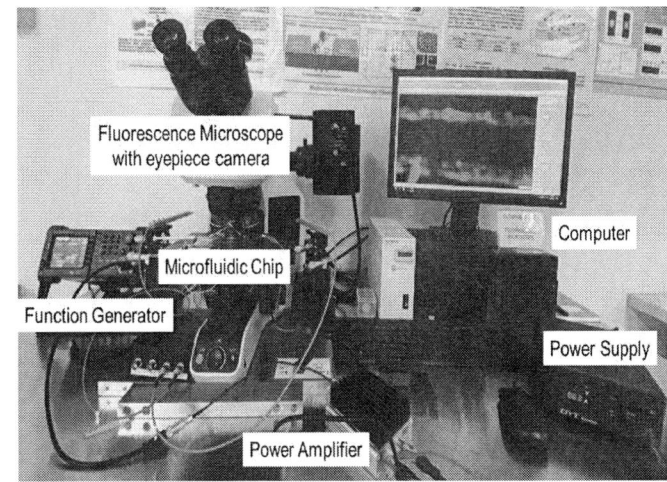

Fig.3. Experimental Setup

III. EXPERIMENTAL RESULTS

In our experimental setup, by utilizing a surface acoustic transducer positioned beneath the microchannel on a piezoelectric substrate, we generated an acoustic effect on the liquids within the microfluidic device. The direction of wave propagation in our experiments was perpendicular to the flow direction in the channel, specifically chosen to avoid additional pumping effects that could interfere with data evaluation. It is worth noting that this configuration offers a promising means of integrating active pumps into such systems. First, we applied this to the solution of polystyrene beads with DI water to visualise the acoustic streaming effect. The streaming effect of polystyrene beads can be seen as shown in fig 4.

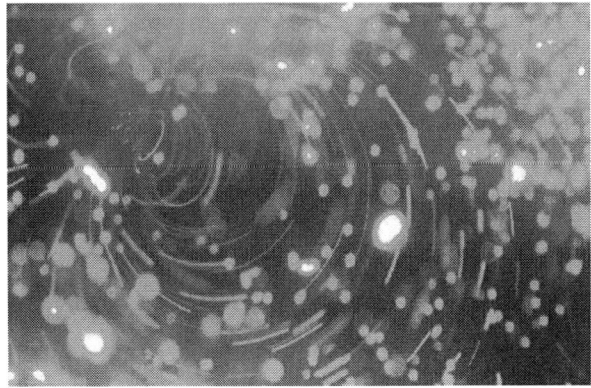

Fig.4 Polystyrene streaming in microfluidic channel

Moreover, we introduced two liquids through separate inlets in a microchannel. One inlet contained a solution of dye diluted in deionized (DI) water in an 2:8 ratio, while the other inlet contained solution of phosphate Buffer saline with DI water (1 tablet in 100ml) as shown in figure.5. Depending on the applied radiofrequency (rf) power, we observed mixing in the fluid as a result of its interaction with the acoustic effect see fig 4 (b).

979-8-3503-2369-6/23 $31.00 © 2023 IEEE

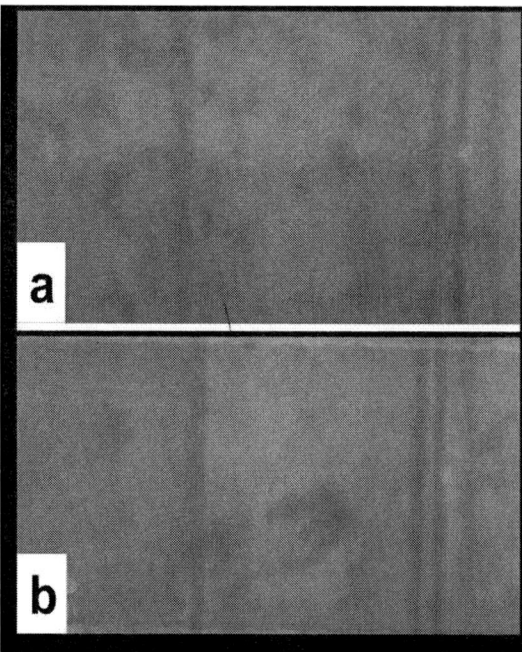

Fig.5. **(a)** Two Solution in Microfluidic channel without Acoustic effect (b) Two Solution in Microfluidic channel with Acoustic effect

The mixing efficiency is calculated with the help of imagJ softwareand and it is noted that without acoustic the mixing index is 56% and after applying acoustic effect it is 91%. So, Acoustofluidics based mixer has really an efficient mixer.

IV. CONCLUSION

In conclusion, we have presented a highly effective approach for achieving mixing in microfluidic channels using high-frequency acoustic waves. By transmitting this effect through the channel material, acoustic streaming is induced, leading to the formation of intricate folding patterns in the fluid. These patterns greatly enhance the mixing efficiency upto 91 % within microfluidic devices. Importantly, the piezoelectric chip can be acoustically coupled to the microfluidic chip without being directly integrated into the fluidic system, providing flexibility in the design and operation of the setup.

V. ACKNOWLEDGMENTS

We thank Dr. Ramdzan Buyong for his support and help. This work is supported by Ministry of Higher Education and

Universiti Kebangsaan Malaysia. Project code: GP-2021-K019988

VI. REFERENCES

[1] Ongaro, A. E., Ndlovu, Z., Sollier, E., Otieno, C., Ondoa, P., Street, A., & Kersaudy-Kerhoas, M. (2022). Engineering a sustainable future for point-of-care diagnostics and single-use microfluidic devices. Lab on a Chip, 22(17), 3122-3137.

[2] Wu, J., Tomsa, D., Zhang, M., Komenda, P., Tangri, N., Rigatto, C., & Lin, F. (2018). A passive mixing microfluidic urinary albumin chip for chronic kidney disease assessment. ACS sensors, 3(10), 2191-2197.

[3] Ward, K., & Fan, Z. H. (2015). Mixing in microfluidic devices and enhancement methods. Journal of Micromechanics and Microengineering, 25(9), 094001.

[4] Lei, J., Cheng, F., & Li, K. (2020). Numerical simulation of boundary-driven acoustic streaming in microfluidic channels with circular cross-sections. Micromachines, 11(3), 240.

[5] Aralekallu, S., Boddula, R., & Singh, V. (2022). Development of glass-based microfluidic devices: A review on its fabrication and biologic applications. Materials & Design, 111517.

[6] Açıkgöz, H. N., Karaman, A., Şahin, M. A., Çaylan, Ö. R., Büke, G. C., Yıldırım, E., ... & Özer, M. B. (2023). Assessment of silicon, glass, FR4, PDMS and PMMA as a chip material for acoustic particle/cell manipulation in microfluidics. Ultrasonics, 129, 106911.

[7] Saylan, Y., & Denizli, A. (2019). Molecularly imprinted polymer-based microfluidic systems for point-of-care applications. Micromachines, 10(11), 766.

[8] Van Assche, D., Reithuber, E., Qiu, W., Laurell, T., Henriques-Normark, B., Mellroth, P., ... & Augustsson, P. (2020). Gradient acoustic focusing of sub-micron particles for separation of bacteria from blood lysate. Scientific reports, 10(1), 1-13.

[9] Lissandrello, C., Dubay, R., Kotz, K. T., & Fiering, J. (2018). Purification of lymphocytes by acoustic separation in plastic microchannels. SLAS TECHNOLOGY: Translating Life Sciences Innovation, 23(4), 352-363.

[10] Mahboubidoust, A., Velisi, A. H., Ramiar, A., & Mosharafi, H. (2023). Development of a hybrid acousto-inertial microfluidic platform for the separation of CTCs from neutrophil. European Journal of Mechanics-B/Fluids, 99, 57-73.

[11] Augustsson, P., Karlsen, J. T., Su, H. W., Bruus, H., & Voldman, J. (2016). Iso-acoustic focusing of cells for size-insensitive acousto-mechanical phenotyping. Nature communications, 7(1), 11556.

[12] Mahboubidoust, A., Velisi, A. H., Ramiar, A., & Mosharafi, H. (2023). Development of a hybrid acousto-inertial microfluidic platform for the separation of CTCs from neutrophil. European Journal of Mechanics-B/Fluids, 99, 57-73.

[13] Hiremath, N., Kumar, V., Motahari, N., & Shukla, D. (2021). An overview of acoustic impedance measurement techniques and future prospects. Metrology, 1(1), 17-38.

Author Index

Name	Page No	Name	Page No
A. Zaidi Asad	134	Azrif Manut	94
Abdelkader Hassein-Bey	78	Burhanuddin Yeop Majlis	78
Abdul Karimi Halim	102	Chien Fat Chau	9
Abdul Manaf Hashim	5 21 44	Chin Fhong Soon	122
Abdur Rahman	66	Dahiru Shu'aibu	90
Abu Hashem	74	Darven Raj Ponnuthurai	36
Affa Rozana Abdul Rashid	58	Deyline Samail	47
Afiq Hamzah	1	Dharma Ram	5
Ahmad Ashrif A. Bakar	32	Faezah Harun	40
Ahmad Muhajer Abdul Aziz	130	Faisal Mohd-Yasin	13
Ahmad Nasrull Mohamed	122	Farhanulhakim Mohd razip wee	142
Ahmad Sabirin Zoolfakar	62 94	Farid Zubir	28
Ahmad Shuhaimi	122	Fazliyatul Azwa Md Rezali	82
Ahmad Wafi Mahmood Zuhdi	9	Ghulam Ali	13
Ahmad Zaki Abu Bakar	86	Guan Kai Oh	47
Aina Syakirah Mohd Masri	54 70	Hanim Hussin	1 50 74 102
Akmal Mustaffa Zulhakim	86	Ili Shairah Abdul Halim	86
Alireza Kalantari	21	Iskandar Yahya	32
Aliza Aini Md Ralib	118 138	Ismail Umar	47
Amit Verma	17	Izzuddin Iskandar	47
Anis Nabilah Mohd Daud	54	Jahariah Sampe	28
Anis Nurashikin Nordin	118 138	Jamila Lamido Sumaila	90
Anjam Waheed	142	Jason Kai Seng Kong	98
Arulampalam Kunaraj	32	Julie Roslita Rusli	114
Arvind Ajoy	17	Jumril Yunas	44
Avinash Kumaresan	32	Lai Ming Lim	118
Azlinda Abu Bakar	90	Leila Sabeha Asmaa Hassein-Bey	78

Name	Page No	Name	Page No
Lutfi Albasha	110	Muhammad Idzdihar Idris	130
M. A. Motalib Hossain	74	Muhammad Izzat Alif Muslan	86
Madhav Ramesh	17	Nafarizal Nayan	122
Maizan Muhamad	50 74 102	Nasir Quadir	110
Maizatul Zolkapli	62 94	Nirmala Kampan	58
Marzaini Rashid	130	Noor Fitrah Abu Bakar	94
Mathangi Ramakrishnan	1	Nor Haslinda Abd Aziz	58
Md Mushfiqur Rahman	28	Norain Sahari	122
Md. Yasir Arafat	126 134	Norazlina Saidin	138
Michael Loong Peng Tan	1 106	Norhayati Soin	50 82
Mohammad Al Mamun	74	Norhazlin Khairudin	94
Mohammad Islam	126	Nowshad Amin	9
Mohd Ambri Mohamed	5	Nur Iffah Irdina Maizal Hairi	118
Mohd Amir Zulkefli	25 47	Nur Mahirah Sallehuddin	102
Moh'd Khier Alshamaileh	110	Nur Shahirah Shaari	58
Mohd Nizar Hamidon	90	Nuradden Magaji	90
Mohd Rafie Johan	74	Nurbahirah Norddin	130
Muhamad Ramdzan Buyong	142	NurSyahirah Kamarozaman	54 66 70
Muhammad Adhwa Fathullah bin Nor Asmadi	138	Nurul Alias	74
		Nurul Ezaila Alias	1 50 106 126
Muhammad Ahmad	122	P. Susthitha Menon	58
Muhammad AlHadi Zulkefle	54	Puvaneswaran Chelvanathan	32
Muhammad Aniq Shazni Mohammad Haniff	5	Robaiah Mamat	86
Muhammad Farhan Affendi Mohamad Yunos	118	Rosminazuin Ab Rahim	118
		Rozina Abdul Rani	62 94
Muhammad Haziq Ilias	94	Sadiq Ur Rehman	134
Muhammad Hilmi Johari	25	Samira Abdelli-Messaci	78

Name	Page No	Name	Page No
Shaharin Fadzli Abd Rahman	21	Zarina Baharudin Zamani	130
Sharifah Fatmadiana Wan Muhamad Hatta	50 82 126 134	Zubaida Yusoff	28
Siti Aisyah Ishak	44	Zul Atfyi Fauzan Mohammed Napiah	130
Siti Nabila Aidit	82	Zulkifli Azman	122
Siti Nasuha Mustaffa	58	Zurita Zulkifli	54 66 70
Slimane Lafane	78		
Subathra Muniandy	130		
Suhana Mohamed Sultan	98		
Sukreen Hana Herman	54 66 70		
Suleiman Babani	90		
Toy Zheng Hong	106		
Vanita Manaoogaran	47		
Vatsala Pithaih	58		
Vikneswary Ravi Kumar	58		
Wan Fazlida Hanim Abdullah	86		
Wan Hidayatulhusna Wan Mohamad Rani	47		
Wan Syakirah Wan Abdullah	9		
Yasmin Abdul Wahab	1 50 74 102 106 126 134		
Yew Weng Ho	9		
Yew Weng Kean	9		
Yusmar Wijaya	122		
Yusof Johan	62		
Zainab Yunusa	90		
Zainal Nurbaya	66 70		
Zainiharyati Mohd Zain	94		